Mathematical Modeling: Analysis and Methodologies

Mathematical Modeling: Analysis and Methodologies

Edited by Andrew Clegg

CLANRYE
INTERNATIONAL
www.clanryeinternational.com

Clanrye International,
750 Third Avenue, 9th Floor,
New York, NY 10017, USA

ISBN: 978-1-63240-654-5

Cataloging-in-Publication Data

Mathematical modeling : analysis and methodologies / edited by Andrew Clegg.
    p. cm.
Includes bibliographical references and index.
ISBN 978-1-63240-654-5
1. Mathematical models. 2. Engineering mathematics.
3. Simulation methods. I. Clegg, Andrew.
TA342 .M38 2018
511.8--dc23

For information on all Clanrye International publications
visit our website at www.clanryeinternational.com

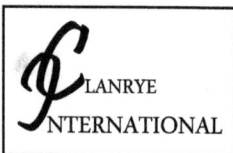

CLANRYE
INTERNATIONAL

Contents

Permissions

List of Contributors

Index

Preface

Mathematical models use mathematical tools to describe the performance and behavior of a system. They can be of different forms, like statistical models, game theoretic models, dynamical systems, etc. Mathematical modeling has significant applications in the diverse areas of science and engineering such as physics, artificial intelligence, economics, operations, research, etc. Through this book, we attempt to further enlighten the readers about the new concepts in this field. For someone with an interest and eye for detail, this book covers the most significant topics in the field of mathematical modeling. It will provide comprehensive knowledge to the readers.

The researches compiled throughout the book are authentic and of high quality, combining several disciplines and from very diverse regions from around the world. Drawing on the contributions of many researchers from diverse countries, the book's objective is to provide the readers with the latest achievements in the area of research. This book will surely be a source of knowledge to all interested and researching the field.

In the end, I would like to express my deep sense of gratitude to all the authors for meeting the set deadlines in completing and submitting their research chapters. I would also like to thank the publisher for the support offered to us throughout the course of the book. Finally, I extend my sincere thanks to my family for being a constant source of inspiration and encouragement.

Editor

Shape optimization of a Timoshenko beam together with an elastic foundation

J. Machalová[a], H. Netuka[a,*], R. Šimeček[a]

[a]*Faculty of Science, Palacký University in Olomouc, 17. listopadu 1192/12, 771 46 Olomouc, Czech Republic*

Abstract

In this article we are going first to aim at the variational formulation of the bending problem for the Timoshenko beam model. Afterwards we will extend this problem to the Timoshenko beam resting on the Winkler foundation, which is firmly connected with the beam. Hereafter a shape optimization for the aforementioned problems is presented. The state problem is here represented by the system of two ordinary differential equations of the second order. The optimization problem is given as a minimization of the so-called compliance functional on the set of all admissible design variables. For our purpose as the design variable we will select the beam thickness. Shape optimization problems have attracted the interest of many applied mathematicians and engineers. The objective of this article is to present a solution method for one of these problems and its demonstration by examples.

Keywords: Timoshenko beam, Winkler foundation, finite element method, shape optimization

1. Introduction

Nowadays it is well known that the classical Euler-Bernoulli beam theory is valid only for long span, equivalently thin, beams. In 1921 S. P. Timoshenko proposed a new beam theory that has been used for short, equivalently thick, beams. Unlike the Euler-Bernoulli hypothesis, the Timoshenko beam theory supposes that the plane section originally normal to the beam middle axis remains plane but not necessarily normal to the deformed axis, as in addition also transverse shear deformations can occur. Thus, using this theory it is possible to analyze thicker beams more accurately than by the classical beam theory.

A variational formulation of the bending problem and a finite element model will be interested us in the first part of this work. But foremost we want to present here some shape optimization of the Timoshenko beam with regard to the beam thickness. The criterion will be done by the compliance functional. Afterwards we are going to deal with the beam resting on a Winkler foundation. For this case it is possible to consider beside the beam thickness also the foundation stiffness optimization, but this opportunity will not be demonstrated in this article.

Several works have been done on this field but none of them is concerning the Timoshenko beam. The thickness optimization for the Euler–Bernoulli beam model was mostly examined, as it is from the theoretical and computational point of view a sort of fundamental and interesting case. Optimization of a beam with a subsoil of Winkler's type was studied in [7]. Object of optimization was the thickness and the subsoil stiffness. Especially it was focused on numerical modeling of the problem using ANSYS software system. Thickness optimization of a beam with a rigid obstacle was treated in [11], where an approach based on sensitivity analysis and

---

*Corresponding author. e-mail: netuka@inf.upol.cz.

nonlinear optimization methods was used. Optimal design of a beam on an unilateral elastic subsoil was presented in [12]. Existence of at least one solution was proved and conditions ensuring the solvability of the state problem were formulated.

2. Timoshenko beam with Winkler foundation

Let us consider a beam of the length L. The displacement field is

$$u_x(x,y) = u(x) - y\theta(x), \quad u_y(x,y) = w(x), \quad u_z(x,y) = 0, \tag{1}$$

where θ denotes the rotation of the cross-section plane about a normal to the middle axis x, u is the axial displacement of this axis and w is its transverse displacement. Little analyzing gives us

$$\theta(x) = w'(x) - \gamma(x), \quad \gamma(x) = \frac{Q(x)}{\kappa GA}. \tag{2}$$

Here Q is the transverse shear force, γ is the angle of shearing, G is the shear modulus, A is the cross-section area and κ is the shear correction factor. This factor is dependent on the cross-section and on the type of problem; one frequently used formula is $\kappa = \frac{10(1+\nu)}{12+11\nu}$ (ν is the Poisson's ratio) or sometimes $\kappa = \frac{5}{6}$ (see e.g. [3]).

Substituting (1) into the Green-St Venant strain tensor we obtain after some rearrangements (for the details see e.g. [3] or [6]) the system of two equations with the unknowns w and θ

$$(EI\theta')' + \kappa GA(w' - \theta) + m = 0, \tag{3}$$
$$(\kappa GA(w' - \theta))' + q = 0, \tag{4}$$

with E denoting the elasticity modulus, I the moment of inertia, $q(x)$ the applied transverse load and $m(x)$ the applied moment. The values of E and G are assumed constant, whereas I and A will be functions of the beam cross-section proportions b and t. Here b denotes the beam width and its height is considered in the interval $[-t, t]$. As we want later to optimize the beam thickness, t will be a function of x and for simplicity referred as the thickness although it is actually only a "half-thickness". For definiteness we will study the beam with a rectangular cross-section, hence we have

$$I(x) = \frac{2}{3}bt^3(x), \quad A(x) = 2bt(x). \tag{5}$$

Now we will deal with the variational formulation of the Timoshenko beam bending problem. First we must define suitable spaces for our unknowns. Let V be the space of kinematically admissible deflections v such that

$$H_0^1((0, \mathrm{L})) \subseteq V \subseteq H^1((0, \mathrm{L})). \tag{6}$$

Let us remember that the *Sobolev space* $H^1((0, \mathrm{L}))$ consists of those functions $v \in L^2((0, \mathrm{L}))$ for which derivatives v' (in the distribution sense) belong to the space $L^2((0, \mathrm{L}))$. The *Lebesgue space* $L^2((0, \mathrm{L}))$ is defined as the space of all measurable functions on $(0, \mathrm{L})$, the squares of which have a finite Lebesgue integral. Finally we define in (6) the space $H_0^1((0, \mathrm{L}))$ by

$$H_0^1((0, \mathrm{L})) = \{v \in H^1((0, \mathrm{L})) : v(0) = v(\mathrm{L}) = 0\}. \tag{7}$$

More information in relation to Sobolev spaces can be found e.g. in [1]. The same we can make also for kinematically admissible rotations η. The respective spaces will be distinguished as V_1 and V_2.

It is well known that the finite element method distinguishes between *natural* and *essential boundary conditions*. The first ones are contained in the space V, the second ones are built into the variational formulation. Let us remark, that the beam fixed at the both ends requires working with the spaces (7). Since the concrete boundary conditions are not important for our next explanation, without a loss of generality we can simply assume that the beam is fixed at both ends, so that we will work with the spaces $V_1 = V_2 = H_0^1((0, \mathrm{L}))$.

Using test functions from the above defined spaces we can obtain from the system (3)–(4) after integration by parts

$$\int_0^L EI\theta'\eta' \, dx - \int_0^L \kappa GA(w' - \theta)\eta \, dx = \int_0^L m\eta \, dx \qquad \forall \eta \in V_2, \tag{8}$$

$$\int_0^L \kappa GA(w' - \theta)v' \, dx = \int_0^L qv \, dx \qquad \forall v \in V_1. \tag{9}$$

Summing these equations together leads to

$$\int_0^L EI\theta'\eta' \, dx + \int_0^L \kappa GA(w' - \theta)(v' - \eta) \, dx = \int_0^L m\eta \, dx + \int_0^L qv \, dx \quad \forall v \in V_1, \eta \in V_2 \tag{10}$$

and this can be interpreted as the equation for a stationary point of the potential energy of the Timoshenko beam and it is possible to write it as follows

$$J_{TB}'(w, \theta; v, \eta) = 0 \qquad \forall v \in V_1, \eta \in V_2. \tag{11}$$

$J_{TB}'(w, \theta; v, \eta)$ denotes the Gâteaux derivative of J_{TB} at the point $\{w, \theta\}$ in the directions v, η (see e.g. [1]). Equations (11) and (10) imply that the functional of potential energy has the form

$$J_{TB}(w, \theta) = \frac{1}{2} \int_0^L EI \, (\theta')^2 \, dx + \frac{1}{2} \int_0^L \kappa GA \, (w' - \theta)^2 dx - \int_0^L m\theta \, dx - \int_0^L qw \, dx. \tag{12}$$

It is easy to prove that this functional is strictly convex. Then the equation (11) can be consequently rewritten as

$$J_{TB}(w, \theta) = \min_{v \in V_1, \eta \in V_2} J_{TB}(v, \eta). \tag{13}$$

The problem of finding a pair $\{w, \theta\} \in V_1 \times V_2$ such that (13) holds we will call the *variational formulation* of the Timoshenko beam bending. The convexity implies the unique solution of the minimization problem (13) and also the fact that (13) can be equivalently represented by the pair of equations (8)–(9).

Now let us go forward to the problem with a Winkler foundation. If k_F is the foundation stiffness, then its potential energy reads as

$$J_{WF}(w) = \frac{1}{2} \int_0^L k_F w^2 \, dx. \tag{14}$$

From here we immediately obtain the functional of total energy for the system beam plus foundation

$$J(w, \theta) = J_{TB}(w, \theta) + J_{WF}(w). \tag{15}$$

The variational formulation of this problem is as follows:

$$\begin{cases} \text{Find functions } \{w, \theta\} \in V_1 \times V_2 \text{ such that} \\ J(w, \theta) = \min_{v \in V_1, \eta \in V_2} J(v, \eta). \end{cases} \tag{16}$$

The functional (15) obviously retains the strict convexity, hence (16) can be equivalently rewritten as

$$J'(w, \theta; v, \eta) = 0 \qquad \forall v \in V_1, \eta \in V_2. \tag{17}$$

It gives us

$$\int_0^L EI\theta'\eta' \, dx - \int_0^L \kappa GA(w' - \theta)\eta \, dx = \int_0^L m\eta \, dx \qquad \forall \eta \in V_2, \tag{18}$$

$$\int_0^L \kappa GA(w' - \theta)v' \, dx + \int_0^L k_F w v \, dx = \int_0^L qv \, dx \qquad \forall v \in V_1 \tag{19}$$

and from here we can deduce the following system of two equations

$$(EI\theta')' + \kappa GA(w' - \theta) + m = 0, \tag{20}$$

$$(\kappa GA(w' - \theta))' + k_F w + q = 0 \tag{21}$$

presenting the extension of the original system (3)–(4) by the foundation term.

3. Finite element model for the Timoshenko beam

Now we proceed to a finite element discretization of our problems. As the problem without the foundation (3)–(4) can be obtained from (20)–(21) by putting $k_F = 0$, we will refer to the last one. For this purpose we have to define some division of the interval $[0, L]$ into subintervals $K_i = [x_{i-1}, x_i]$, where we have generated *nodes* $0 = x_0 < x_1 < \ldots < x_n = L$. Without loss of generality, we will restrict ourself to an equidistant division, i.e. $x_i - x_{i-1} = h$ for all i. Formally, the *discrete problem* reads as follows:

$$\begin{cases} \text{Find } \{w_h, \theta_h\} \in V_{1,h} \times V_{2,h} \text{ such that} \\ J(w_h, \theta_h) = \min_{v_h \in V_{1,h}, \eta_h \in V_{2,h}} J(v_h, \eta_h) \, , \end{cases} \tag{22}$$

which is equivalent to

$$\int_0^L EI\theta_h'\eta_h' dx - \int_0^L \kappa GA(w_h' - \theta_h)\eta_h dx = \int_0^L m\eta_h dx \qquad \forall \eta_h \in V_{2,h}, \tag{23}$$

$$\int_0^L \kappa GA(w_h' - \theta_h)v_h' dx + \int_0^L k_F w_h v_h dx = \int_0^L qv_h dx \qquad \forall v_h \in V_{1,h}. \tag{24}$$

$V_{k,h}$ is a finite-dimensional subspace of the given space V_k, $k = 1, 2$. Because, in contrast to the Euler-Bernoulli beam, we need not C^1-continuity, it is quite natural to choose the simplest approximation method, i.e.

$$V_{k,h} = \{v_h \in V_k : v_h|_{K_i} \in P_1(K_i) \quad \forall i = 1, \ldots, n\} \qquad k = 1, 2, \tag{25}$$

where $P_1(K_i)$ denotes the set of linear polynomials defined on K_i and hence $V_{k,h}$ contains continuous piecewise linear functions.

Now we can continue as it is usual for the standard finite element method. We define the *Lagrange basis functions* for our space (25) and afterwards the *shape functions* on a single element, which are beneficial from the practical computation point of view. Finally we obtain a system of linear equations (see e.g. [2] or [9]).

Unfortunately, there is a serious difficulty in this procedure — the phenomenon called the *shear locking* (see e.g. [10]). A brief explanation is as follows. Let us consider (2) after the finite element discretization. It results in

$$\gamma_h = w_h' - \theta_h. \tag{26}$$

We expect that γ_h converges to zero as the thickness $t \to 0$ (so-called Euler-Bernoulli limit). But according to (25) on the right side of (26) we have the difference of a constant and a linear function on every element and it will never give zero. Hence the shear strains, which are equal to γ_h, cannot be arbitrary small and in practice the computed deflections can be much smaller than the exact solution.

There are several possibilities how to handle this problem. We have chosen the way that is mathematically completely correct. Therefore we define a new approximation for the unknown w so that w_h' will have the same polynomial degree as θ_h. For this purpose we put

$$V_{1,h} = \{v_h \in V_1 : v_h|_{K_i} \in P_2(K_i) \quad \forall i = 1, \ldots, n\} \tag{27}$$

and $P_2(K_i)$ denotes the set of quadratic polynomials defined on K_i. Of course, we must add an extra node in the middle of each interval K_i. The space $V_{2,h}$ remains the same as in (25).

Now we are able to evaluate the element matrix for an element of the length h. Let $x_i = 0$, $x_{i+1/2} = \frac{h}{2}$, $x_{i+1} = h$ and let us denote

$$w_i = w_h(x_i), \ w_{i+1/2} = w_h(x_{i+1/2}), \ w_{i+1} = w_h(x_{i+1}), \tag{28}$$

$$\theta_i = \theta_h(x_i), \ \theta_{i+1} = \theta_h(x_{i+1}). \tag{29}$$

Then we have for $x \in [0, h]$

$$w_h(x) = \frac{1}{h^2}[(2x^2 - 3hx + h^2)w_i + (2x^2 - hx)w_{i+1} + (-4x^2 + 4hx)w_{i+1/2}], \tag{30}$$

$$\theta_h(x) = \frac{1}{h}[(-x + h)\theta_i + x\theta_{i+1}]. \tag{31}$$

Substituting these relations into (23)–(24) gets after some integrations the beam element matrix

$$
\begin{pmatrix}
\frac{7\kappa GA}{3h} & \frac{5\kappa GA}{6} & \frac{\kappa GA}{3h} & \frac{\kappa GA}{6} & -\frac{8\kappa GA}{3h} \\
\frac{5\kappa GA}{6} & \frac{EI}{h} + \frac{\kappa GAh}{3} & -\frac{\kappa GA}{6} & -\frac{EI}{h} + \frac{\kappa GAh}{6} & -\frac{2\kappa GA}{3} \\
\frac{\kappa GA}{3h} & -\frac{\kappa GA}{6} & \frac{7\kappa GA}{3h} & -\frac{5\kappa GA}{6} & -\frac{8\kappa GA}{3h} \\
\frac{\kappa GA}{6} & -\frac{EI}{h} + \frac{\kappa GAh}{6} & -\frac{5\kappa GA}{6} & \frac{EI}{h} + \frac{\kappa GAh}{3} & \frac{2\kappa GA}{3} \\
-\frac{8\kappa GA}{3h} & -\frac{2\kappa GA}{3} & -\frac{8\kappa GA}{3h} & \frac{2\kappa GA}{3} & \frac{16\kappa GA}{3h}
\end{pmatrix}
\begin{pmatrix}
w_i \\
\theta_i \\
w_{i+1} \\
\theta_{i+1} \\
w_{i+1/2}
\end{pmatrix} . \tag{32}
$$

As we have internal nodes in every element, we can use the *static condensation* technique (see e.g. [2]) to eliminate the unknowns associated with these nodes. After that we obtain

$$
\begin{pmatrix}
\frac{\kappa GA}{h} & \frac{\kappa GA}{2} & -\frac{\kappa GA}{h} & \frac{\kappa GA}{2} \\
\frac{\kappa GA}{2} & \frac{EI}{h} + \frac{\kappa GAh}{4} & -\frac{\kappa GA}{2} & -\frac{EI}{h} + \frac{\kappa GAh}{4} \\
-\frac{\kappa GA}{h} & -\frac{\kappa GA}{2} & \frac{\kappa GA}{h} & -\frac{\kappa GA}{2} \\
\frac{\kappa GA}{2} & -\frac{EI}{h} + \frac{\kappa GAh}{4} & -\frac{\kappa GA}{2} & \frac{EI}{h} + \frac{\kappa GAh}{4}
\end{pmatrix}
\begin{pmatrix}
w_i \\ \theta_i \\ w_{i+1} \\ \theta_{i+1}
\end{pmatrix}. \tag{33}
$$

The same can be done also for the foundation element matrix. For example [13] contains more relevant details.

4. Optimization

In this section we shall formulate a shape optimization problem for the Timoshenko beam model presented in the previous sections. We will optimize the thickness of the beam with respect to a compliance cost functional. Let us note, that here can be used also the name *sizing optimization*, as only typical size of a structure is optimized.

Let the thickness t be a function depending on x and occurring in the Timoshenko beam model as a part of the cross-section area A and its inertia moment I as it is in (5). To define an optimization problem we have to specify the class U_{ad} of *admissible thicknesses*

$$
U_{ad} = \left\{ t \in C^{0,1}([0, \mathrm{L}]) \; : \; 0 < T_0 \leq t(x) \leq T_1 \quad x \in [0, \mathrm{L}], \right.
$$

$$
\left. \int_0^{\mathrm{L}} t(x) \, \mathrm{d}x = T_2, \quad |t'(x)| \leq T_3 \quad \forall x \in [0, \mathrm{L}] \right\}, \tag{34}
$$

where the positive constants T_0, T_1, T_2 and T_3 are chosen in such a way that U_{ad} is nonempty. The set U_{ad} consists of all Lipschitz continuous functions that are uniformly bounded together with the absolute value of their first derivatives in $[0, \mathrm{L}]$. Moreover the volume of the beam is preserved and fixed during the optimization.

For an arbitrary but fixed $t \in U_{ad}$ the state problem is defined by (8), (9) and (18), (19), respectively. It can be proved that there is a continuous dependence between the design variable t and the state problem solution $\{w, \theta\}$. Further let us define the *compliance cost functional*

$$
\mathcal{J}(t, w, \theta) = \int_0^{\mathrm{L}} q(x) \, w(x) \, \mathrm{d}x. \tag{35}
$$

Functional $\mathcal{J}$ corresponds to the compliance of the transversally loaded beam. The compliance cost functional is not explicitly dependent on the design variable t, but generally the cost functional is a mapping $\mathcal{J} : U_{ad} \times V_1 \times V_2 \to \mathbb{R}$, see e.g. [4, 8]. Now we are ready to formulate the *shape optimization problem*:

$$
\begin{cases}
\text{Find } t^* \in U_{ad} \text{ such that} \\
\mathcal{J}(t^*, w^*, \theta^*) \leq \mathcal{J}(t, w, \theta) \quad \forall t \in U_{ad},
\end{cases} \tag{36}
$$

where $\{w, \theta\} = \{w(t), \theta(t)\}$ is a solution of the state problem for corresponding $t \in U_{ad}$. It can be proved that the optimization problem has at least one solution. We can describe the optimization problem by the following scheme:

$$
t \longmapsto \{w, \theta\} \longmapsto \mathcal{J}(t, w, \theta). \tag{37}
$$

Optimization problem in this form is not suitable for a numerical realization. Now we proceed to a discretization of our problem. The problem will be transformed to a new one defined by finite number of parameters. We start with the discretization of U_{ad}. Instead of general thickness $t(x)$ we will consider only those functions from U_{ad} that are Lipschitz continuous and piecewise linear on the partition $0 = x_0 < x_1 < \ldots < x_n = L$, i.e., we define

$$U_{ad}^h = \left\{ t_h \in C^{0,1}([0,L]) : t_h|_{K_i} \in P_1(K_i), \ \forall i = 1, \ldots, n \right\} \cap U_{ad}. \tag{38}$$

There is also an option to consider a stepped beam. It means to use piecewise constant thickness distribution instead of piecewise linear, see e.g. [4]. We can associate the design variable $t_h \in U_{ad}^h$ with an $(n+1)$-dimensional vector. Components of this vector are nodal values of t_h; i.e., $t_i = t_h(x_i)$, $i = 0, \ldots, n$. Then it is easy to see that U_{ad}^h can be identified with the finite dimensional set

$$\mathcal{U}^h = \left\{ t = (t_0, t_1, \ldots, t_n) \in \mathbb{R}^{n+1} : T_0 \leq t_i \leq T_1, \ i = 0, \ldots, n, \right.$$

$$\left. \sum_{i=1}^{n} \frac{h}{2}(t_{i-1} + t_i) = T_2, \ |t_{i-1} - t_i| \leq h\,T_3, \ i = 1, \ldots, n \right\}. \tag{39}$$

Using the finite element approach presented in the previous section the state problem transforms into a system of linear algebraic equations

$$\boldsymbol{K}(\boldsymbol{t})\,\boldsymbol{w}(\boldsymbol{t}) = \boldsymbol{F}, \tag{40}$$

where $\boldsymbol{K} = \boldsymbol{K}_b + \boldsymbol{K}_s$. Stiffness matrices $\boldsymbol{K}_b$, $\boldsymbol{K}_s \in \mathbb{R}^{(2n+2)\times(2n+2)}$ correspond to the beam and its foundation, respectively. These matrices are assembled from the element matrices (33). The vector $\boldsymbol{w} \in \mathbb{R}^{2n+2}$ consists of two parts. Coefficients w_i and θ_i, $i = 0, \ldots, n$, correspond to the transversal displacement and rotation of the cross section, respectively. These values are arranged as follows

$$\boldsymbol{w} = (w_0, \ \theta_0, \ w_1, \ \theta_1, \ \ldots, \ w_n, \ \theta_n) \in \mathbb{R}^{2n+2}. \tag{41}$$

Finally we can approach to a discretization of the cost functional. We use the trapezoid formula for numerical integration:

$$\mathcal{J}_h(\boldsymbol{t}, \boldsymbol{w}) = \sum_{i=1}^{n} \frac{h}{2} \left(w_{i-1}\,q(x_{i-1}) + w_i\,q(x_i) \right) = \boldsymbol{w}^T \boldsymbol{B}\,\boldsymbol{q}, \tag{42}$$

where

$$\boldsymbol{B} = h\,\mathrm{diag}\left(1/2, 0, \ 1, 0, \ 1, 0, \ \ldots, \ 1, 0, \ 1/2, 0\right) \in \mathbb{R}^{(2n+2)\times(2n+2)},$$

$$\boldsymbol{q} = (q(x_0), \ 0, \ q(x_1), \ 0, \ \ldots, \ q(x_n), \ 0) \in \mathbb{R}^{2n+2}.$$

Therefore the discrete optimization problem leads to the following *nonlinear programming problem*:

$$\begin{cases} \text{Find } \boldsymbol{t}^* \in \mathcal{U}^h \text{ such that} \\ \mathcal{J}_h(\boldsymbol{t}^*, \boldsymbol{w}^*) \leq \mathcal{J}_h(\boldsymbol{t}, \boldsymbol{w}) \quad \forall \boldsymbol{t} \in \mathcal{U}^h, \end{cases} \tag{43}$$

where $\boldsymbol{w} = \boldsymbol{w}(\boldsymbol{t})$ is a solution of the linear system (40) for corresponding $\boldsymbol{t} \in \mathcal{U}^h$. It can be proved that if we let $h \to 0+$ then solutions $(t_h, \{w_h, \theta_h\})$ of the approximate optimization problem will converge to the solution of the original problem (36).

The evaluation of the cost functional involves a solving of the linear state problem. Consequently, the optimization algorithm should use as few function evaluations as possible. Thus some gradient information is needed. In what follows we shall evaluate the gradient of $\mathcal{J}_h$ with respect to t. By using the classical chain rule of differentiation we can compute the derivative of the cost functional at point $t \in \mathcal{U}^h$ and in direction $s \in \mathbb{R}^{n+1}$:

$$\mathcal{J}_h'(t;\, s) = \nabla_w^T \mathcal{J}_h(t,\, w)\, w'(t;\, s). \tag{44}$$

By differentiating (40) we obtain

$$K(t)\, w'(t;\, s) = -K'(t;\, s)\, w(t). \tag{45}$$

To get full information on the gradient $\nabla_t \mathcal{J}_h$, we need to compute the directional derivative (44) in $n + 1$ linearly independent directions. We use the adjoint state technique to overcome this difficulty. Let us define the adjoint state problem

$$K(t)\, p(t) = \nabla_w \mathcal{J}_h(t,\, w) = B\, q. \tag{46}$$

Then multiplying (46) by $w'(t;\, s)$ we have

$$-p^T(t)\, K'(t;\, s)\, w(t) = p^T(t)\, K(t)\, w'(t;\, s) = \nabla_w^T \mathcal{J}_h(t,\, w)\, w'(t;\, s), \tag{47}$$

where we used (45). Making use of (44) and (47) we obtain the final form of the directional derivative:

$$\mathcal{J}_h'(t;\, s) = \nabla_w^T \mathcal{J}_h(t,\, w)\, w'(t;\, s) = -p^T(t)\, K'(t;\, s)\, w(t). \tag{48}$$

For more detailed treatment of the sensitivity analysis approach we refer to [4, 5].

5. Computational examples

In the *first example* we consider the beam of length L $= 10$. The load function q is piecewise constant and given by

$$q(x) = \begin{cases} 100 & x, < 5, \\ 1\,000 & x \geq 5. \end{cases} \tag{49}$$

The parameters related to the material properties and the cross sectional area of the beam are defined as follows: $b = 0.2$, $E = 2.19\,e6$, $G = E/[2(1 + \nu)]$, $\nu = 0.3$. The beam is not supported by a foundation, thus $k_F = 0$. Let the set $\mathcal{U}^h$ be defined by the following parameters: $T_0 = 0.5$, $T_1 = 1$, $T_2 = 7.5$, $T_3 = 0.2$ and let the initial guess be $t_i^0 = 0.75$ for $i = 0, \ldots, n$. We used 32 finite elements in discretization; i.e., $n = 32$ and $h = 10/32$. The following boundary conditions are prescribed: $w(0) = 0$, $\theta(0) = 0$ and $w(\mathrm{L}) = 0$. The beam is clamped at the left end and simply supported at the right end. No moments are applied.

The nonlinear mathematical programming problem (43) has been solved by the *sequential quadratic programming method* implemented in Matlab function *fmincon*. The state problems were solved by the Cholesky method.

In fig. 1 the optimal thickness of the beam is shown. We can compare the optimal shapes when both the Timoshenko model and the Euler-Bernoulli model for the state problem were used. We can also compare the deflection of the beam for both models. The results are shown in fig. 2. The cost functional values are summarized in tab. 1.

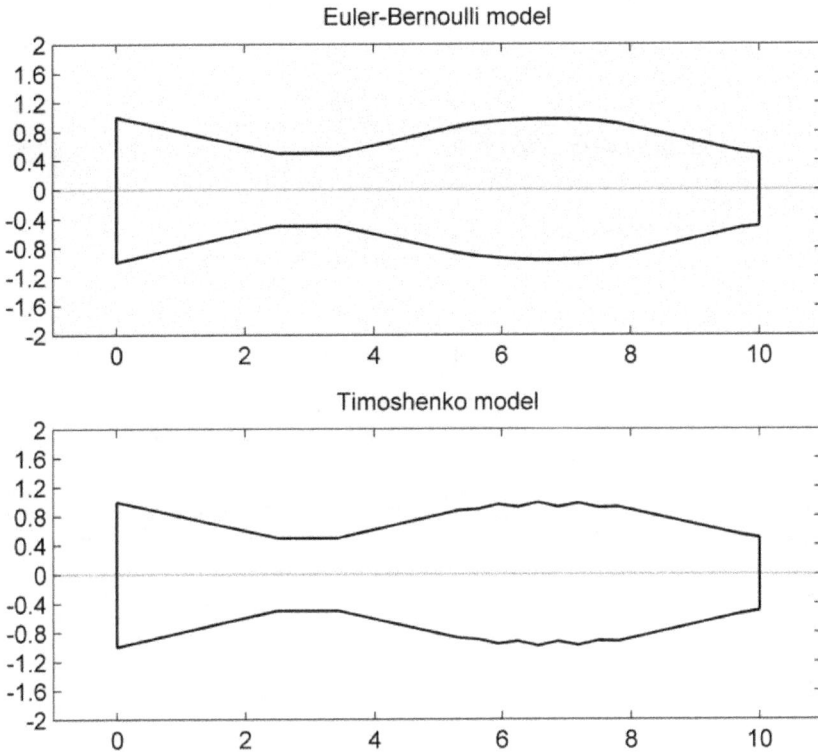

Fig. 1. An optimal thickness with respect to the compliance cost functional

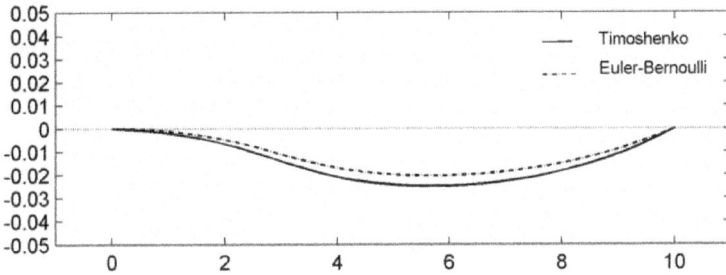

Fig. 2. A deflection of the optimal beam

Table 1. Cost functional values and number of iterations

Model	Initial	Final	Iter
Euler-Bernoulli	109.993 961 88	78.278 316 31	11
Timoshenko	125.507 121 96	96.189 756 66	16

The following abbreviations are used: *Model* = mathematical model used for the state problem, *Initial* = initial value of the cost functional, *Final* = final value of the cost functional, *Iter* = number of iterations.

In the *second example* we consider a beam of length $L = 10$ that is supported by a foundation with the piecewise constant stiffness coefficient given by:

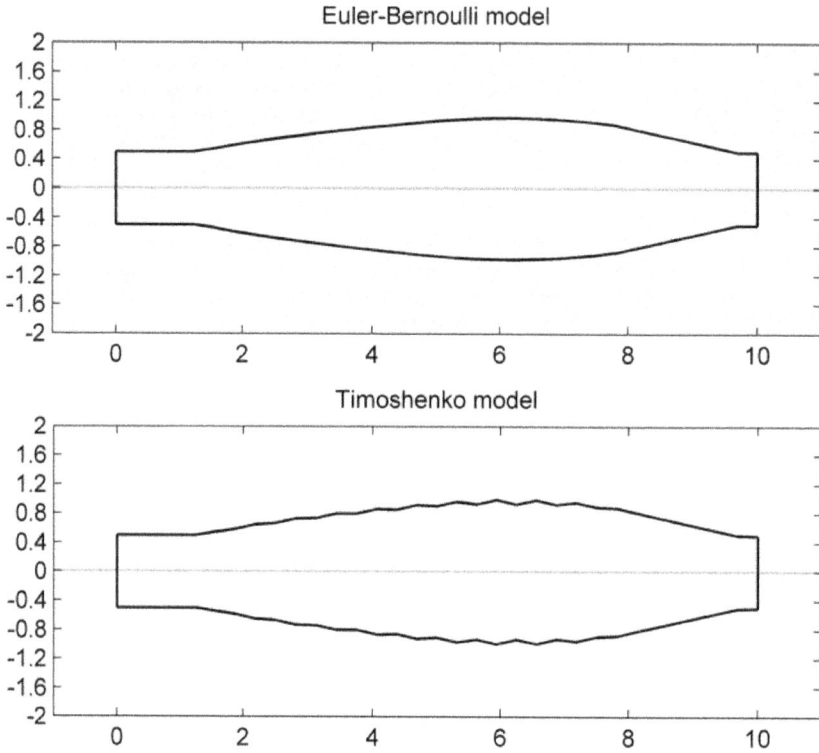

Fig. 3. A beam minimizing the compliance

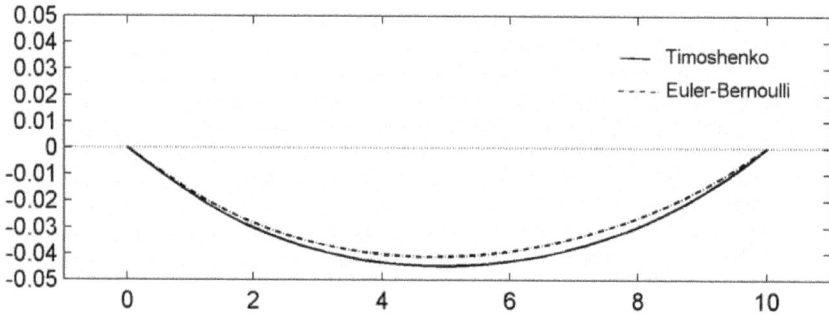

Fig. 4. A deflection of the beam with optimal thickness

$$k_F(x) = \begin{cases} 1\,000 & x < \frac{50}{16}, \\ 100 & x \geq \frac{50}{16}. \end{cases}$$

Let the beam be loaded by a piecewise constant load $q(x)$ given by (49). The parameters in the definition of $\mathcal{U}^h$ and the parameters related to the material and the cross section of the beam are the same as in the first example. We used 32 finite elements in discretization; i.e., $n = 32$ and $h = 10/32$. Boundary conditions defining a simply supported beam are prescribed as $w(0) = 0$ and $w(L) = 0$. No moments are applied.

Optimal shapes and beam deflections reached for the second example are presented in fig. 3 and fig. 4. Decrease of the cost functional for both models is shown in tab. 2. From both

Table 2. Cost functional values and number of iterations

Model	Initial	Final	Iter
Euler-Bernoulli	209.187 411 08	150.600 022 27	18
Timoshenko	222.556 647 52	166.877 166 35	21

examples it follows that the Timoshenko model is more sensitive to changes of parameter T_3. In practice the constraint $|t'(x)| \leq T_3$ prevents thickness oscillation. If we drop this constraint or set $T_3 > 1$, the shape can oscillate wildly. For the Timoshenko model this phenomena starting to be apparent for $T_3 > 0.2$. Further, it can be seen that the optimal solutions for both models produce a significant decrease of the compliance in comparison with a reference design.

6. Conclusion

We presented the shape optimization problem of the transversally loaded elastic beam with a foundation of Winkler's type. The Timoshenko beam theory was used for modeling of the state problem. The variational formulation and the finite element approximation of the beam bending problem were demonstrated. The objective of the optimization was the thickness of the beam. The optimization problem was formulated as a minimization of the compliance cost functional over a set of admissible thicknesses. Several numerical experiments were done, where optimal shapes for a beam with and without foundation were shown and compared with results attained for the Euler-Bernoulli beam model. An apparent decrease of the compliance for optimal thickness in comparison to the reference design has been obtained in all examples. From these results it follows that the Timoshenko model is more sensitive to changes of the parameter T_3 defining the bounds of the thickness first derivative.

Acknowledgements

The work has been supported by the Council of Czech Government MSM 6198959214 and by PrF_2010_00 - Mathematical and informatical models and structures of Internal Grant Agency of Palacký University in Olomouc.

References

[1] Atkinson, K., Han, W., Elementary numerical analysis: A functional analysis framework, Springer, New York, 2001.

[2] Bathe, K.-J., Finite element procedures, Prentice-Hall, New York, 1996.

[3] Dym, C. L., Shames, I. H., Energy and finite element methods in structural mechanics, Taylor & Francis, New York, 1991.

[4] Haslinger, J., Mäkinen, R. A. E., Introduction to shape optimization: Theory, approximaion and computation, SIAM, Philadelphia, 2003.

[5] Haslinger, J., Neittaanmäki, P., Finite element approximation for optimal shape, material and topology design, Second edition, John Wiley and Sons, Chichester, 1997.

[6] Hjelmstad, K. D., Fundamentals of structural mechanics, Second edition, Springer, New York, 2005.

[7] Horák, J. V., Šimeček, R., ANSYS implementation of shape design optimization problems, 1. AN-SYS conference 2008, 16. ANSYS FEM Users' Meeting, Luhačovice 5.–7. November 2008, 32 pages, released on CD, published by SVS-FEM, Brno, 2008.

[8] Chleboun, J., Optimal design of an elastic beam on an elastic basis, Applications of Mathematics, 31 (2) (1986) 118–140.

[9] Reddy, J. N., An introduction to the finite element method, 2nd edition, McGraw-Hill Book Co., New York, 1993.

[10] Reddy, J. N., On locking free shear deformable beam finite element, Computational Methods in Applied Mechanics and Engineering, 149 (1997), 113–132.

[11] Šimeček, R., Sizing optimization of an elastic beam with a rigid obstacle: Numerical realization, Proceedings of SVOC competition, Olomouc, 2009.

[12] Šimeček, R., Optimalizace nosníku na jednostranném podloží: Existence řešení, str. 68—84, Sborník konference Olomoucké dny aplikované matematiky ODAM2009, KMAaAM, PřF UP Olomouc, 2009 (in Czech).

[13] Shrikhande, M., Finite element method and computational structural dynamics, textbook in prepa-ration, Indian Institute of Technology Roorkee, Roorkee, 2008.

2

Complex model of the lower urinary tract

M. Brandner[a], J. Egermaier[b], H. Kopincová[a], J. Rosenberg[c,*]

[a]NTIS -- New Technologies for Information Society, University of West Bohemia in Pilsen, Univerzitni 8, 306 14 Pilsen, Czech Republic
[b]Department of Mathematics, University of West Bohemia in Pilsen, Univerzitni 8, 306 14 Pilsen, Czech Republic
[c]New Technologies-Research Centre, University of West Bohemia in Pilsen, Univerzitni 8, 306 14 Pilsen, Czech Republic

Abstract

The complex model of the lower part of the urinary tract is introduced. It consists of the detrusor smooth muscle cell model and the detailed 1D model of the urethra flow. The nerve control is taken into account. In future this model will allow to simulate the influence of different drugs and mechanical obstructions in the bladder neck and urethra. A general muscle model involving the calcium dynamics in the smooth muscle cell and the growth and remodelling theory will be shortly introduced. For the modelling calcium dynamics the approach of Koenigsberger published in Biophysical Journal (Koenigsberger, M., Sauser, R., Seppey, D., Beny, J.-L., Meister, J.-J., Calcium dynamics and vosomotion in arteries subject to isometric, isobaric and isotonic conditions, Biophysical Journal 95 (2008) 2 728–2 738.) was adopted. The model includes the ATP consumption calculation according to Hai et al. (Hai, C. M., Murphy, R. A., Adenosine 5'-triphosphate consumption by smooth muscle as predicted by the coupled four-state crossbridge model, Biophysical Journal 61(2) (1992) 530–541.). The main part is devoted to the development of a simple bladder model and the detrusor contraction during voiding together with the detailed model of the urethra flow.

Keywords: urinary tract, bladder, urethra fluid flow, steady state preserving

1. Introduction

The voiding is a very complex process. As we can see from Fig. 1, it consists of the transfer of information about the state of the bladder filling in to the spinal cord. Next part is the sending of the action potentials to the smooth muscle cells of the bladder. Even this process is not simple and includes the spreading of the action potential along the nerve axon and the transmission of the mediator (Ach – acetylcholine) in the synapse. The action potential starts the process of the smooth muscle contraction.

The smooth muscles have a lot of different forms in contradiction with the striated muscles. They are present in vesicles, arteries and others hollow organs.

Although the own biological motor – sliding between actin and myosin fueled by hydrolysis of ATP – is the same here as well as in striated and heart muscles, there are important differences between these basic types of muscles and also between smooth muscles in different organs. The sliding between actin and myosin causing the change of the form (length) of the muscle cell and its stiffness can be observed as a kind of growth and remodeling. This approach described e.g. in [15] and [14] is used in this model. It should be mentioned that a lot of different smooth muscle cells (SMC) models exist. They are based either on Huxley model where the calcium dynamics is not taken into account in details or on the contrary the calcium dynamics and the

---

*Corresponding author. e-mail: rosen@kme.zcu.cz.

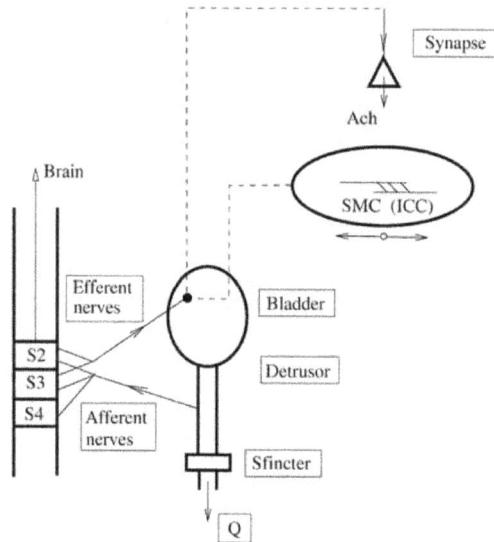

Fig. 1. The simplified scheme of the lower part of the urinary tract. SMC – smooth muscle cell, ICC – interstitial cell of Cajal (in this contribution are not taken in account), Ach – acetocholine, Q – flux of the urine

phosphorilation is modeled very precisely but the mechanochemical coupling is based on the work on [6] where the stress in the muscle cell depends linearly on the amount of the bonded crossbridges either phosphorilized or unphosphorilized (e.g., [7, 10], where the model is applied to the SMC in the vessels).

To be able to describe the very complex processes in the SMC in the efficient form it is necessary to use the irreversible thermodynamics. This approach was described in [16].

Using all these approaches the algorithm published in [13] was developed. In this contribution we join on the results of this paper. The simple model of the whole bladder and the detailed 1D model of the urethra flow is added. Some examples of the numerical experiments are shown.

2. Bladder contraction

As it was already mentioned, the whole model of the bladder contraction is described in [12]. It consists of the following parts:

- Model of the time evolution of the Ca2++ concentration – five equations [7]. The Ca2++ intracellular concentration is the main control parameter for the next processes and finally for the smooth muscle contraction. Its increase depends on the flux $J_{agonist}$ of the mediator (in this case acetylcholine) via the nerve synapse.

- Model of the time evolution of the phosporilation of the light myosin chain – three equations [6]. The muscle cell contraction is caused by the relative movement of the myosin and actin filaments. For this it is necessary that the phophorilation of the mentioned light myosin chain on the heads of the myosin occurs. Knowing this process also the time evolution of the ATP consumption (J_{cycl}) can be determined. The ATP (adenosintriphosphate) is the main energy source for the muscle contraction.

3. Model of the own contraction based on the GRT and the irreversible thermodynamics

The growth and remodelling theory [4] together with the laws of irreversible thermodynamics with internal variables was applied in [16] to describe the mechano-chemical coupling of the smooth muscle cell contraction. The product of the chemical reaction affinity (the ATP hydrolysis) with its rate plays an important role in the discussed model. Further it can be assumed that the rate of the ATP hydrolysis depends on the ATP consumption. The corresponding equations in the non-dimensional form are following:

$$\dot{x} = \frac{g}{h}\left[\tau' - z(x-1)\right] = k_1\left[\tau' - z(x-1)\right], \tag{1}$$

$$\dot{y} = \frac{y}{k_2}\left[x\tau' - \frac{1}{2}z(x-1)^2 + C'\right], \tag{2}$$

where

$$\dot{z} = \mathrm{sgn}\,(m) \cdot \left[r - \frac{1}{2}z(x-1)^2\right], \quad x = \frac{l'}{l'_r}, \quad y = l'_r, \quad z = k, \tag{3}$$

$$\frac{l}{l_0} = x \cdot y, \quad k' = k\sqrt{\frac{|m|}{g}}, \quad l'_r = \frac{l_r}{l_0}, \quad t' = \frac{t}{\sqrt{g|m|}}.$$

l_0 is the initial length of the muscle fibre, l_r its length after stimulation when the fibre is unloaded (s. c. resting length), l the actual length (when the contraction is isometric this is the input value), t' the stress and k is the fibre stiffness. The non-dimensional values are labeled with the single quote mark. The others symbols are the parameters. The most important parameter is C'. Using the irreversible thermodynamics we can obtain the following relations

$$C' = p \cdot (C - a_{chem}Y)\sqrt{\frac{|m|}{g}},$$

$$C\sqrt{\frac{|m|}{g}} = C_0 + C_t e^{q\left(\frac{l}{l_0} - \frac{l}{l_0}|opt\right)^2}, \tag{4}$$

$$p = p_0 e^{s\left(\frac{l}{l_0} - \frac{l}{l_0}|opt\right)^2},$$

where for the afinity of the chemical reactions especially for the hydrolysis of the ATP gilt

$$a_{chem} = -Q \cdot Y. \tag{5}$$

Q is the constant and Y is the concentration of ATP. For its time evolution gilts [11]

$$\dot{Y} = -QQ \cdot Y + L \cdot J_{cycle}. \tag{6}$$

Here QQ is the damping parameter. Than the whole model is finished because the ATP consumption J_{cycl} as a function of the Ca2+ concentration in the cytoplasm was already determined.

4. Bladder and voiding model

To model the contraction of the bladder during the voiding process we will use the very simple model according [9] and [1]. The bladder is modelled as a hollow sphere with the output corresponding to the input into urethra.

For the pressure in the bladder the following formula is introduced in [9]

$$p = \frac{V_{sh}}{3V} \cdot \tau, \qquad \tau = \frac{F}{S}, \tag{7}$$

where V_{sh} is the volume of the wall, V the inner volume, τ stress in the muscle fibre, S the inner surface and F the force in the muscle cell.

For the flux q gilts

$$q = \frac{dV}{dt}, \tag{8}$$

where ρ is the density of the fluid.

Using the formulas for the isotonic contraction, we can at first obtain the relation for the volume. It gilts

$$l' = \frac{l}{l_0} = x \cdot y \tag{9}$$

and then

$$V = \kappa \cdot (x \cdot y)^3, \tag{10}$$

where κ is the constant which in the theoretical case if only one cell will occupy the circumference of the spherical bladder will be $1/6\pi^2$. Putting this formula into the equation for q and using the e quations for the derivatives of x and y mentioned before we obtain the equation, from which we can calculate τ:

$$\tau = \frac{\frac{q}{3\kappa(x \cdot y)^2} + \left[k_1 z y^2 (x-1) + \frac{z y^2 x}{2 k_2}(x-1)^2 - \frac{x y}{k_2} C' \right]}{k_1 y + \frac{x^2 y}{k_2}}. \tag{11}$$

For the pressure gilts then

$$p = \frac{V_{sh}}{3\kappa \cdot (x \cdot y)^3} \cdot \frac{\frac{q}{3\kappa(x \cdot y)^2} + \left[k_1 z y^2 (x-1) + \frac{z y^2 x}{2 k_2}(x-1)^2 - \frac{x y}{k_2} C' \right]}{k_1 y + \frac{x^2 y}{k_2}}. \tag{12}$$

This will be putted into the equations for the isotonic contraction.

5. Urethral flow

We now briefly introduce a problem describing fluid flow through the elastic tube represented by hyperbolic partial differential equations with the source term. In the case of the male urethra, the system based on model in [17] has the following form

$$a_t + q_x = 0,$$
$$q_t + \left(\frac{q^2}{a} + \frac{a^2}{2\rho\beta} \right)_x = \frac{a}{\rho} \left(\frac{a_0}{\beta} \right)_x + \frac{a^2}{2\rho\beta^2}\beta_x - \frac{q^2}{4a^2}\sqrt{\frac{\pi}{a}}\lambda(Re), \tag{13}$$

where $a = a(x,t)$ is the unknown cross-section area, $q = q(x,t)$ is the unknown flow rate (we also denote $v = v(x,t)$ as the fluid velocity, $v = \frac{q}{a}$), ρ is the fluid density, $a_0 = a_0(x)$ is the cross-section of the tube under no pressure, $\beta = \beta(x,t)$ is the coefficient describing tube compliance and $\lambda(Re)$ is the Mooney-Darcy friction factor ($\lambda(Re) = 64/Re$ for laminar flow). Re is the Reynolds number defined by

$$Re = \frac{\rho q}{\mu a} \sqrt{\frac{4a}{\pi}}, \tag{14}$$

where μ is fluid viscosity. This model contains constitutive relation between the pressure and the cross section of the tube

$$p = \frac{a - a_0}{\beta} + p_e, \tag{15}$$

where p_e is surrounding pressure.

Presented system (13) can be written in the compact matrix form

$$\mathbf{u}_t + [\mathbf{f}(\mathbf{u}, x)]_x = \boldsymbol{\psi}(\mathbf{u}, x), \tag{16}$$

with $\mathbf{q}(x, t)$ being the vector of conserved quantities, $\mathbf{f}(\mathbf{q}, x)$ the flux function and $\boldsymbol{\psi}(\mathbf{q}, x)$ the source term. This relation represents the balance laws. For the following consideration, we reformulate this problem to the nonconservative form.

5.1. Nonconservative problems

We consider the nonlinear hyperbolic problem in nonconservative form

$$\mathbf{u}_t + \mathbf{A}(\mathbf{u})\mathbf{u}_x = 0, \ x \in \mathbf{R}, \ t \in (0, T), \tag{17}$$
$$\mathbf{u}(x, 0) = \mathbf{u}_0(x), \ x \in \mathbf{R}.$$

The numerical schemes for solving problems (17) can be written in fluctuation form

$$\frac{\partial \mathbf{U_j}}{\partial t} = -\frac{1}{\Delta x}[\mathbf{A}^-(\mathbf{U}_{j+1/2}^-, \mathbf{U}_{j+1/2}^+) + \mathbf{A}(\mathbf{U}_{j+1/2}^-, \mathbf{U}_{j-1/2}^+) + \mathbf{A}^+(\mathbf{U}_{j-1/2}^-, \mathbf{U}_{j-1/2}^+)], \tag{18}$$

where $\mathbf{A}^\pm(\mathbf{U}_{j+1/2}^-, \mathbf{U}_{j+1/2}^+)$ are so called fluctuations. They can be defined by the sum of waves moving to the right or to the left. The directions are dependent on the signs of the speeds of these waves, which are related to the eigenvalues of matrix $\mathbf{A}(\mathbf{u})$. In what follows, we use the notation $\mathbf{U}_{j+1/2}^+$ and $\mathbf{U}_{j+1/2}^-$ for the reconstructed values of unknown function. Reconstructed values represent the approximations of limit values at the points $x_{j+1/2}$. The most common reconstructions are based on the minmod function (see for example [8]) or ENO and WENO techniques [3].

The reconstruction can be applied to each component of $\mathbf{u}$. But this approach does not work well in general. It is better to apply the reconstruction to the characteristic field of $\mathbf{u}$. It means that each jump is decomposed to the eigenvectors $\mathbf{r}$ of Jacobian matrix $\mathbf{A}(\mathbf{u})$.

$$\mathbf{U}_{j+1} - \mathbf{U}_j = \sum_{p=1}^{m} \alpha_{j+1/2}^p \mathbf{r}_{j+1/2}^p. \tag{19}$$

Then the reconstruction based on minmod function can be defined by following

$$\mathbf{U}_{j+1/2}^+ = \mathbf{U}_{j+1} + \sum_p \phi_{I+1/2}^{p,+} \alpha_{j+1/2}^p \mathbf{r}_{j+1/2}^p, \tag{20}$$

$$\mathbf{U}_{j+1/2}^- = \mathbf{U}_j + \sum_p \phi_{I+1/2}^{p,-} \alpha_{j+1/2}^p \mathbf{r}_{j+1/2}^p,$$

where

$$\phi_{I+1/2}^{p,\pm} = \mp \frac{1}{2}\left(1 + \text{sgn}(\theta_{I+1/2}^p)\right) \min(1, |\theta_{I+1/2}^p|) \tag{21}$$

and

$$I = \begin{cases} j - 1/2, & \text{if } s^p_{j+1/2} \geq 0, \\ j + 3/2, & \text{if } s^p_{j+1/2} < 0. \end{cases} \tag{22}$$

The function $\theta^p_{j+1/2}$ can be determined by the following way

$$\theta^p_{j+1/2} = \frac{\alpha^p_{j+1/2} \mathbf{r}^p_{j+1/2} \cdot \mathbf{r}^p_{I+1/2}}{\alpha^p_{I+1/2} \mathbf{r}^p_{I+1/2} \cdot \mathbf{r}^p_{I+1/2}}. \tag{23}$$

When the problem (17) is derived from the conservation form (16), i.e. $\mathbf{f}'(\mathbf{u}) = \mathbf{A}(\mathbf{u})$ is the Jacobi matrix of the system, fluctuations can be defined as follows

$$\begin{aligned}
\mathbf{A}(\mathbf{U}^-_{j+1/2}, \mathbf{U}^+_{j-1/2}) &= \mathbf{f}(\mathbf{U}^-_{j+1/2}) - \mathbf{f}(\mathbf{U}^+_{j-1/2}), \\
\mathbf{A}^-(\mathbf{U}^-_{j+1/2}, \mathbf{U}^+_{j+1/2}) &= \mathbf{F}^-_{j+1/2} - \mathbf{f}(\mathbf{U}^-_{j+1/2}), \\
\mathbf{A}^+(\mathbf{U}^-_{j-1/2}, \mathbf{U}^+_{j-1/2}) &= \mathbf{f}(\mathbf{U}^+_{j-1/2}) - \mathbf{F}^+_{j-1/2}.
\end{aligned} \tag{24}$$

5.2. Decompositions based on augmented system

This procedure is based on the extension of the system (13) by other equations (for simplicity we omit viscous term). This was derived in [5] for the shallow water flow. The advantage of this step is in the conversion of the nonhomogeneous system to the homogeneous one. In the case of urethra flow we obtain the system of four equations, where the augmented vector of unknown functions is $\mathbf{w} = [a, q, \frac{a_0}{\beta}, \beta]^T$. Furthermore we formally augment this system by adding components of the flux function $\mathbf{f}(\mathbf{u})$ to the vector of the unknown functions. We multiply balance law (16) by Jacobian matrix $\mathbf{f}'(\mathbf{u})$ and obtain following relation

$$\mathbf{f}'(\mathbf{u})\mathbf{u}_t + \mathbf{f}'(\mathbf{u})[\mathbf{f}(\mathbf{u})]_x = \mathbf{f}'(\mathbf{u})\psi(\mathbf{u}, x). \tag{25}$$

Because of $\mathbf{f}'(\mathbf{u})\mathbf{u}_t = [\mathbf{f}(\mathbf{u})]_t$ we obtain hyperbolic system for the flux function

$$[\mathbf{f}(\mathbf{u})]_t + \mathbf{f}'(\mathbf{u})[\mathbf{f}(\mathbf{u})]_x = \mathbf{f}'(\mathbf{u})\psi(\mathbf{u}, x). \tag{26}$$

In the case of the urethra fluid flow modelling we add only one equation for the second component of the flux function i.e. $\phi = av^2 + \frac{a^2}{2\rho\beta}$ (the first component q is unknown function of the original balance law), which has the form

$$\phi_t + \left(-v^2 + \frac{a}{2\rho\beta}\right)(av)_x + 2v\phi_x - \frac{2av}{\rho}\left(\frac{a_0}{\beta}\right)_x - \frac{a^2 v}{\rho\beta^2}\beta_x = 0. \tag{27}$$

Finally augmented system can be written in the nonconservative form

$$\begin{bmatrix} a \\ q \\ \phi \\ \frac{a_0}{\beta} \\ \beta \end{bmatrix}_t + \begin{bmatrix} 0 & 1 & 0 & 0 & 0 \\ -\frac{q^2}{a^2} + \frac{a}{\rho\beta} & 2\frac{q}{a} & 0 & -\frac{a}{\rho} & -\frac{a^2}{\rho\beta^2} \\ 0 & -\frac{q^2}{a^2} + \frac{a}{\rho\beta} & 2\frac{q}{a} & 2\frac{q}{a} & -\frac{a q}{\rho} & -\frac{aq}{\rho\beta^2} \\ 0 & 0 & 0 & 0 & 0 \\ 0 & 0 & 0 & 0 & 0 \end{bmatrix} \begin{bmatrix} a \\ q \\ \phi \\ \frac{a_0}{\beta} \\ \beta \end{bmatrix}_x = \mathbf{0}, \tag{28}$$

briefly $\mathbf{w}_t + \mathbf{B}(\mathbf{w})\mathbf{w}_x = \mathbf{0}$, where matrix $\mathbf{B}(\mathbf{w})$ has following eigenvalues

$$\lambda^1 = v - \sqrt{\frac{a}{\rho\beta}}, \lambda^2 = v + \sqrt{\frac{a}{\rho\beta}}, \lambda^3 = 2v, \lambda^4 = \lambda^5 = 0 \tag{29}$$

and corresponding eigenvectors

$$\mathbf{r}^1 = \begin{bmatrix} 1 \\ \lambda^1 \\ (\lambda^1)^2 \\ 0 \\ 0 \end{bmatrix}, \mathbf{r}^2 = \begin{bmatrix} 1 \\ \lambda^2 \\ (\lambda^2)^2 \\ 0 \\ 0 \end{bmatrix}, \mathbf{r}^3 = \begin{bmatrix} 0 \\ 0 \\ 1 \\ 0 \\ 0 \end{bmatrix}, \mathbf{r}^4 = \begin{bmatrix} \frac{-a}{\rho \lambda^1 \lambda^2} \\ 0 \\ \frac{a}{\rho} \\ 1 \\ 0 \end{bmatrix}, \mathbf{r}^5 = \begin{bmatrix} \frac{-a^2}{\rho \beta^2 \lambda^1 \lambda^2} \\ 0 \\ \frac{a^2}{2\rho \beta^2} \\ 0 \\ 1 \end{bmatrix}. \tag{30}$$

We have five linearly independent eigenvectors. The approximation is chosen to be able to prove the consistency and provide the stability of the algorithm. In some special cases this scheme is conservative and we can guarantee the positive semidefiniteness, but only under the additional assumptions (see [2]).

The fluctuations are then defined by

$$\mathbf{A}^-(\mathbf{U}^-_{j+1/2}, \mathbf{U}^+_{j+1/2}) = \begin{bmatrix} 0 & 1 & 0 & 0 & 1 \\ 0 & 1 & 0 & 0 & 1 \end{bmatrix} \cdot \sum_{p=1, s^{p,n}_{j+1/2}<0}^{m} \gamma^p_{j+1/2} \mathbf{r}^p_{j+1/2},$$

$$\mathbf{A}^+(\mathbf{U}^-_{j+1/2}, \mathbf{U}^+_{j+1/2}) = \begin{bmatrix} 0 & 1 & 0 & 0 & 1 \\ 0 & 1 & 0 & 0 & 1 \end{bmatrix} \cdot \sum_{p=1, s^{p,n}_{j+1/2}>0}^{m} \gamma^p_{j+1/2} \mathbf{r}^p_{j+1/2}, \tag{31}$$

$$\mathbf{A}(\mathbf{U}^+_{j-1/2}, \mathbf{U}^-_{j+1/2}) = \mathbf{f}(\mathbf{U}^-_{j+1/2}) - \mathbf{f}(\mathbf{U}^+_{j-1/2}) - \mathbf{\Psi}(\mathbf{U}^-_{j+1/2}, \mathbf{U}^+_{j-1/2}),$$

where $\mathbf{\Psi}(\mathbf{U}^-_{j+1/2}, \mathbf{U}^+_{j-1/2})$ is a suitable approximation of the source term and $\mathbf{r}^p_{j+1/2}$ are suitable approximations of the eigenvectors (30).

5.3. Steady states

The steady state for the augmented system means $\mathbf{B}(\mathbf{w})\mathbf{w}_x = 0$, therefore $\mathbf{w}_x$ is a linear combination of the eigenvectors corresponding to the zero eigenvalues. The discrete form of the vector $\Delta \mathbf{w}$ corresponds to the certain approximation of these eigenvectors. It can be shown [2] that

$$\Delta \begin{bmatrix} A \\ Q \\ \Phi \\ \frac{a_0}{\beta} \\ \beta \end{bmatrix} = \begin{bmatrix} \frac{\bar{A}}{\rho} \frac{1}{\widetilde{\lambda^1 \lambda^2}} \\ 0 \\ \frac{\bar{A}}{\rho} \frac{\widetilde{\lambda^1 \lambda^2}}{\widetilde{\lambda^1 \lambda^2}} \\ 1 \\ 0 \end{bmatrix} \Delta \left(\frac{a_0}{\beta} \right) + \begin{bmatrix} \frac{\tilde{A}^2}{\rho \beta_{j+1} \beta_j} \frac{1}{\widetilde{\lambda^1 \lambda^2}} \\ 0 \\ \frac{\tilde{A}^2}{\rho \beta_{j+1} \beta_j} \frac{\widetilde{\lambda^1 \lambda^2}}{\widetilde{\lambda^1 \lambda^2}} - \frac{\tilde{A}^2}{2\rho \beta_{j+1} \beta_j} \\ 0 \\ 1 \end{bmatrix} \Delta \beta, \tag{32}$$

where $\bar{A} = \frac{A_j + A_{j+1}}{2}, \bar{\beta} = \frac{\beta_j + \beta_{j+1}}{2}, \tilde{A}^2 = \frac{A_j^2 + A_{j+1}^2}{2}, \tilde{V}^2 = |V_j V_{j+1}|, \bar{V}^2 = \left(\frac{V_j + V_{j+1}}{2} \right)^2$ and

$$\widetilde{\lambda^1 \lambda^2} = -\tilde{V}^2 + \frac{\bar{A}\bar{\beta}}{\rho \beta_{j+1} \beta_j}, \qquad \overline{\lambda^1 \lambda^2} = -\bar{V}^2 + \frac{\bar{A}\bar{\beta}}{\rho \beta_{j+1} \beta_j}. \tag{33}$$

Therefore we use vectors on the RHS of (32) as approximations of the fourth and fifth eigenvectors of the matrix $\mathbf{B}(\mathbf{w})$ to preserve general steady state.

5.4. Positive semidefiniteness

Positive semidefiniteness of this scheme is shown in [5] for the case of shallow water equation. It is based on a special choice of approximations of the eigenvectors (30). This, in the case of urethra flow, is more complicated because of the structure of the eigenvectors. Some necessary conditions for approximation of these eigenvectors are presented in [2].

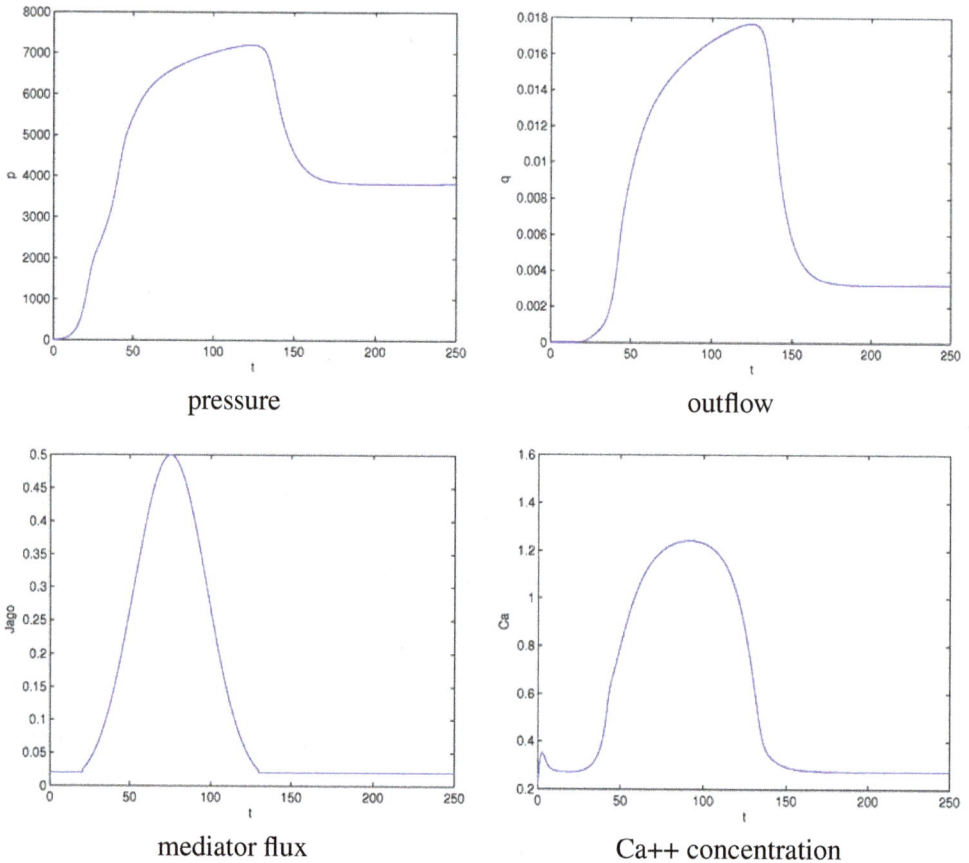

pressure

outflow

mediator flux

Ca++ concentration

Fig. 2. Time evolution of the quantities at the bladder neck

6. Numerical experiment

Now we present numerical experiment based on the system of differential equations described detrusor smooth muscle cell model (12 equations) and urethral flow (30 equations). The equations describing urethral flow are based on spatial high-resolution discretization of the urethra (15 finite volumes) described in section 5.2. The parameters used in this experiment are the same as in [12]. The Fig. 2 illustrate time evolution of the quantities at the bladder neck. For the further application it is necessary to fit the parameters because of non-dimensionality of the equations describing the muscle contraction.

1. For the simplicity the precious modelling of the synapse is neglected and the mediator flux $J_{agonist}$ is chosen – see Fig. 2. The IC units are used although in the medical paper are used for intravesical pressure cmH_2O ($1\,cmH_2O = 0.1$ kPa) and for the outflow ml/s. The concentration is measured in μM where $M = mol/l$.

2. At the Fig. 3 there are shown the cross-section area, velocity and flow rate along the whole urethra in two different times after beginning of voiding.

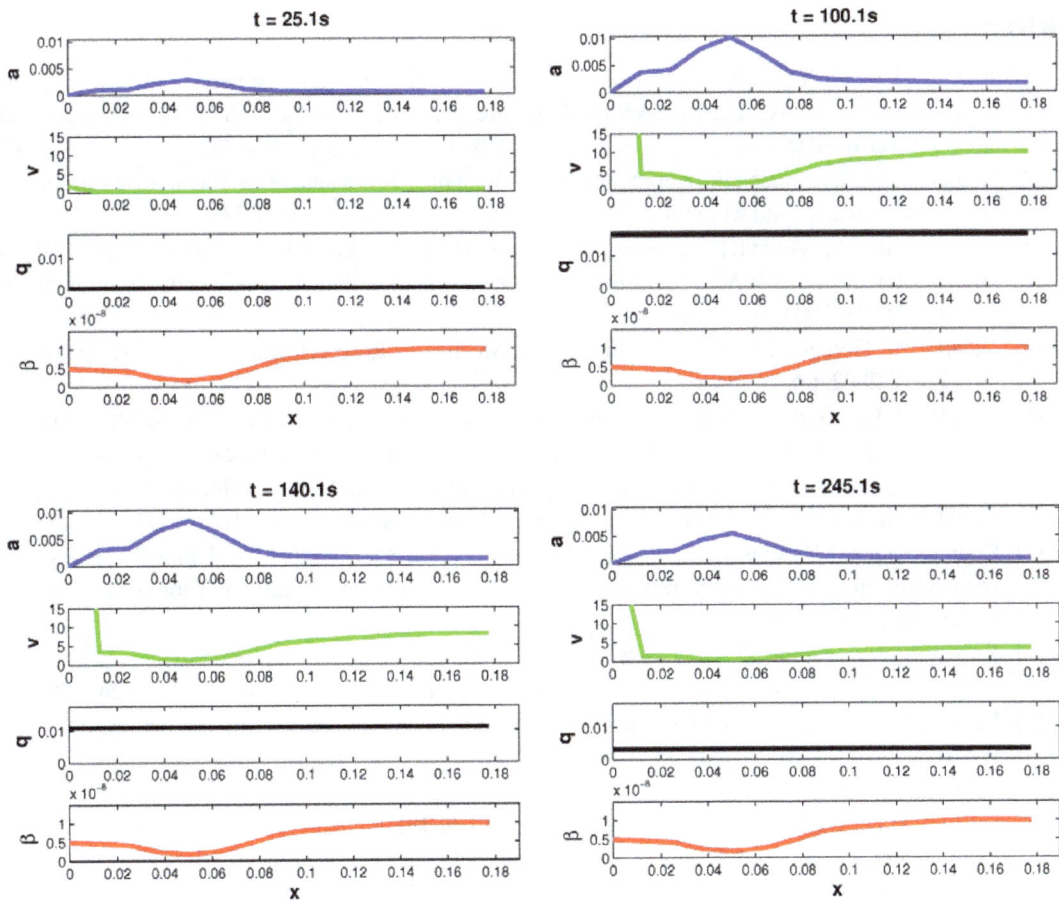

Fig. 3. Time evolution of the quantities through the urethra (cross section area, velocity, flow rate, tube compliance)

7. Conclusion

We presented the complex model of the lower part of the urinary tract. A simple bladder model and the detrusor contraction model were developed during voiding together with the detailed model of urethra flow. The urethra flow was described by the high-resolution positive semidefiniteness method, which preserves general steady states. For the practical application the identification of the parameters is necessary.

Acknowledgements

This work was supported by the European Regional Development Fund (ERDF), project "NTIS – New Technologies for Information Society", European Centre of Excellence, CZ.1.05/1.1.00/02.0090 and the CENTEM project, reg. no. CZ.1.05/2.1.00/03.0088.

References

[1] Arts, T., Bovendeerd, P. H. M., Prinzen, F. W., Reneman, R. S., Relation between left ventricular cavity pressure and volume and systolic fiber stress and strain in the wall, Biophysical Journal 59(1) (1991) 93–102.

[2] Brandner, M., Egermaier, J., Kopincová, H., Augmented Riemann solver for urethra flow modelling, Mathematics and Computers in Simulations 80(6) (2009) 1 222–1 231.

[3] Črnjarič-Zič, N., Vukovič, S., Sopta, L., Balanced finite volume WENO and central WENO schemes for the shallow water and the open-channel flow equations, Journal of Computational Physics 200(2) (2004) 512–548.

[4] Dicarlo, A., Quiligotti, S., Growth and balance, Mechanics Research Communications 29, Pergamon Press, 2002, pp. 449–456.

[5] George, D. L., Augmented Riemann solvers for the shallow water equations over variable topography with steady states and inundation, Journal of Computational Physics 227 (2008) 3 089–3 113.

[6] Hai, C. M., Murphy, R. A., Adenosine 5'-triphosphate consumption by smooth muscle as predicted by the coupled four-state crossbridge model, Biophysical Journal 61(2) (1992) 530–541.

[7] Koenigsberger, M., Sauser, R., Seppey, D., Beny, J.-L., Meister, J.-J., Calcium dynamics and vosomotion in arteries subject to isometric, isobaric and isotonic conditions, Biophysical Journal 95 (2008) 2 728–2 738.

[8] Kurganov, A., Tadmor, E., New high-resolution central schemes for nonlinear conservation laws and convection-diffusion equations, Journal of Computational Physics 160(1) (2000) 241–282.

[9] Laforet, J., Guiraud, D., Smooth muscle model for functional electric stimulation applications, Proceedings of the 29th Annual International Conference of the IEEE EMBS, Cite Internationale, Lyon, France, 2007.

[10] Parthimos, D., Edwards, D. H., Hill, C. E., Griffith, T. M., Dynamics of a three-variable nonlinear model of vasomotion: Comparison of theory and experiment, Biophysical Journal 93 (2007) 1 534–1 556.

[11] Pokrovski, V. N., Extended thermodynamics in a discrete/system approach, European Journal of Physics 26 (2005) 769–781.

[12] Rosenberg, J., Smooth muscle model applied to bladder, Proceeding of the 4th international conference Modelling of Mechanical and Mechatronic Systems – MMaMS 2011, Herlany, Slovakia, 2011.

[13] Rosenberg, J., Modelling of the voiding process, Proceedings of the 27th conference with international participation Computational Mechanics 2011, Plzeň, Czech Republic, 2011.

[14] Rosenberg, J., Hynčík, L., Contribution to the simulation of growth and remodelling applied to muscle fibre stimulation, Short communications of the 1st IMACS International Conference on Computational Biomechanice and Biology – ICCBB 2007, Plzeň, Czech Republic, 2007, pp. 1–4.

[15] Rosenberg, J., Hynčík, L., Modelling of the influence of the stiffness evolution on the behaviour of the muscle fibre, Proceedings of the conference Human Biomechanics 2008, Praha, Czech Republic, 2008.

[16] Rosenberg, J., Svobodová, M., Comments on the thermodynamical background to the growth and remodelling theory applied to the model of muscle fibre contraction, Applied and Computational Mechanics 4(1) (2010) 101–112.

[17] Stergiopulos, N., Tardy, Y., Meister, J.-J., Nonlinear separation of forward and backward running waves in elastic conduits, Journal of Biomechanics 26 (1993) 201–209.

3

Determination of principal residual stresses' directions by incremental strain method

A. Civín[a,*], M. Vlk[a]

[a] *Institute of Solid Mechanics, Mechatronics and Biomechanics, Brno University of Technology, Technická 2896/2, 616 69 Brno, Czech Republic*

Abstract

The ring-core method is the semi-destructive experimental method used for evaluation of the homogeneous and non-homogeneous residual stresses, acting over depth of drilled core. By using incremental strain method (ISM) for the residual state of stress determination, this article describes procedure how unknown directions and magnitudes of principal residual stresses can be determined. Finite element method (FEM) is used for the numerical simulation of homogenous residual state of stress and for subsequent strain determination. Relieved strains on the top of the model's core are measured by simulated three-element strain gauge, turned by the axis of strain gauge "a" from the direction of the principal stress σ_1 about unknown angle α. Depth dependent magnitudes of relieved strains, their differences and set of known values of calibration coefficients K_1 and K_2 or relaxation coefficients A and B are used together for determination of the angle α and for re-calculation of principal stresses.

Keywords: ring-core method, incremental strain method, residual stress, calibration coefficients, strain gauge

1. Introduction

The ring-core method (RCM) is a semi-destructive experimental method used for the evaluation of homogeneous and non-homogeneous residual stresses, acting over depth of drilled core. Therefore, the specimen is not totally destroyed during measurement and it could be used for further application in many cases.

One of the applicable theories, based on the procedure of evaluating magnitude of the residual stress, is called the incremental strain method (ISM). It is still used quite often, despite its numerous theoretical shortcomings. On the one hand, ISM assumes that the measured deformations $d\varepsilon_a$, $d\varepsilon_b$ and $d\varepsilon_c$ are functions only of the residual stresses acting in the current depth "z" of the drilled hole and they do not depend on the previous increments "dz" including another residual stresses, see Fig. 1. On the other hand, relieved strains do not depend only on the stress acting within the drilled layer, but also on the geometric changes of the ring groove during deepening. Consequently, strain relaxations are still continuing and grooving with drilled depth, even though the next step's increment is stress free. Therefore, the proposed theory purveys only approximate information about the real state of stress and RCM method is not suitable for the types of measurements with a steep gradient of residual state of stress.

By using incremental strain method for the residual state of stress determination and FEM, this article describes procedure how directions of the principal residual stresses can be determined. Finite element method is used for the numerical simulation of homogenous residual state of stress and relieved strains on the top of the model's core are measured by simulated

*Corresponding author. e-mail: civin.adam@seznam.cz.

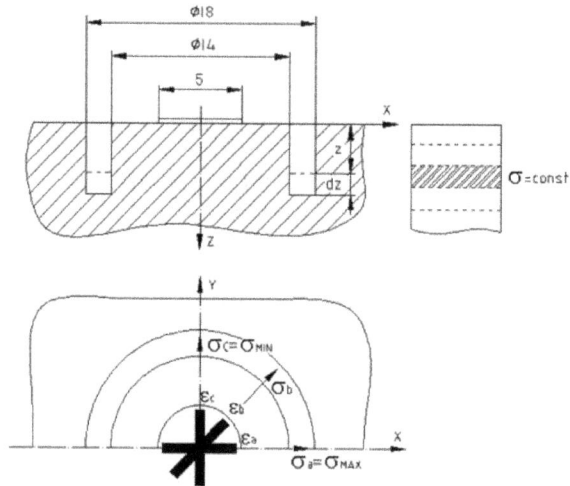

Fig. 1. Principle of ISM with known directions of principal stresses ($\alpha = 0°$)

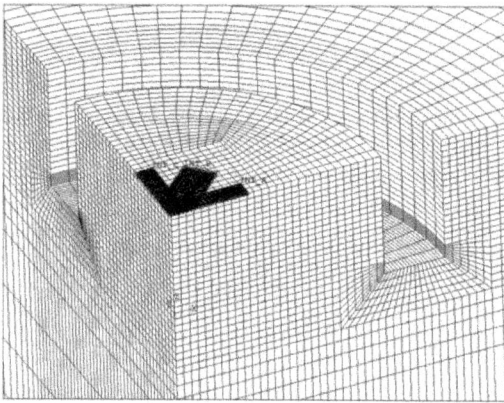

Fig. 2. Strain gauges "a, b, c" placed in principal stress directions

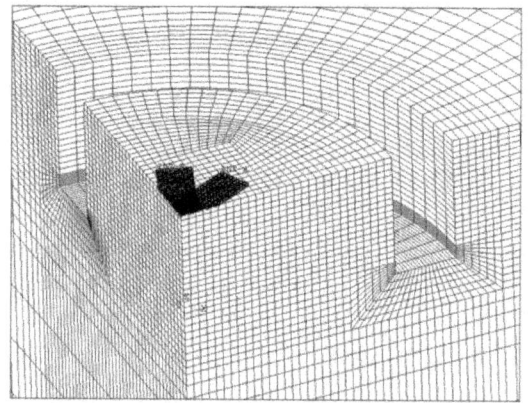

Fig. 3. Strain gauges "a, b, c" placed in general direction

three-element strain gauge rosette, turned by the axis of the strain gauge "a" from the direction of the principal stress σ_1 about unknown angle α (Fig. 3).

2. Problem description

Like the integral method, the incremental strain method requires a set of depth-dependent coefficients, which are necessary for further residual stress determination, carried out by the ring-core method in this case. Values of calibration coefficients K_1 and K_2 have been already determined by the simulation under various types of uniaxial and biaxial state of stress conditions and published in articles [2, 3] and [1, 7]. Their dependence on the depth of drilled hole and on the disposition of the homogenous residual state of stress as well as geometry changes of the annular groove and finite element model's dimensions have been considered too. Another way, how to determine residual state of stress between two specifics depths of drilled groove, is possible by calculation of relaxation coefficients A and B.

    This paper deals with results obtained by the FE-measurement of relieved strains ε_a, ε_b and ε_c, by generally placed strain gauge rosette on the top of the core, where their differences and set of known values of calibration coefficients K_1 and K_2 or relaxation coefficients A and B

is used for the proper determination of the principal stresses σ_1 and σ_2. Placing of the three-element strain gauge rosette on the top of the ring-core is shown in case of known direction of principal residual stress (Fig. 2) and in case of general direction (Fig. 3).

3. Basic equations

Like each method, incremental strain method has its own theoretical background to define certain relations between known and unknown parameters. Residual state of stress can be determined either by differentials or differences of relieved strains.

3.1. Using calibration coefficients K_1 and K_2

Equations (2)–(4) describe strain differentials, used to express determination of the principal stress σ_1 and σ_2 by the known set of calibration coefficients K_1, K_2, calculated from principal strains ε_1 and ε_2 on the top surface of the core, where the three-element ring-core rosette is placed ([1–3] and [5]).

Relieved general strains ε_a, ε_b and ε_c are measured every i-th step of drilled depth z_i and size of step's difference Δz is always referred to the previous step's size (z_{i-1}). Magnitude of each step used in FEM simulation (1) is $\Delta z = \text{const.} = 0.2$ mm

$$\Delta z = z_i - z_{i-1} = 0.2 \text{ mm, for } i = 1 \div 40. \tag{1}$$

With known magnitude of the calibration coefficient K_1, K_2 (Fig. 4 and 5, Table 1) and relevant derivation of principal strains $d\varepsilon_1/dz$ and $d\varepsilon_2/dz$ in dependence on the specific magnitude of step's increment dz could by principal stresses of *homogenous* residual state of stress obtained by following equations:

$$\sigma_1 = \frac{E}{K_1^2 - \mu^2 K_2^2} \cdot \left(K_1 \frac{d\varepsilon_1}{dz} + \mu K_2 \frac{d\varepsilon_2}{dz} \right), \tag{2}$$

$$\sigma_2 = \frac{E}{K_1^2 - \mu^2 K_2^2} \cdot \left(K_1 \frac{d\varepsilon_2}{dz} + \mu K_2 \frac{d\varepsilon_1}{dz} \right), \tag{3}$$

$$\frac{d\varepsilon_1}{dz} = \varepsilon_1', \quad \frac{d\varepsilon_2}{dz} = \varepsilon_2', \tag{4}$$

where E is Young's modulus, μ is Poisson's ratio and ε_1', ε_2' are numerical derivations of relieved strains.

Attention should be paid to formulations suggested in (2), (3). If the denominator $K_1^2 - \mu^2 K_2^2$ becomes zero for certain values of K_1 and K_2, the stress will become infinite, i.e. for steel material with $0.3 \cong \mu = K_1/K_2$. Further, expressions of (2)–(4) could be modified into equations used for determination of calibration coefficients under uniaxial and biaxial state of stress conditions.

In case of the uniaxial state of stress, $(\sigma_1 \neq 0, \sigma_2 = 0)$, equations for the calibration coefficients K_1, K_2 are described by:

$$K_1 = \frac{E}{\sigma_1} \cdot \varepsilon_1', \qquad K_2 = -\frac{E}{\mu \sigma_1} \cdot \varepsilon_2'. \tag{5}$$

In case of the biaxial state of stress $(\sigma_1 \neq 0, \sigma_2 \neq 0)$, equations for calibration coefficients K_1 and K_2 are described by:

$$K_1 = \frac{E}{\sigma_1(1-\kappa^2)} \cdot (\varepsilon_1' - \kappa \cdot \varepsilon_2'), \tag{6}$$

$$K_2 = \frac{E}{\mu\sigma_1(1-\kappa^2)} \cdot (\kappa \cdot \varepsilon_1' - \varepsilon_2'), \tag{7}$$

$$\kappa = \frac{\sigma_2}{\sigma_1}. \tag{8}$$

Formulations suggested by (6) and (7) have a problem with the denominator too. If $\sigma_1 = \sigma_2$ or $\sigma_1 = -\sigma_2$, then $(1 - \kappa^2)$ becomes zero and magnitude of calibration coefficient will become infinite.

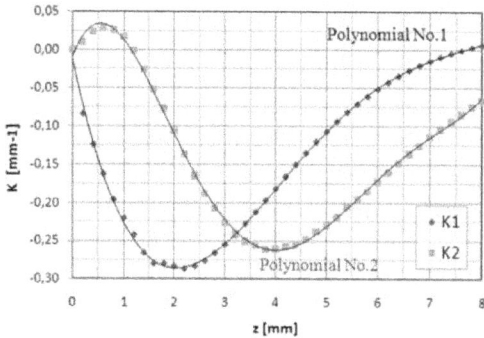

Fig. 4. Calibration coefficients determined under uniaxial residual state of stress simulation

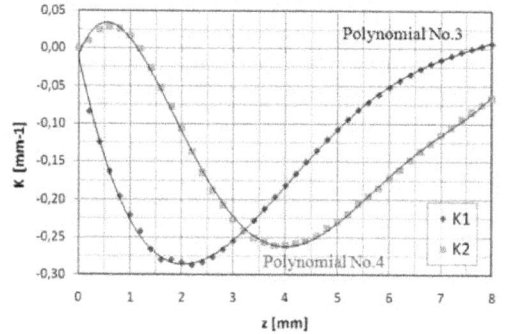

Fig. 5. Calibration coefficients determined under biaxial residual state of stress simulation

Calculated points of calibration coefficients K_1, K_2 in dependence on the drilled depth of the ring-groove are plotted in Figs. 4 and 5. Appropriate polynomial functions of the sixth degree with constants reproduced by (9) are written in Table 1.

Table 1. Coefficients of polynomial functions

Polyno-	Coefficients [–]						
mial No.:	a_0	a_1	a_2	a_3	a_4	a_5	a_6
1	−0,010 670 4	−0,314 649 7	0,111 879 8	−0,011 659 6	−0,000 084 8	0,000 075 7	−0,000 003 0
2	−0,010 055 6	0,173 809 1	−0,205 010 2	0,057 055 6	−0,005 849 3	0,000 150 4	0,000 005 7
3	−0,010 676 9	−0,314 756 6	0,112 016 7	−0,011 715 9	−0,000 074 6	0,000 074 8	−0,000 003 0
4	−0,010 116 6	0,173 977 5	−0,204 814 0	0,056 843 4	−0,005 784 4	0,000 142 2	0,000 006 1

Entire hole was made by 40 increments of step's size $\Delta z = 0.2$ mm. In Figs. 4 and 5 is obvious, that behavior of K_1 and K_2 polynomial functions still remains the same for various magnitudes of simulated uniaxial and biaxial states of stress [3]. Therefore, no change in the numerical evaluation of calibration coefficients K_1 and K_2 is observed, because modification of homogenous state of stress have no influence on calibration coefficients determination. For this reason, only one universal set of calibration coefficients K_1 and K_2 is applicable.

$$K_i = a_0 + a_1 z^1 + a_2 z^2 + a_3 z^3 + a_4 z^4 + a_5 z^5 + a_6 z^6. \tag{9}$$

Polynomial constants in Table 1 prove the fact that the functions of calibration coefficients K_1 and K_2 are the same for various types of simulated homogenous residual states of stress, i.e. only one universal set of calibration coefficients K_1 and K_2 is applicable.

Dependence of the calibration coefficient K_2 on type of the residual state of stress was published by Hwang [4], but this contention was disproved.

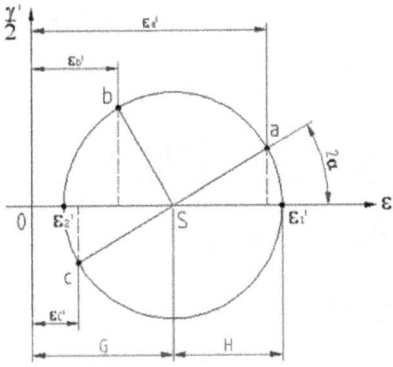

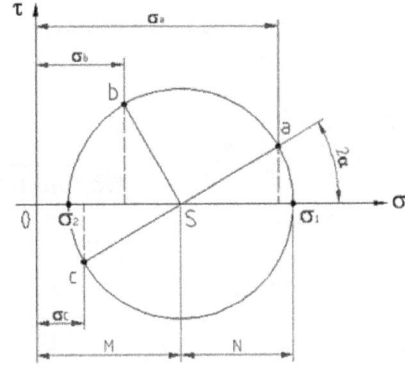

Fig. 6. Modified Mohr's circle for strain   Fig. 7. Mohr's circle for stress

3.2. Determination of principal stresses with unknown principal directions

Relationship between principal strains $\varepsilon'_1, \varepsilon'_2$ and general strains $\varepsilon'_a, \varepsilon'_b, \varepsilon'_c$ measured in unknown angle α between direction of principal stress σ_1 and axis of the strain gauge's measuring grid:

$$\varepsilon'_a = G + H \cdot \cos 2\alpha = \frac{\varepsilon'_1 + \varepsilon'_2}{2} + \frac{\varepsilon'_1 - \varepsilon'_2}{2} \cdot \cos 2\alpha, \tag{10}$$

$$\varepsilon'_b = G + H \cdot \cos(2\alpha + 90°) = \frac{\varepsilon'_1 + \varepsilon'_2}{2} - \frac{\varepsilon'_1 - \varepsilon'_2}{2} \cdot \sin 2\alpha, \tag{11}$$

$$\varepsilon'_c = G + H \cdot \cos(2\alpha + 180°) = \frac{\varepsilon'_1 + \varepsilon'_2}{2} - \frac{\varepsilon'_1 - \varepsilon'_2}{2} \cdot \cos 2\alpha. \tag{12}$$

According to modified Mohr's circle in Fig. 6, relationship between principal strains $\varepsilon'_1, \varepsilon'_2$ and generally measured relieved strains $\varepsilon'_a, \varepsilon'_b, \varepsilon'_c$ is:

$$G = \frac{\varepsilon'_1 + \varepsilon'_2}{2} = \frac{\varepsilon'_a + \varepsilon'_c}{2}, \tag{13}$$

$$H = \frac{\varepsilon'_1 - \varepsilon'_2}{2} = \frac{1}{2}\sqrt{(\varepsilon'_a - \varepsilon'_c)^2 + (\varepsilon'_a + \varepsilon'_c - \varepsilon'b)^2}. \tag{14}$$

Angle between direction of principal residual stress σ_1 and axis of strain gauge's measuring grid "a":

$$\tan 2\alpha = \frac{\varepsilon'_b - G}{G - \varepsilon'_a} = \frac{2\varepsilon'_b - \varepsilon'_a - \varepsilon'_c}{\varepsilon'_c - \varepsilon'_a} \rightarrow \alpha = \arctan\left(\frac{2\varepsilon'_b - \varepsilon'_a - \varepsilon'_c}{\varepsilon'_c - \varepsilon'_a}\right). \tag{15}$$

Table 2. Specified quadrants

Numerator: $2\varepsilon'_b - \varepsilon'_a - \varepsilon'_c$	Denominator: $\varepsilon'_c - \varepsilon'_a$	2α [°]
$+$	$+$	$0 \div 90$
$+$	$-$	$90 \div 180$
$-$	$-$	$180 \div 270$
$-$	$+$	$270 \div 360$

Similarly, derivation of strains ε_1', ε_2' and ε_a', ε_b', ε_c' could be used in (10)–(15) instead of ε_1, ε_2 and ε_a, ε_b, ε_c strains in order to evaluate stress σ_a, σ_b and σ_c in direction of strain gauge's axis:

$$\sigma_a = \frac{E}{K_1^2 - \mu^2 K_2^2} \cdot (K_1 \varepsilon_a' + \mu K_2 \varepsilon_c'), \tag{16}$$

$$\sigma_b = \frac{E}{K_1^2 - \mu^2 K_2^2} \cdot [K_1 \varepsilon_b' + \mu K_2 (\varepsilon_a' - \varepsilon_b' + \varepsilon_c')], \tag{17}$$

$$\sigma_c = \frac{E}{K_1^2 - \mu^2 K_2^2} \cdot (K_1 \varepsilon_c' + \mu K_2 \varepsilon_a'). \tag{18}$$

According to Mohr's circle in Fig. 7, principal stress σ_1 and σ_2 could be recalculated by using known magnitudes of non-principal stresses σ_a, σ_b, σ_c measured by generally turned strain gauge rosette:

$$M = \frac{\sigma_a + \sigma_c}{2}, \qquad N = \frac{1}{2}\sqrt{(\sigma_a - \sigma_c)^2 + (\sigma_a + \sigma_c - 2\sigma_b)^2}, \tag{19}$$

$$\sigma_1 = M + N, \qquad \sigma_2 = M - N. \tag{20}$$

3.3. Using relaxation coefficients A and B

Magnitude of principal residual stresses, acting within two drilled depths, can be determined by using relaxation coefficients too. Therefore, relieved strains are measured only at two different depths and step's difference Δz consist of two particular depths z_i and $2z_i$, described by

$$\Delta z = 2z_i - z_i = z_i, \text{ for } z_i = 1, 2, 3, 4 \text{ [mm]}. \tag{21}$$

Assuming that $\frac{d\varepsilon_i}{dz} \approx \frac{\Delta\varepsilon_i}{\Delta z}$, equations of principal strains (2) and (3) can be rewritten:

$$\sigma_1 = \frac{E}{K_1^2 - \mu^2 K_2^2} \cdot \frac{1}{\Delta z} \cdot (K_1 \Delta\varepsilon_1 + \mu K_2 \Delta\varepsilon_2), \tag{22}$$

$$\sigma_2 = \frac{E}{K_1^2 - \mu^2 K_2^2} \cdot \frac{1}{\Delta z} \cdot (K_1 \Delta\varepsilon_2 + \mu K_2 \Delta\varepsilon_1), \tag{23}$$

$$\Delta\varepsilon_1 = (\varepsilon_1)_{2z_i} - (\varepsilon_1)_{z_i}, \qquad \Delta\varepsilon_2 = (\varepsilon_2)_{2z_i} - (\varepsilon_2)_{z_i}. \tag{24}$$

Confrontation of the calibration coefficients K_1, K_2 and relaxation coefficients A, B:

$$A = \frac{E \cdot K_1}{K_1^2 - \mu^2 K_2^2} \cdot \frac{1}{\Delta z}, \qquad B = \frac{E \cdot K_2}{K_1^2 - \mu^2 K_2^2} \cdot \frac{1}{\Delta z}. \tag{25}$$

If $\bar{\varepsilon}_1 = \frac{\sigma_1}{E}$ and $\Delta\varepsilon_1^* = \frac{\Delta\varepsilon_1}{\bar{\varepsilon}_1}$; $\Delta\varepsilon_2^* = \frac{\Delta\varepsilon_2}{\mu \cdot \bar{\varepsilon}_1}$ then relaxation coefficients A, B are determined:

$$A = \frac{E\frac{\Delta\varepsilon_1^*}{\Delta z}}{\frac{1}{(\Delta z)^2}[(\Delta\varepsilon_1^*)^2 - (\mu\Delta\varepsilon_2^*)^2] \cdot \Delta z} = \frac{E \cdot \Delta\varepsilon_1^*}{(\Delta\varepsilon_1^*)^2 - (\mu\Delta\varepsilon_2^*)^2}, \tag{26}$$

$$B = -\frac{E \cdot \mu \cdot \frac{\Delta\varepsilon_2^*}{\Delta z}}{\frac{1}{(\Delta z)^2}[(\Delta\varepsilon_1^*)^2 - (\mu\Delta\varepsilon_2^*)^2] \cdot \Delta z} = -\frac{E \cdot \mu \cdot \Delta\varepsilon_2^*}{(\Delta\varepsilon_1^*)^2 - (\mu\Delta\varepsilon_2^*)^2}. \tag{27}$$

Finally, equations for residual stress determination, which are based on differences of relieved strains and relaxation coefficients A, B are:

$$\sigma_1 = A \cdot \Delta\varepsilon_1 - B \cdot \Delta\varepsilon_2, \qquad \sigma_2 = A \cdot \Delta\varepsilon_2 - B \cdot \Delta\varepsilon_1. \tag{28}$$

4. FEM simulation

A prerequisite for correct and accurate measurement of relieved strains on the top of the core is to use FEM simulation. It is the only reasonable way to obtain desired information or simulate real experiment. The ANSYS analysis system is used for the FE-simulation.

FE-analysis is based on a specimen volume with dimensions of $a \times a = 50 \times 50$ mm and thickness of $t = 50$ mm. Due to symmetry, only a quarter of the model has been modeled with centre of the core on the surface as the origin. The shape of the model is simply represented by a block with planar faces, with a quarter of the annular groove drilled away (Figs. 8 and 9). The annular groove has been made by $n = 40$ increments with the step size of $\Delta z = 0.2$ mm in case of approach described by using calibration coefficients K_1 and K_2. The maximum depth of drilled groove is $z = 8$ mm. Dimension of outer diameter is $D = 2r_i = 18$ mm and groove width is $h = 2$ mm.

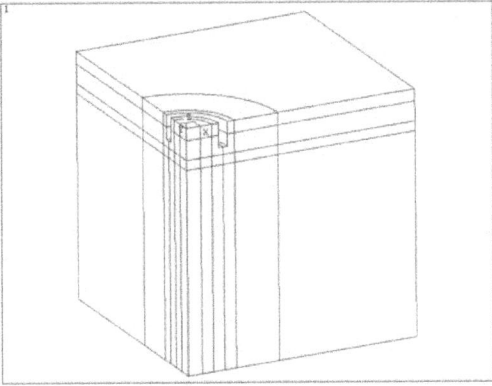

Fig. 8. Quarter of global solid model

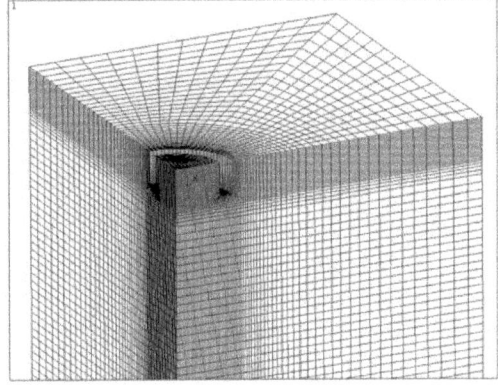

Fig. 9. Finite element model

Linear, elastic and isotropic material model is used with material properties of Young's modulus $E = 210$ GPa and Poisson's ratio $\mu \cong 0.3$. Relaxed strains ε_1, ε_2 and ε_3 have been measured at real positions of strain gauge rosettes' measuring grids by integration across its surface. Type of considered strain gauge rosette is FR-5-11-3LT, with length and width of each measuring grid 5 mm and 1.9 mm, respectively [6].

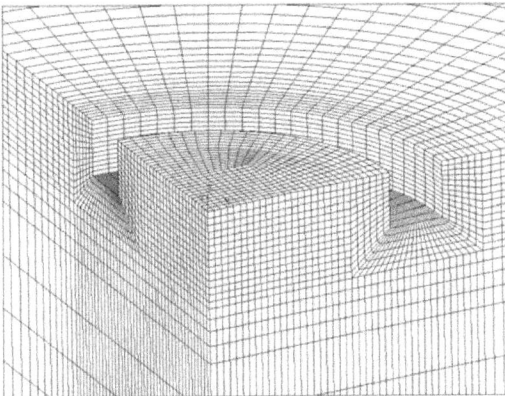

Fig. 10. Depth of drilled groove for $z = 2$ mm

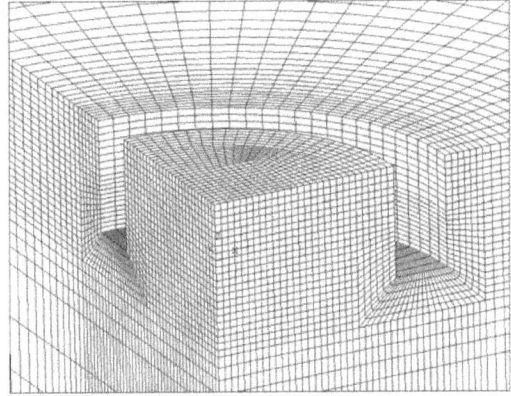

Fig. 11. Depth of drilled groove for $z = 4$ mm

5. Results

5.1. *Using calibration coefficients K_1 and K_2*

Released strains on the top of the core are obtained by the FE-analysis (Fig. 12). Application of the general-purposed finite element model in order to simulate homogenous uniaxial state of stress with magnitude of principal stress $\sigma_1 = 60$ MPa, $\sigma_2 = 0$ MPa has been used to verify basic equations (10)–(20) and theoretical approach proposed by ISM.

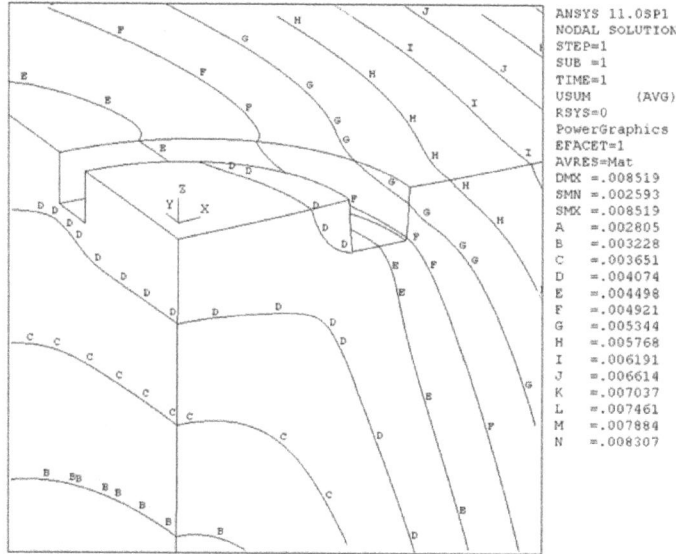

```
ANSYS 11.0SP1
NODAL SOLUTION
STEP=1
SUB =1
TIME=1
USUM        (AVG)
RSYS=0
PowerGraphics
EFACET=1
AVRES=Mat
DMX =.008519
SMN =.002593
SMX =.008519
A   =.002805
B   =.003228
C   =.003651
D   =.004074
E   =.004498
F   =.004921
G   =.005344
H   =.005768
I   =.006191
J   =.006614
K   =.007037
L   =.007461
M   =.007884
N   =.008307
```

Fig. 12. Plot of total displacement [mm] — uniaxial state of stress, depth of drilled groove $z = 2$ mm

Graphs of relaxed strains calculated by integration across strain gauge's measuring grid [3] are plotted in Fig. 13 and their numerical derivations are plotted in Fig. 14. Axis of strain gauge's measuring gird "a" was for this simulation turned from the direction of principal stress σ_1 about angle $\alpha = 30°$.

Non-principal residual stresses σ_a, σ_b, σ_c acting in axis direction of turned strain gauge rosette's measuring grids "a, b, c", are calculated by (16)–(18) and plotted in Fig. 15 in dependence on the depth of drilled hole. The set of calibration coefficients K_1 and K_2 (Figs. 4 and 5), determined under uniaxial or biaxial state of stress conditions, needs to be used for this reason.

Fig. 16 shows angle α between direction of principal residual stress σ_1 and axis of strain gauge's measuring grid "a" determined in each drilled depth by (15). Table 2 gives an advice how to consider signs of numerator and denominator of (15) in order to determine correct quadrant of strain gauge grid's position.

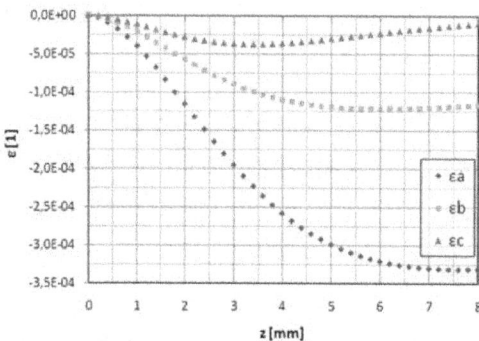

Fig. 13. Measured strains on the top of the core

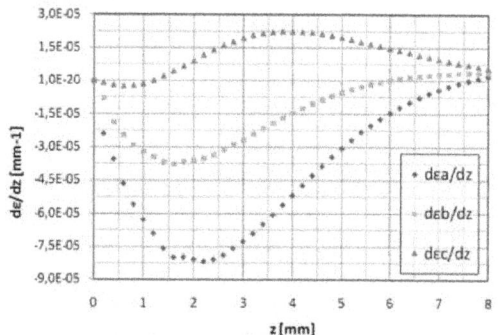

Fig. 14. Derivations of relieved strains

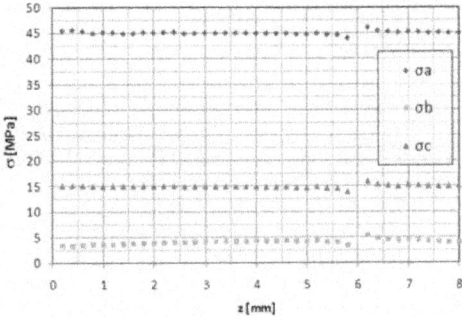

Fig. 15. Measured stress in general direction of strain gauges measuring grid's axis

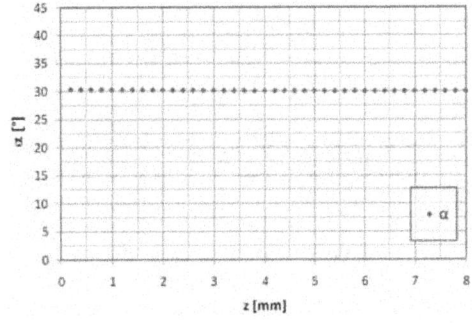

Fig. 16. Determined angle α

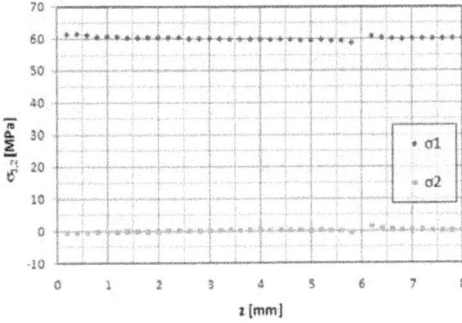

Fig. 17. Re-calculated principal stresses

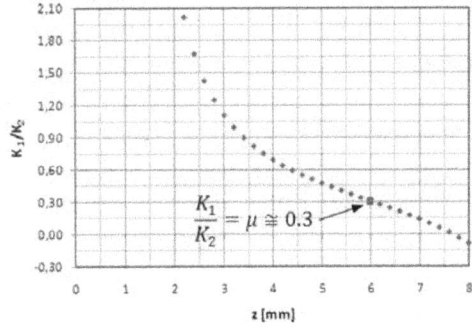

Fig. 18. Ratio of calibration coefficients

Re-calculated magnitudes of principal residual stresses σ_1, σ_2 by (19)–(20) are plotted in Fig. 17 and their magnitudes correctly correspond with simulated homogenous state of stress with principal stresses $\sigma_1 = 60$ MPa, $\sigma_2 = 0$ MPa.

Shortcoming of ISM is obvious in Figs. 15, 17 and 18 where values of results are missing in depth of drilled hole $z = 6$ mm. This problem is caused by denominator $K_1^2 - \mu^2 K_2^2$ in all equations where it appears. Only one case is possible when denominator $K_1^2 - \mu^2 K_2^2$ becomes zero for certain values of K_1 and K_2, and this condition is met in case of Poisson's ration $K_1/K_2 = \mu \cong 0.3$ (steel material) exactly in depth of $z = 6$ mm (Fig. 18). For this reason, magnitude of stress is non-numerable in this depth.

5.2. Using relaxation coefficients A and B

Magnitudes of residual stresses, acting between two specific depths z_i and $2z_i$ (Figs. 10 and 11) of drilled groove can be determined by the method using differences $\Delta\varepsilon/\Delta z$ too (21). Values of general strains, used for determination of relaxation coefficients A, B by simulation of homogenous uniaxial stress state ($\sigma_1 = 60$ MPa, $\sigma_2 = 0$ MPa), are measured across strain gauge's measuring grid.

Unknown angle α can be determined for set of strains $\varepsilon_a, \varepsilon_b$ and ε_c in each depth z_i by (15). Principal strains ε_1 and ε_2 can be re-calculated by (10)–(12). After that, calibration coefficients A and B can be determined by (26) and (27), using normalized strains $\Delta\varepsilon_1^*, \Delta\varepsilon_2^*$ of differentials $\Delta\varepsilon_1, \Delta\varepsilon_2$ (24). All necessary constants are written in Table 3 for specific variations of drilled depths.

Incontestable advantage of residual stress determination by relaxation functions A and B is independency on determination of depth-dependent calibration coefficients like K_1 and K_2, which are possible to obtain, either by FEM simulation or experimental measurement.

Table 3. Residual stress determination by relaxation constants

z_i [mm]	$\Delta\varepsilon_1$ [1]	$\Delta\varepsilon_2$ [1]	$\Delta\varepsilon_1^*$ [1]	$\Delta\varepsilon_2^*$ [1]	A [MPa]	B [MPa]	σ [MPa]	α [°]
1 / 2	3.671E–06	–7.599E–05	1.285E–02	–8.866E–01	**–3.823E+05**	**–7.914E+05**	60.00	30
2 / 4	3.617E–05	3.753E–04	1.266E–01	4.375E–00	**–1.558E+04**	**1.615E+05**	60.00	30
3 / 6	5.940E–05	–1.248E–04	2.079E–01	–1.456E–00	**–2.958E+05**	**–6.216E+05**	60.00	30
4 / 8	5.791E–05	–7.200E–05	2.027E–01	–8.400E–01	**–1.898E+06**	**–2.360E+06**	60.00	30

6. Conclusions

This paper provided basic information about semi-destructive ring-core method. By using incremental strain method for residual state of stress determination by the finite element method, this article gives additional information about homogenous residual stress measurement. By using slightly turned strain gauge rosettes' measuring grids from the directions of acting principal stresses about general angle α, magnitudes and directions of principal stresses need to be re-calculated.

Theoretical background described by basic differential or difference equations and application of universal set of the depth-dependent calibration coefficients K_1, K_2 or relaxation functions A, B in order to determine principal residual stresses and their orientation, has been presented.

One of the shortcomings of the ISM, such as impossibility of stress measurement in specified depth in dependence on the Poisson's ratio, has been clarified. Another shortcoming of this method is inaccurate non-homogenous stress evaluation and measuring of more than full released strains in depth greater than $z = 5$ mm [2, 3]. Where the steep gradients of residual state of stress are occurred, measurement is not suitable in this case too.

Incremental strain method had been used frequently until the integral method has overcome its shortcomings. By concentrating the research on the observed weaknesses and the ambiguous details the ring-core method can be made an accurate and reliable method for residual stress measurement.

Acknowledgements

This work has been supported by the specific research FSI-S-11-11.

References

[1] Bohdan, P., Holý, S., Jankovec, J., Jaroš, P., Václavík, J., Weinberg, O., Residual Stress Measurement Using Ring-Core Method. In Experimental Stress Analysis 2008, Horní Bečva, 2008.

[2] Civín, A., Vlk, M., Analysis of Calibration Coefficients for Incremental Strain Method Used for Residual Stress Measurement by Ring-Core Method. In Applied Mechanics 2010, Jablonec nad Nisou, 2010, pp. 25–28.

[3] Civín, A., Vlk M., Assessment of Incremental Strain Method Used for Residual Stress Measurement by Ring-Core Method. In Experimental Stress Analysis 2010, Velké Losiny, 2010, pp. 27–34.

[4] Hwang, B–W., Suh, C–M., Kim, S–H., Finite element analysis of calibration factors for the modified incremental strain method, J. Strain Analysis, Vol. 38, 2003, No. 1, pp. 45–51.

[5] Keil, S., On-line evaluation of measurement results during the determination of residual stress using strain gages, HBM — Reports in Applied Measurement, Vol. 9, 1995, No. 1, pp. 15–20.

[6] Preusser Messtechnik GmbH [online] URL: <http://www.dms-technik.de/> [cit. 2010–10–28]

[7] Václavík, J., Weinberg, O., Bohdan, P., Jankovec, J., Holý, S., Evaluation of Residual Stress using Ring Core Method. In ICEM 14 — 14th International Conference on Experimental Mechanics, France, 2010.

4

Flexural analysis of deep beam subjected to parabolic load using refined shear deformation theory

Y. M. Ghugal[a,*], A. G. Dahake[b]

[a]Applied Mechanics Department, Govt. College of Engineering, Karad – 415 124, MS, India

[b]Applied Mechanics Department, Govt. College of Engineering, Aurangabad – 431 005, MS, India

Abstract

A trigonometric shear deformation theory for flexure of thick or deep beams, taking into account transverse shear deformation effects, is developed. The number of variables in the present theory is same as that in the first order shear deformation theory. The sinusoidal function is used in displacement field in terms of thickness coordinate to represent the shear deformation effects. The noteworthy feature of this theory is that the transverse shear stresses can be obtained directly from the use of constitutive relations with excellent accuracy, satisfying the shear stress free conditions on the top and bottom surfaces of the beam. Hence, the theory obviates the need of shear correction factor. Governing differential equations and boundary conditions are obtained by using the principle of virtual work. The thick isotropic beams are considered for the numerical studies to demonstrate the efficiency of the theory. It has been shown that the theory is capable of predicting the local effect of stress concentration due to fixity of support. The fixed isotropic beams subjected to parabolic loads are examined using the present theory. Results obtained are discussed critically with those of other theories.

Keywords: thick beam, trigonometric shear deformation, principle of virtual work, equilibrium equations, displacement, stress

1. Introduction

It is well-known that elementary theory of bending of beam based on Euler-Bernoulli hypothesis disregards the effects of the shear deformation and stress concentration. The theory is suitable for slender beams and is not suitable for thick or deep beams since it is based on the assumption that the sections normal to neutral axis before bending remain so during bending and after bending, implying that the transverse shear strain is zero. Since theory neglects the transverse shear deformation, it underestimates deflections in case of thick beams where shear deformation effects are significant.

Bresse [5], Rayleigh [16] and Timoshenko [20] were the pioneer investigators to include refined effects such as rotatory inertia and shear deformation in the beam theory. Timoshenko showed that the effect of transverse shear is much greater than that of rotatory inertia on the response of transverse vibration of prismatic bars. This theory is now widely referred to as Timoshenko beam theory or first order shear deformation theory (FSDT) in the literature. In this theory transverse shear strain distribution is assumed to be constant through the beam thickness and thus requires shear correction factor to appropriately represent the strain energy of deformation. Cowper [6] has given refined expression for the shear correction factor for different

---

*Corresponding author. e-mail: ghugal@rediffmail.com.

cross-sections of beam. The accuracy of Timoshenko beam theory for transverse vibrations of simply supported beam in respect of the fundamental frequency is verified by Cowper [7] with a plane stress exact elasticity solution. To remove the discrepancies in classical and first order shear deformation theories, higher order or refined shear deformation theories were developed and are available in the open literature for static and vibration analysis of beam.

Levinson [15], Bickford [4], Rehfield and Murty [18], Krishna Murty [14], Baluch et al. [2], Bhimaraddi and Chandrashekhara [3] presented parabolic shear deformation theories assuming a higher variation of axial displacement in terms of thickness coordinate. These theories satisfy shear stress free boundary conditions on top and bottom surfaces of beam and thus obviate the need of shear correction factor. Irretier [12] studied the refined dynamical effects in linear, homogenous beam according to theories, which exceed the limits of the Euler-Bernoulli beam theory. These effects are rotary inertia, shear deformation, axial pre-stress, twist and coupling between bending and torsion.

Hilderbrand and Reissner [11] have given the distribution of stress in built-in beam of narrow rectangular cross section using Airy's stress function and the principle of least work. Timoshenko and Goodier [21] presented the elasticity solutions for simply supported and cantilever beams using Airy's stress polynomial functions and using stress functions in the form of a Fourier series.

Kant and Gupta [13], Heyliger and Reddy [10] presented finite element models based on higher order shear deformation uniform rectangular beams. However, these displacement based finite element models are not free from phenomenon of shear locking (Averill and Reddy [1]; Reddy [17]).

There is another class of refined theories, which includes trigonometric functions to represent the shear deformation effects through the thickness. Vlasov and Leont'ev [22], Stein [19] developed refined shear deformation theories for thick beams including sinusoidal function in terms of thickness coordinate in displacement field. However, with these theories shear stress free boundary conditions are not satisfied at top and bottom surfaces of the beam. A study of literature by Ghugal and Shimpi [8] indicates that the research work dealing with flexural analysis of thick beams using refined trigonometric and hyperbolic shear deformation theories is very scarce and is still in infancy.

In this paper development of theory and its application to thick fixed beams is presented.

2. Development of theory

The beam under consideration as shown in Fig. 1 occupies in $0 - x - y - z$ Cartesian coordinate system the region:

$$0 \leq x \leq L, \qquad 0 \leq y \leq b, \qquad -\frac{h}{2} \leq z \leq \frac{h}{2},$$

where x, y, z are Cartesian coordinates, L and b are the length and width of beam in the x and y directions respectively, and h is the thickness of the beam in the z-direction. The beam is made up of homogeneous, linearly elastic isotropic material.

2.1. The displacement field

The displacement field of the present beam theory is of the form:

$$u(x, z) = -z\frac{\mathrm{d}w}{\mathrm{d}x} + \frac{h}{\pi}\sin\frac{\pi z}{h}\phi(x), \tag{1}$$

$$w(x, z) = w(x),$$

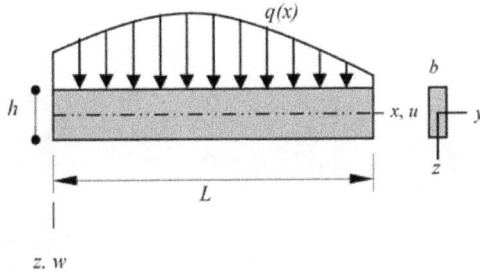

Fig. 1. Beam under bending in x–z plane

where u is the axial displacement in x direction and w is the transverse displacement in z direction of the beam. The sinusoidal function is assigned according to the shear stress distribution through the thickness of the beam. The function ϕ represents rotation of the beam at neutral axis, which is an unknown function to be determined. The normal and shear strains obtained within the framework of linear theory of elasticity using displacement field given by Eq. (1) are as follows:

$$\text{Normal strain: } \varepsilon_x = \frac{\partial u}{\partial x} = -z\frac{\mathrm{d}^2 w}{\mathrm{d}x^2} + \frac{h}{\pi}\sin\frac{\pi z}{h}\frac{\mathrm{d}\phi}{\mathrm{d}x}, \tag{2}$$

$$\text{Shear strain: } \gamma_{zx} = \frac{\partial u}{\partial z} + \frac{\mathrm{d}w}{\mathrm{d}x} = \cos\frac{\pi z}{h}\phi. \tag{3}$$

The stress-strain relationships used are as follows:

$$\sigma_x = E\varepsilon_x, \qquad \tau_{zx} = G\gamma_{zx}. \tag{4}$$

2.2. Governing equations and boundary conditions

Using the expressions for strains and stresses (2) through (4) and using the principle of virtual work, variationally consistent governing differential equations and boundary conditions for the beam under consideration can be obtained. The principle of virtual work when applied to the beam leads to:

$$b\int_{x=0}^{x=L}\int_{z=-h/2}^{z=+h/2}(\sigma_x\delta\varepsilon_x + \tau_{zx}\delta\gamma_{zx})\mathrm{d}x\,\mathrm{d}z - \int_{x=0}^{x=L}q(x)\delta w\,\mathrm{d}x = 0, \tag{5}$$

where the symbol δ denotes the variational operator. Employing Green's theorem in Eq. (4) successively, we obtain the coupled Euler-Lagrange equations which are the governing differential equations and associated boundary conditions of the beam. The governing differential equations obtained are as follows:

$$EI\frac{\mathrm{d}^4 w}{\mathrm{d}x^4} - \frac{24}{\pi^3}EI\frac{\mathrm{d}^3\phi}{\mathrm{d}x^3} = q(x), \tag{6}$$

$$\frac{24}{\pi^3}EI\frac{\mathrm{d}^3 w}{\mathrm{d}x^3} - \frac{6}{\pi^2}EI\frac{\mathrm{d}^2\phi}{\mathrm{d}x^2} + \frac{GA}{2}\phi = 0. \tag{7}$$

The associated consistent natural boundary conditions obtained are of following form:

At the ends $x = 0$ and $x = L$

$$V_x = EI\frac{\mathrm{d}^3w}{\mathrm{d}x^3} - \frac{24}{\pi^3}EI\frac{\mathrm{d}^2\phi}{\mathrm{d}x^2} = 0 \quad \text{or } w \text{ is prescribed,} \tag{8}$$

$$M_x = EI\frac{\mathrm{d}^2w}{\mathrm{d}x^2} - \frac{24}{\pi^3}EI\frac{\mathrm{d}\phi}{\mathrm{d}x} = 0 \quad \text{or } \frac{\mathrm{d}w}{\mathrm{d}x} \text{ is prescribed,} \tag{9}$$

$$M_a = EI\frac{24}{\pi^3}\frac{\mathrm{d}^2w}{\mathrm{d}x^2} - \frac{6}{\pi^2}EI\frac{\mathrm{d}\phi}{\mathrm{d}x} = 0 \quad \text{or } \phi \text{ is prescribed.} \tag{10}$$

Thus the boundary value problem of the beam bending is given by the above variationally consistent governing differential equations and boundary conditions.

2.3. The general solution of governing equilibrium equations of the beam

The general solution for transverse displacement $w(x)$ and warping function $\phi(x)$ is obtained using Eqs. (6) and (7) using method of solution of linear differential equations with constant coefficients. Integrating and rearranging the first governing Eq. (6), we obtain the following equation

$$\frac{\mathrm{d}^3w}{\mathrm{d}x^3} = \frac{24}{\pi^3}\frac{\mathrm{d}^2\phi}{\mathrm{d}x^2} + \frac{Q(x)}{EI}, \tag{11}$$

where $Q(x)$ is the generalized shear force for beam and it is given by $Q(x) = \int_0^x q\,\mathrm{d}x + C_1$.

Now the second governing Eq. (7) is rearranged in the following form:

$$\frac{\mathrm{d}^3w}{\mathrm{d}x^3} = \frac{\pi}{4}\frac{\mathrm{d}^2\phi}{\mathrm{d}x^2} - \beta\phi. \tag{12}$$

A single equation in terms of ϕ is now obtained using Eqs. (11) and (12) as

$$\frac{\mathrm{d}^2\phi}{\mathrm{d}x^2} - \lambda^2\phi = \frac{Q(x)}{\alpha EI}, \tag{13}$$

where constants α, β and λ in Eqs. (12) and (13) are as follows

$$\alpha = \left(\frac{\pi}{4} - \frac{24}{\pi^3}\right), \qquad \beta = \left(\frac{\pi^3}{48}\frac{GA}{EI}\right) \quad \text{and} \quad \lambda^2 = \frac{\beta}{\alpha}.$$

The general solution of Eq. (13) is as follows:

$$\phi(x) = C_2\cosh\lambda x + C_3\sinh\lambda x - \frac{Q(x)}{\beta EI}. \tag{14}$$

The equation of transverse displacement $w(x)$ is obtained by substituting the expression of $\phi(x)$ in Eq. (12) and then integrating it thrice with respect to x. The general solution for $w(x)$ is obtained as follows:

$$EIw(x) = \iiiint q\,\mathrm{d}x\,\mathrm{d}x\,\mathrm{d}x\,\mathrm{d}x + \frac{C_1 x^3}{6} + \tag{15}$$

$$\left(\frac{\pi}{4}\lambda^2 - \beta\right)\frac{EI}{\lambda^3}(C_2\sinh\lambda x + C_3\cosh\lambda x) + C_4\frac{x^2}{2} + C_5 x + C_6,$$

where C_1, C_2, C_3, C_4, C_5 and C_6 are arbitrary constants and can be obtained by imposing boundary conditions of beam.

3. Illustrative example

In order to prove the efficacy of the present theory, the following numerical example is considered. The material properties for beam used are: $E = 210$ GPa, $\mu = 0.3$ and $\rho = 7\,800$ kg/m^3, where E is the Young's modulus, ρ is the density, and μ is the Poisson's ratio of beam material.

A fixed-fixed beam has its origin at left hand side support and is fixed at $x = 0$ and L. The beam is subjected to parabolic load $q(x) = q_0 \left(\frac{x}{L}\right)^2$ on surface $z = -h/2$ acting in the downward z direction with maximum intensity of load q_0 as shown in Fig. 2. The boundary conditions associated with this beam at fixed ends are: $\frac{dw}{dx} = \phi = w = 0$ at $x = 0$ and L.

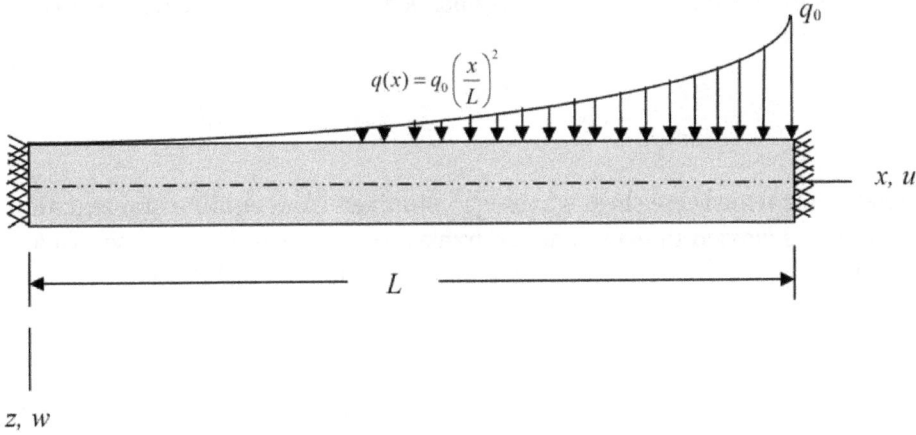

Fig. 2. Fixed beam with parabolic load

General expressions obtained for $w(x)$ and $\phi(x)$ are as follows:

$$
w(x) = \frac{q_0 L^4}{120 EI} \left[\frac{1}{3}\frac{x^6}{L^6} + \frac{x^2}{L^2} - \frac{4}{3}\frac{x^3}{L^3} - \frac{12}{\pi^2}\frac{E}{G}\frac{h^2}{L^2}\left(\frac{5}{6}\frac{x^4}{L^4} - \frac{5}{3}\frac{x^2}{L^2}\right) - \right. \tag{16}
$$
$$
\left. \frac{4}{5}\frac{E}{G}\frac{h^2}{L^2}\left(-\frac{x}{L} + \frac{1}{2}\frac{x^2}{L^2} + \frac{\sinh \lambda x - \cosh \lambda x + 1}{\lambda L}\right)\right],
$$

$$
\phi(x) = \frac{1}{15}\frac{q_0 L}{\beta EI}\left(1 + 5\frac{x^3}{L^3} + \sinh \lambda x - \cosh \lambda x\right). \tag{17}
$$

The expression for axial displacement u is obtained by substituting Eqs. (16) and (17) into the first equation in (1) and it is as follows:

$$
u = \frac{q_0 h}{Eb}\left[-\frac{1}{10}\frac{z}{h}\frac{L^3}{h^3}\left(2\frac{x^5}{L^5} + 2\frac{x}{L} - 4\frac{x^2}{L^2} - \frac{40}{\pi^2}\frac{E}{G}\frac{h^2}{L^2}\left(\frac{x^3}{L^3} - \frac{x}{L}\right) - \right.\right.
$$
$$
\left. \frac{4}{5}\frac{E}{G}\frac{h^2}{L^2}\left(-1 + \frac{x}{L} + \cosh \lambda x - \sinh \lambda x\right)\right) - \tag{18}
$$
$$
\left. \frac{16}{5\pi^4}\sin\frac{\pi z}{h}\frac{E}{G}\frac{L}{h}\left(-1 + 5\frac{x^3}{L^3} + \cosh \lambda x - \sinh \lambda x\right)\right].
$$

The expression for axial stress is obtained using Eqs. (2), (4), (16) and (17) as follows:

$$\sigma_x = \frac{q_0}{b} \left\{ -\frac{1}{10} \frac{z}{h} \frac{L^2}{h^2} \left[10\frac{x^4}{L^4} + 2 - 4\frac{x}{L} - \frac{120}{\pi^2} \frac{E}{G} \frac{h^2}{L^2} \left(\frac{x^2}{L^2} - \frac{1}{3} \right) - \right. \right.$$
$$\left. \frac{4}{5} \frac{E}{G} \frac{h^2}{L^2} (1 + \lambda L(\sinh \lambda x - \cosh \lambda x)) \right] - \qquad (19)$$
$$\frac{16}{5\pi^4} \sin \frac{\pi z}{h} \frac{E}{G} \left(15\frac{x^2}{L^2} + \lambda L(\sinh \lambda x - \cosh \lambda x) \right) \right\}.$$

The expressions for transverse shear stress is obtained using constitutive relation (4) and using Eq. (17) as follows:

$$\tau_{zx}^{CR} = \frac{16}{5\pi^3} \frac{q_0}{b} \frac{L}{h} \cos \frac{\pi z}{h} \left(1 - 5\frac{x^3}{L^3} + \sinh \lambda x - \cosh \lambda x \right). \qquad (20)$$

Expression for transverse shear stress τ_{zx}^{EE} obtained from equilibrium equation

The alternate approach to determine the transverse shear stress is the use of equilibrium equations. The first stress equilibrium equation of two dimensional theory of elasticity is as follows:

$$\frac{\partial \sigma_x}{\partial x} + \frac{\partial \tau_{zx}}{\partial z} = 0. \qquad (21)$$

Substituting expression for σ_x into Eq. (21) and integrating it with respect to the thickness coordinate z and imposing the boundary condition $\tau_{zx} = 0$ at the bounding surfaces $z = \pm h/2$ of the beam one can obtain the final expression of transverse shear stress, which is follows:

$$\tau_{zx}^{EE} = \frac{q_0 L}{80bh} \left(4\frac{z^2}{h^2} - 1 \right) \left[40\frac{x^3}{L^3} - 4 - \frac{240}{\pi^2} \frac{x}{L} - \frac{4}{5} \frac{E}{G} \frac{h^2}{L^2} \lambda^2 L^2 (\cosh \lambda x - \sinh \lambda x) \right] - \qquad (22)$$
$$\frac{16}{5\pi^5} \cos \frac{\pi z}{h} \frac{E}{G} \frac{q_0 h}{bL} \left(30\frac{x}{L} + \lambda^2 L^2 (\cosh \lambda x - \sinh \lambda x) \right).$$

Results are obtained using expressions (16) through (22) for displacements and stresses. The numerical results are presented in Table 1 and graphically presented in Figs. 3 – 11.

Table 1. Non-dimensional axial displacement ($\bar{u}$) at ($x = 0.75L$, $z = h/2$), transverse deflection ($\bar{w}$) at ($x = 0.75L$, $z = 0.0$), axial stress ($\bar{\sigma}_x$) at ($x = 0$, $z = h/2$), maximum transverse shear stresses $\bar{\tau}_{zx}^{CR}$ and $\bar{\tau}_{zx}^{EE}$ ($x = 0.01L$, $z = 0.0$) of the beam for slenderness ratio (S) 4 and 10

Source	S	$\bar{u}$	$\bar{w}$	$\bar{\sigma}_x$	$\bar{\tau}_{zx}^{CR}$	$\bar{\tau}_{zx}^{EE}$
Present		0.293 2	0.251 3	3.227 3	0.196 9	−0.442 1
Ghugal and Sharma [9]		0.295 5	0.251 1	3.527 7	0.232 5	−0.455 4
Krishna Murthy [14]	4	0.297 9	0.251 4	3.270 2	0.205 3	−0.283 4
Timoshenko [20]		−0.881 2	0.110 7	1.600 0	0.048 2	0.399 9
Bernoulli-Euler		−0.881 2	0.059 3	1.600 0	—	0.399 9
Present		−10.833 1	0.090 2	13.433 9	0.827 8	−0.087 3
Ghugal and Sharma [9]		−10.827 5	0.090 2	14.189 1	0.885 1	0.425 1
Krishna Murthy [14]	10	−10.821 5	0.090 2	13.542 2	0.834 7	0.419 7
Timoshenko [20]		−13.769 5	0.067 5	10.000 0	0.753 8	0.999 9
Bernoulli-Euler		−13.769 5	0.059 3	10.000 0	—	0.999 9

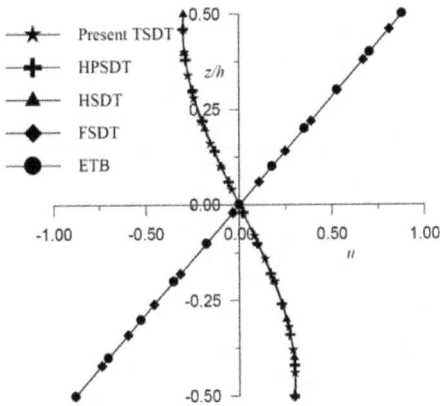

Fig. 3. Variation of axial displacement ($\bar{u}$) through the thickness of fixed-fixed beam at ($x = 0.75L, z$) for slenderness ratio 4

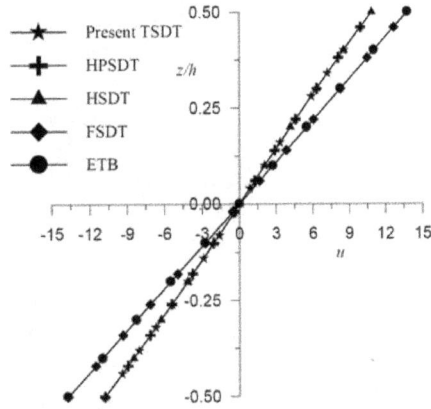

Fig. 4. Variation of axial displacement ($\bar{u}$) through the thickness of fixed-fixed beam at ($x = 0.75L, z$) for slenderness ratio 10

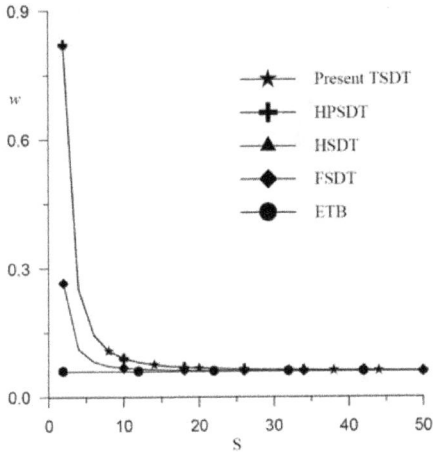

Fig. 5. Variation of maximum transverse displacement ($\bar{w}$) of fixed-fixed beam at ($x = 0.75L$, $z = 0$) with slenderness ratio S

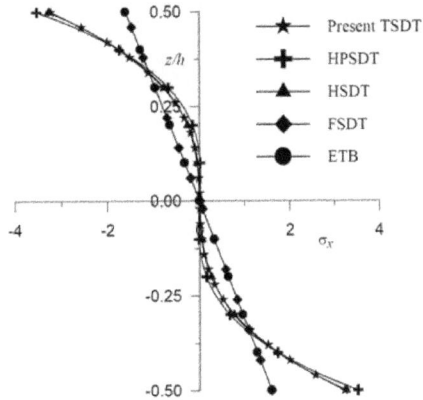

Fig. 6. Variation of axial stress ($\bar{\sigma}_x$) through the thickness of fixed-fixed beam at ($x = 0, z$) for slenderness ratio 4

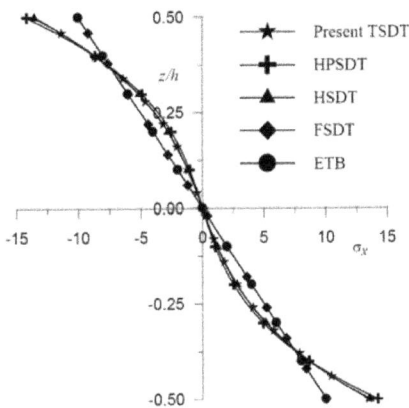

Fig. 7. Variation of axial stress ($\bar{\sigma}_x$) through the thickness of fixed-fixed beam at ($x = 0, z$) for slenderness ratio 10

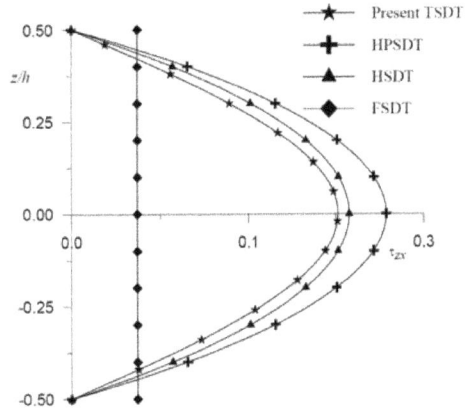

Fig. 8. Variation of transverse shear stress ($\bar{\tau}_{zx}$) through the thickness of fixed-fixed beam at ($x = 0.01L, z$) obtained using constitutive relation for slenderness ratio 4

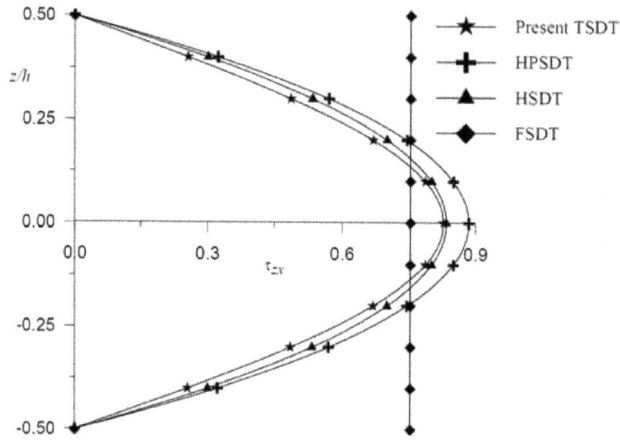

Fig. 9. Variation of transverse shear stress ($\bar{\tau}_{zx}$) through the thickness of fixed-fixed beam at ($x = 0.01L$, z) obtained using constitutive relation for slenderness ratio 4

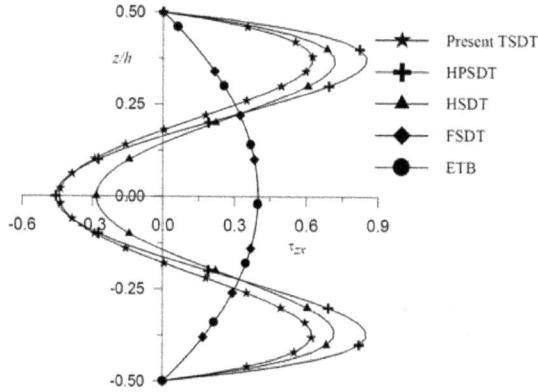

Fig. 10. Variation of transverse shear stress ($\bar{\tau}_{zx}$) through the thickness of fixed-fixed beam at ($x = 0.01L$, z) obtained using equilibrium equation for slenderness ratio 4

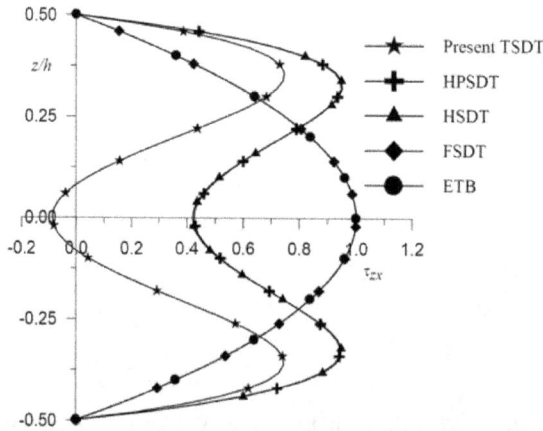

Fig. 11. Variation of transverse shear stress ($\bar{\tau}_{zx}$) through the thickness of fixed-fixed beam at ($x = 0.01L$, z) obtained using equilibrium equation for slenderness ratio 10

4. Results

The results for inplane displacement, transverse displacement, axial and transverse stresses are presented in the following non dimensional form for the purpose of presenting the results in this paper:

$$\bar{u} = \frac{Ebu}{q_0 h}, \qquad \bar{w} = \frac{10Ebh^3 w}{q_0 L^4}, \qquad \bar{\sigma}_x = \frac{b\sigma_x}{q_0}, \qquad \bar{\tau}_{zx} = \frac{b\tau_{zx}}{q_0}, \qquad S = \frac{L}{h}.$$

The numerical results for displacements and stresses are obtained using FORTRAN programs developed based on the non-dimensional expressions for these quantities.

5. Discussion and conclusion

The variationally consistent theoretical formulation of the theory with general solution technique of governing differential equations is presented. The general solutions for beam with parabolic load is obtained in case of thick fixed beams. The displacements and stresses obtained by present theory are in excellent agreement with those of other equivalent refined and higher order theories. The present theory yields the realistic variation of axial displacement and stresses through the thickness of beam. The theory is shown to be capable of predicting the effects of stress concentration on the axial and transverse stresses in the vicinity of the built-in end of the beam which is the region of heavy stress concentration. Thus the validity of the present theory is established.

References

[1] Averill, R. C., Reddy, J. N., An assessment of four-noded plate finite elements based on a generalized third order theory, International Journal of Numerical Methods in Engineering 33 (1992) 1 553–1 572.

[2] Baluch, M. H., Azad, A. K., Khidir, M. A., Technical theory of beams with normal strain, ASCE Journal of Engineering Mechanics 110 (8) (1984) 1 233–1 237.

[3] Bhimaraddi, A., Chandrashekhara, K., Observations on higher order beam theory, ASCE Journal of Aerospace Engineering 6 (4) (1993) 408–413.

[4] Bickford, W. B., A consistent higher order beam theory, Development in Theoretical Applied Mechanics (SECTAM).11 (1982) 137–150.

[5] Bresse, J. A. C., Course of applied mechanics, Mallet-Bachelier, Paris, 1859. (in French)

[6] Cowper, G. R., The shear coefficients in Timoshenko beam theory, ASME Journal of Applied Mechanic 33(2) (1966) 335–340.

[7] Cowper, G. R., On the accuracy of Timoshenko beam theory, ASCE Journal of Engineering Mechanics Division 94 (EM6) (1968) 1 447–1 453.

[8] Ghugal, Y. M., Shmipi, R. P., A review of refined shear deformation theories for isotropic and anisotropic laminated beams, Journal of Reinforced Plastics And Composites 20(3) (2001) 255–272.

[9] Ghugal, Y. M., Sharma, R., A hyperbolic shear deformation theory for flexure and vibration of thick isotropic beams, International Journal of Computational Methods 6(4) (2009) 585–604.

[10] Heyliger, P. R., Reddy, J. N., A higher order beam finite element for bending and vibration problems, Journal of Sound and Vibration 126(2) (1988) 309–326.

[11] Hildebrand, F. B., Reissner, E. C., Distribution of stress in built-in beam of narrow rectangular cross section, Journal of Applied Mechanics 64 (1942) 109–116.

[12] Irretier, H., Refined effects in beam theories and their influence on natural frequencies of beam, International Proceeding of Euromech Colloquium 219 – Refined Dynamical Theories of Beam, Plates and Shells and Their Applications, Edited by I. Elishakoff and H. Irretier, Springer-Verlag, Berlin, (1986), pp. 163–179.

[13] Kant, T., Gupta, A., A finite element model for higher order shears deformable beam theory, Journal of Sound and Vibration 125(2) (1988) 193–202.

[14] Krishna Murthy, A. V., Towards a consistent beam theory, AIAA Journal 22(6) (1984) 811–816.

[15] Levinson, M., A new rectangular beam theory, Journal of Sound and Vibration 74(1) (1981) 81–87.

[16] Lord Rayleigh, J. W. S., The theory of sound, Macmillan Publishers, London, 1877.

[17] Reddy, J. N., An Introduction to finite element method. Second edition, McGraw-Hill, Inc., New York, 1993.

[18] Rehfield, L. W., Murthy, P. L. N., Toward a new engineering theory of bending: fundamentals, AIAA Journal 20(5) (1982) 693–699.

[19] Stein, M., Vibration of beams and plate strips with three dimensional flexibility, ASME Journal of Applied Mechanics 56(1) (1989) 228–231.

[20] Timoshenko, S. P., On the correction for shear of the differential equation for transverse vibrations of prismatic bars, Philosophical Magazine 41(6) (1921) 742–746.

[21] Timoshenko, S. P., Goodier, J. N., Theory of elasticity. Third international edition, McGraw-Hill, Singapore. 1970.

[22] Vlasov, V. Z., Leont'ev, U. N., Beams, plates and shells on elastic foundations, Moskva, Chapter 1, 1–8. Translated from the Russian by Barouch A. and Plez T., Israel Program for Scientific Translation Ltd., Jerusalem, 1966.

Nomenclature

A	Cross sectional area of beam $= bh$
b	Width of beam in y-direction
E, G, μ	Elastic constants of the beam material
h	Thickness of beam
I	Moment of inertia of cross-section of beam
L	Span of the beam
q_0	Intensity of parabolic transverse load
S	Slenderness ratio of the beam $= L/h$
w	Transverse displacement in z-direction
$\bar{w}$	Non-dimensional transverse displacement
$\bar{u}$	Non-dimensional axial displacement
x, y, z	Rectangular Cartesian coordinates
$\bar{\sigma}_x$	Non-dimensional axial stress in x-direction
$\bar{\tau}_{zx}^{CR}$	Non-dimensional transverse shear stress via constitutive relation
$\bar{\tau}_{zx}^{EE}$	Non-dimensional transverse shear stress via equilibrium equation
$\phi(x)$	Unknown function associated with the shear slope

List of abbreviations

CR	Constitutive Relations
EE	Equilibrium Equations
TSDT	Trigonometric Shear Deformation Theory
HPSDT	Hyperbolic Shear Deformation Theory
HSDT	Third Order Shear Deformation Theory
FSDT	First Order Shear Deformation Theory
ETB	Elementary Theory of Beam

5

Dynamic response of nuclear fuel assembly excited by pressure pulsations

V. Zeman[a,*], Z. Hlaváč[a]

[a] *Faculty of Applied Sciences, University of West Bohemia, Univerzitní 22, 306 14 Plzeň, Czech Republic*

Abstract

The paper deals with dynamic load calculation of the hexagonal type nuclear fuel assembly caused by spatial motion of the support plates in the reactor core. The support plate motion is excited by pressure pulsations generated by main circulation pumps in the coolant loops of the primary circuit of the nuclear power plant. Slightly different pumps revolutions generate the beat vibrations which causes an amplification of fuel assembly component dynamic deformations and fuel rods coating abrasion. The cyclic and central symmetry of the fuel assembly makes it possible the system decomposition into six identical revolved fuel rod segments which are linked with central tube and skeleton by several spacer grids in horizontal planes.

The modal synthesis method with condensation of the fuel rod segments is used for calculation of the normal and friction forces transmitted between fuel rods and spacer grids cells.

Keywords: vibrations, nuclear fuel assembly, pressure pulsations, modal synthesis method, dynamic load

1. Introduction

An assessment of nuclear fuel assemblies (FA) behaviour at standard operating conditions of the nuclear reactors belongs to important safety and reliability audits. A significant part of FA assessment plays dynamic deformations and load of FA components and abrasion of fuel rods coating [6]. Dynamic properties of FA are usually investigated experimentally using their physical models [4, 7]. Frequency lowest modal values (eigenfrequencies and eigenvectors) investigated by measurement in the air, serve as initial data for parametric identification of the FA global model of beam type [9] used in dynamic analyses of the VVER1000 type reactors [2, 11]. These FA global models do not enable investigation of dynamic deformations of FA components and dynamic coupling forces between FA components taken to fuel road coating assessment.

The goal of this paper, in direct sequence at an interpretation of FA modelling and FA modal analysis published in the paper [13], is a presentation of the newly developed method for calculation of the hexagonal type FA component dynamic load caused by pressure pulsations generated by main circulation pumps in coolant loops [5] of the primary circuit of the nuclear power plants (NPP). The method is applied to Russian TVSA-T FA [1, 8] installed in reactor VVER1000 in NPP Temelín.

---

*Corresponding author. e-mail: zemanv@kme.zcu.cz.

2. Mathematical model of the FA kinematical excited vibration

In order to model the hexagonal FA can be divided into six rod segments (S), centre tube (CT) and load-bearing skeleton (LS) (see Fig. 1). Each rod segment (on Fig. 2 drawn in lateral cross-section and circumscribed by triangles) is composed of many fuel rods with fixed bottom ends in the lower tailpiece and several more guide thimbles fully restrained in lower (LP) and head pieces (HP). The centre tube is fully restrained into these pieces. The skeleton is created of six angle pieces (AP) fast linked with LP and mutually coupled by divided grid rims (GR). All FA components are linked by transverse spacer grids of three types (SG_1–SG_3) which elastic properties are expressed by linear springs placed on several level spacings $g = 1, \dots, G$ (see Fig. 1). The fuel rods are embedded into spacer grids with small initial tension, which would not fall below zero during core operation.

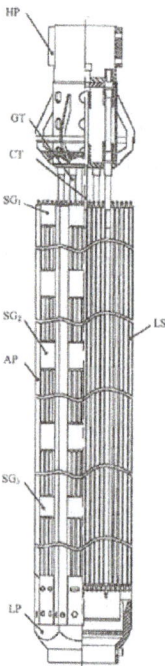

Fig. 1. Scheme of the fuel assembly   Fig. 2. The FA cross-section

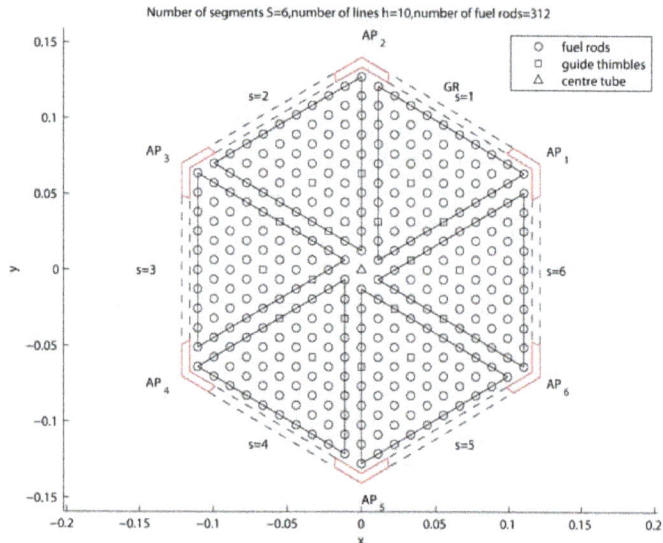

Each FA is fixed by means of lower tailpiece (LP) into mounting plate in core barrel bottom and by means of head piece (HP) into lower supporting plate of the block of protection tubes. These support plates with pieces can be considered in transverse direction as rigid bodies.

Let use consider the spatial motion of the support plates described in coordinate systems x_X, y_X, z_X ($X = L, U$) with origins in plate gravity centres L, U by displacement vectors (see Fig. 3)

$$\boldsymbol{q}_X = [x_X, y_X, z_X, \varphi_{x,X}, \varphi_{y,X}, \varphi_{z,X}]^T, \quad X = L, U. \tag{1}$$

The lateral $\xi_{r,X}^{(s)}$, $\eta_{r,X}^{(s)}$ and bending $\vartheta_{r,X}^{(s)}$, $\psi_{r,X}^{(s)}$ displacements in the end-nodes of the fuel rod or guide thimbles r in segment s (in Fig. 3 illustrated for $s = 1$) coupled with plates can be expressed by the displacements of the lower ($X = L$) and upper ($X = U$) plates in the form

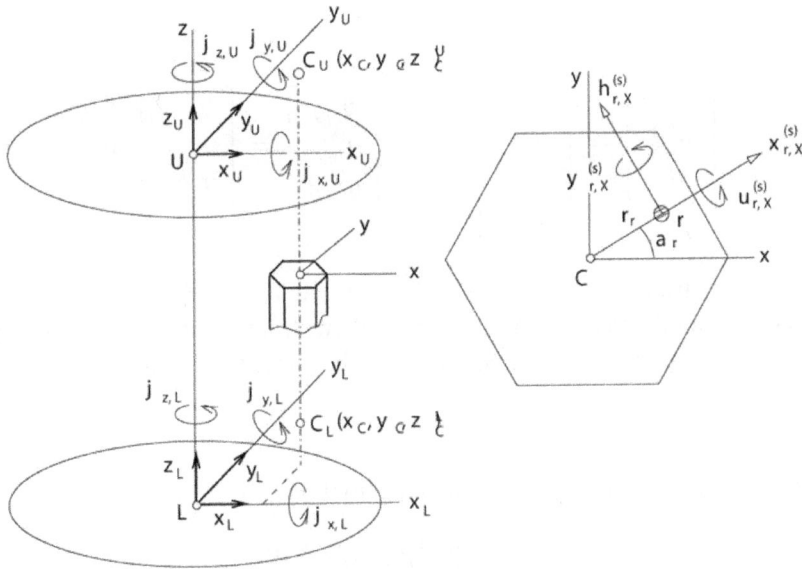

Fig. 3. Spatial motion of the FA support plates

$$
\begin{bmatrix} \xi_{r,X}^{(s)} \\ \eta_{r,X}^{(s)} \\ \vartheta_{r,X}^{(s)} \\ \psi_{r,X}^{(s)} \end{bmatrix} = \begin{bmatrix} C_r^{(s)} & S_r^{(s)} & 0 & -z_C^X S_r^{(s)} & z_C^X C_r^{(s)} & x_C S_r^{(s)} - y_C C_r^{(s)} \\ -S_r^{(s)} & C_r^{(s)} & 0 & -z_C^X C_r^{(s)} & -z_C^X S_r^{(s)} & x_C C_r^{(s)} + y_C S_r^{(s)} + r_r \\ 0 & 0 & 0 & C_r^{(s)} & S_r^{(s)} & 0 \\ 0 & 0 & 0 & -S_r^{(s)} & C_r^{(s)} & 0 \end{bmatrix} \begin{bmatrix} x_X \\ y_X \\ z_X \\ \varphi_{x,X} \\ \varphi_{y,X} \\ \varphi_{z,X} \end{bmatrix}, \quad (2)
$$

shortly

$$
q_{r,X}^{(s)} = T_{r,X}^{(s)} q_X, \quad X = L, U, \tag{3}
$$

where $x_C, y_C, z_C^X, X = L, U$ are coordinates of the FA lower C_L and upper C_U piece centres in the coordinate systems of the support plates. Values $C_r^{(s)}$ and $S_r^{(s)}$ corresponding to r−th fuel rod (guide thimble) in segment s for the hexagonal type FA are

$$
C_r^{(s)} = \cos\left[\alpha_r + (s-1)\frac{\pi}{3}\right], \quad S_r^{(s)} = \sin\left[\alpha_r + (s-1)\frac{\pi}{3}\right] \tag{4}
$$

and r_r, α_r are polar coordinates of the fuel rod (guide thimble) centre in transverse plane (Fig. 3). The total transformations between displacements of the all kinematical excited nods of the subsystem components and lower (L) or upper (U) support plate displacements can be expressed according to (2) and (3) in the global matrix form

$$
q_L^{(s)} = T_L^{(s)} q_L, \, s = 1, \ldots, 6, CT, LS; \quad q_U^{(s)} = T_U^{(s)} q_U, \, s = 1, \ldots, 6, CT. \tag{5}
$$

The Russian TVSA-T FA (Fig. 1) in NPP Temelín contains in each segment $s \in \{1, \ldots, 6\}$ 52 fuel rods and 3 guide thimbles at the positions $5, 20, 30$. Therefore the transformation relations (5) for these subsystems have the form

$$
\begin{bmatrix} q_{1,L}^{(s)} \\ \vdots \\ q_{r,L}^{(s)} \\ \vdots \\ q_{55,L}^{(s)} \end{bmatrix} = \begin{bmatrix} T_{1,L}^{(s)} \\ \vdots \\ T_{r,L}^{(s)} \\ \vdots \\ T_{55,L}^{(s)} \end{bmatrix} q_L , \qquad \begin{bmatrix} q_{5,U}^{(s)} \\ q_{20,U}^{(s)} \\ q_{30,U}^{(s)} \end{bmatrix} = \begin{bmatrix} T_{5,U}^{(s)} \\ T_{20,U}^{(s)} \\ T_{30,U}^{(s)} \end{bmatrix} q_U \tag{6}
$$

and transformation matrices are of type $T_L^{(s)} \in R^{220,6}$ and $T_U^{(s)} \in R^{12,6}$.

The vectors of generalized coordinates of the fully restrained subsystems (rod segments and centre tube) loosed in kinematical excited nodes can be partitioned in the form

$$
q_s = [(q_L^{(s)})^T, (q_F^{(s)})^T, (q_U^{(s)})^T]^T , \ s = 1, \ldots, 6, CT \tag{7}
$$

and the skeleton $s = LS$ fixed only in bottom ends in the form

$$
q_{LS} = [(q_L^{(LS)})^T, (q_F^{(LS)})^T]^T . \tag{8}
$$

The displacements of free system nodes (uncoupled with support plates) are integrated in vectors $q_F^{(s)} \in R^{n_s}$. The conservative mathematical models of the loosed subsystems in the decomposed block form corresponding to partitioned vectors can be written as

$$
\begin{bmatrix} M_L^{(s)} & M_{L,F}^{(s)} & 0 \\ M_{F,L}^{(s)} & M_F^{(s)} & M_{F,U}^{(s)} \\ 0 & M_{U,F}^{(s)} & M_U^{(s)} \end{bmatrix} \begin{bmatrix} \ddot{q}_L^{(s)} \\ \ddot{q}_F^{(s)} \\ \ddot{q}_U^{(s)} \end{bmatrix} + \begin{bmatrix} K_L^{(s)} & K_{L,F}^{(s)} & 0 \\ K_{F,L}^{(s)} & K_F^{(s)} & K_{F,U}^{(s)} \\ 0 & K_{U,F}^{(s)} & K_U^{(s)} \end{bmatrix} \begin{bmatrix} q_L^{(s)} \\ q_F^{(s)} \\ q_U^{(s)} \end{bmatrix} = \begin{bmatrix} f_L^{(s)} \\ f_C^{(s)} \\ f_U^{(s)} \end{bmatrix} \tag{9}
$$

for the $s = 1, \ldots, 6, CT$ and for the skeleton as

$$
\begin{bmatrix} M_L^{(LS)} & M_{L,F}^{(LS)} \\ M_{F,L}^{(LS)} & M_F^{(LS)} \end{bmatrix} \begin{bmatrix} \ddot{q}_L^{(LS)} \\ \ddot{q}_F^{(LS)} \end{bmatrix} + \begin{bmatrix} K_L^{(LS)} & K_{L,F}^{(LS)} \\ K_{F,L}^{(LS)} & K_F^{(LS)} \end{bmatrix} \begin{bmatrix} q_L^{(LS)} \\ q_F^{(LS)} \end{bmatrix} = \begin{bmatrix} f_L^{(LS)} \\ f_C^{(LS)} \end{bmatrix} , \tag{10}
$$

where letters M (K) correspond to mass (stiffness) submatrices of the subsystems. The force subvectors $f_C^{(s)}$ express the coupling forces between subsystem s and adjacent subsystems transmitted by spacer grids. The second set of equations extracted from (9) and (10) for each subsystem is

$$
M_F^{(s)} \ddot{q}_F^{(s)} + K_F^{(s)} q_F^{(s)} = -M_{F,L}^{(s)} T_L^{(s)} \ddot{q}_L - M_{F,U}^{(s)} T_U^{(s)} \ddot{q}_U - \tag{11}
$$
$$
K_{F,L}^{(s)} T_L^{(s)} q_L - K_{F,U}^{(s)} T_U^{(s)} q_U + f_C^{(s)} ,
$$

where $M_{F,U}^{(LS)} = 0$, $K_{F,U}^{(LS)} = 0$ because the skeleton (LS) is fixed only with lower support plate.

The global model of the FA has to large DOF number for calculation of dynamic response excited by support plate motion. Therefore we assemble the condensed model using the modal synthesis method presented in the paper [12]. Let the modal properties of the conservative models of the mutually uncoupled subsystems with the strengthened end-nodes coupled with immovable support plates be characterized by spectral Λ_s and modal V_s matrices of order n_s, suitable to orthonormality conditions

$$
V_s^T M_F^{(s)} V_s = E , \quad V_s^T K_F^{(s)} V_s = \Lambda_s, \quad s = 1, \ldots, 6, CT, LS . \tag{12}
$$

The vectors $q_F^{(s)}$ of dimension n_s, corresponding to free nodes of subsystems, can be approximately transformed in the form

$$q_F^{(s)} = {}^m V_s x_s, \quad x_s \in R^{m_s}, \quad s = 1, \ldots, 6, CT, LS, \tag{13}$$

where ${}^m V_s \in R^{n_s, m_s}$ are modal submatrices compound out of chosen m_s master eigenvectors of fixed subsystems. The equations (11) can be rewritten using (12) and (13) in the form

$$\ddot{x}_s + {}^m \Lambda_s x_s = -{}^m V_s^T (M_{F,L}^{(s)} T_L^{(s)} \ddot{q}_L + M_{F,U}^{(s)} T_U^{(s)} \ddot{q}_U + K_{F,L}^{(s)} T_L^{(s)} q_L + \tag{14}$$
$$K_{F,U}^{(s)} T_U^{(s)} q_U) + {}^m V_s^T f_C^{(s)}, \quad s = 1, \ldots, 6, CT, LS,$$

where spectral submatrices ${}^m \Lambda_s \in R^{m_s, m_s}$ correspond to chosen master eigenvectors in ${}^m V_s$. The models (14) of all subsystems can be written in the configuration space $x = [x_s]$, $s = 1, \ldots, 6, CT, LS$ of dimension $m = \sum_s m_s$ as

$$\ddot{x}(t) + \Lambda x(t) = -V^T (M_L \ddot{Q}_L + M_U \ddot{Q}_U + K_L Q_L + K_U Q_U) + V^T f_C, \tag{15}$$

where $f_C = [f_C^{(s)}] \in R^n$, $n = \sum_s n_s$ is global vector of coupling forces between subsystems and matrices

$$\Lambda = \mathrm{diag}[{}^m \Lambda_s] \in R^{m,m}; \quad V = \mathrm{diag}[{}^m V_s] \in R^{n,m}; \quad X_X = \mathrm{diag}[X_{F,X}^{(s)} T_X^{(s)}] \in R^{n,48}$$
$$X = M, K; \quad X = L, U; s = 1, \ldots, 6, CT, LS$$

are block diagonal, composed from corresponding matrices of subsystems. Vectors $Q_X = [q_X^T, \ldots, q_X^T]^T \in R^{48}$, $X = L, U$ are assembled for eight FA subsystems from eight times repeating support plate displacement vectors. The global vector of coupling forces between subsystems can be calculated from identity

$$f_C = -\frac{\partial E_p}{\partial q_F} = -K_C q_F, \quad q_F = [q_F^{(s)}], \tag{16}$$

where E_p is potential (deformation) energy of the all spacer grids (springs) between subsystems. The stiffness matrix K_C of all couplings between subsystems was derived in [13] for Russian TVSA-T fuel assembly. The expressions (16) can be substituted in (15) and then we get the condensed model of the nuclear fuel assembly of order m

$$\ddot{x}(t) + (\Lambda + V^T K_C V) x(t) = -V^T (M_L \ddot{Q}_L(t) + M_U \ddot{Q}_U(t) + K_L Q_L(t) + K_U Q_U(t)). \tag{17}$$

3. Fuel assembly steady vibration and fuel rod coating abrasion caused by pressure pulsations

The steady vibrations of the reactor VVER1000 excited by coolant pressure pulsations in the gap between core barrel and reactor pressure vessel walls generated by the main circulation pumps were investigated in co-operation with NRI Řež [10] and published in [11]. The force effect can be expressed in the global model of the reactor by excitation vector in the complex form [11]

$$f(t) = \sum_j \sum_k f_j^{(k)} e^{ik\omega_j t}, \tag{18}$$

where $\boldsymbol{f}_j^{(k)}$ is vector of complex amplitudes of k−th excitation harmonic component caused by hydrodynamic forces generated in one j−th circulation pump. Corresponding angular rotational frequency of the j−th pump $\omega_j = 2\pi f_j$ is defined by pump revolutions per minute n_j [rpm], where can be for particular pumps slightly different. Steady dynamic response of the reactor in generalized coordinates is given by identical form [11]

$$q(t) = \sum_j \sum_k q_j^{(k)} \mathrm{e}^{\mathrm{i}k\omega_j t} .\tag{19}$$

The vectors of complex amplitudes $q_j^{(k)}$ must be transformed into vectors of $\boldsymbol{Q}_{X,j}^{(k)}$ $(X = L, U)$ describing steady vibration of the support plates caused by k−th harmonic of j−th pump [13, 15].

In consequence of slightly damped fuel assembly components we consider modal damping of the subsystems characterized in the space of modal coordinates $\boldsymbol{x}_s$ by diagonal matrices $\boldsymbol{D}_s = \mathrm{diag}[2D_\nu^{(s)}\Omega_\nu^{(s)}]$, where $D_\nu^{(s)}$ are damping factors of natural modes and $\Omega_\nu^{(s)}$ are eigenfrequencies of the mutually uncoupled subsystems. The damping of spacer grids can be approximately expressed by damping matrix $\boldsymbol{B}_C = \beta\boldsymbol{K}_C$ proportional to stiffness matrix $\boldsymbol{K}_C$ by coefficient β.

That being simplifying supposed and the polyharmonic excitation (18) the conservative condensed model (17) will be completed in the complex form

$$\ddot{\boldsymbol{x}}(t) + (\boldsymbol{D} + \beta\boldsymbol{V}^T\boldsymbol{K}_C\boldsymbol{V})\dot{\boldsymbol{x}}(t) + (\boldsymbol{\Lambda} + \boldsymbol{V}^T\boldsymbol{K}_C\boldsymbol{V})\boldsymbol{x}(t) =$$
$$= -\boldsymbol{V}^T \sum_j \sum_k \left[(\boldsymbol{K}_L - k^2\omega_j^2\boldsymbol{M}_L)\boldsymbol{Q}_{L,j}^{(k)} + (\boldsymbol{K}_U - k^2\omega_j^2\boldsymbol{M}_U)\boldsymbol{Q}_{U,j}^{(k)} \right] \mathrm{e}^{\mathrm{i}k\omega_j t} .\tag{20}$$

Steady response of the fuel assembly subsystems in the complex form according to (13) is

$$\widetilde{\boldsymbol{q}}_F^{(s)}(t) = \sum_j \sum_k {}^m\boldsymbol{V}_s \widetilde{\boldsymbol{x}}_{s,j}^{(k)} \mathrm{e}^{\mathrm{i}k\omega_j t} , \quad s = 1, \dots, 6, CT, LS ,\tag{21}$$

where $\widetilde{\boldsymbol{x}}_{s,j}^{(k)}$ are subvectors of the global vector $\widetilde{\boldsymbol{x}}_j^{(k)}$ of the complex amplitudes

$$\widetilde{\boldsymbol{x}}_j^{(k)} = -[\boldsymbol{\Lambda} + (1 + \mathrm{i}\beta k\omega_j)\boldsymbol{V}^T\boldsymbol{K}_C\boldsymbol{V} + \mathrm{i}k\omega_j\boldsymbol{D}]^{-1} \cdot$$
$$\boldsymbol{V}^T \sum_j \sum_k \left[(\boldsymbol{K}_L - k^2\omega_j^2\boldsymbol{M}_L)\boldsymbol{Q}_{L,j}^{(k)} + (\boldsymbol{K}_U - k^2\omega_j^2\boldsymbol{M}_U)\boldsymbol{Q}_{U,j}^{(k)} \right]\tag{22}$$

corresponding to subsystem s. Subscript $j \in \{1, 2, 3, 4\}$ is assigned to the operating circulation pump and subscript k to the harmonic component of pressure pulsations. The real steady dynamic response expressed by the generalized coordinates vector of the FA subsystems s in dependence on time according to (21) and (22) is

$$\boldsymbol{q}_F^{(s)}(t) = \sum_j \sum_k {}^m\boldsymbol{V}_s \left(\mathrm{Re}[\widetilde{\boldsymbol{x}}_{s,j}^{(k)}] \cos k\omega_j t - \mathrm{Im}[\widetilde{\boldsymbol{x}}_{s,j}^{(k)}] \sin k\omega_j t \right) .\tag{23}$$

The components of of the vector generalized coordinates $\boldsymbol{q}_F^{(s)}(t)$ corresponding to rod segment s are nodal points displacements of particular fuel rod or guide thimble on the level all spacer grids g [13] in the form

$$\boldsymbol{q}_F^{(s)} = [\dots, \xi_{r,g}^{(s)}, \eta_{r,g}^{(s)}, \vartheta_{r,g}^{(s)}, \psi_{r,g}^{(s)}, \dots], \quad r = 1, \dots, R; \quad g = 1, \dots, G,\tag{24}$$

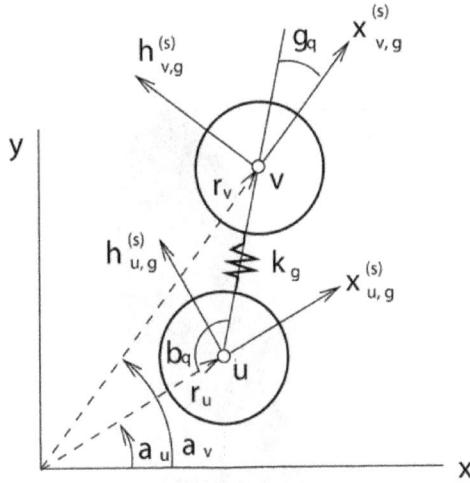

Fig. 4. The coupling between two fuel rods

where R is number of fuel rods and guide thimbles in one rod segment $s \in \{1, \ldots, 6\}$ and G is number of spacer grids.

The dynamic normal force transmitted by spacer grid between two adjacent fuel rods u and v (see Fig. 4) of the segment s on the level spacer grid g can be expressed in the form

$$N_{q,g}^{(s)} = -k_g \left[\xi_{v,g}^{(s)} \cos \gamma_q + \eta_{v,g}^{(s)} \sin \gamma_q + \xi_{u,g}^{(s)} \cos \beta_q - \eta_{u,g}^{(s)} \sin \beta_q \right], \tag{25}$$

where k_g is stiffness of the transverse spring expressing the spacer grid cell stiffness between two adjacent fuel rods. Angles β_q, γ_q correspond to fuel rod couple u and v that is assigned coupling q. The fuel rod positions in the rod segment are determined by polar coordinates r_u, α_u and r_v, α_v of the linked fuel rods [3]. The slip speeds between transverse vibrating spacer grid on the level g and bending vibrating fuel rods u and v due to fuel rods bending inside of spacer grid cell are

$$c_{u,g}^{(s)} = r(\sin \beta_q \dot{\vartheta}_{u,g}^{(s)} + \cos \beta_q \dot{\psi}_{u,g}^{(s)}), \quad c_{v,g}^{(s)} = r(-\sin \gamma_q \dot{\vartheta}_{v,g}^{(s)} + \cos \gamma_q \dot{\psi}_{v,g}^{(s)}), \tag{26}$$

where r is outside diameter of the fuel rod coating. Bending angular velocities of fuel rod cross-section are expressed by corresponding components of the vector

$$\dot{q}_F^{(s)}(t) = -\sum_j \sum_k k\omega_j V_s \left(\mathrm{Re}[\widetilde{x}_{s,j}^{(k)}] \sin k\omega_j t + \mathrm{Im}[\widetilde{x}_{s,j}^{(k)}] \cos k\omega_j t \right) \tag{27}$$

obtained by the derivative of generalized coordinate vector (23) with respect to time. The power of the friction forces in the contact of the fuel rod coating and spacer grid cell is

$$P_{u,g}^{(s)} = f N_{q,g}^{(s)} c_{u,g}^{(s)} \text{ and } P_{v,g}^{(s)} = f N_{q,g}^{(s)} c_{v,g}^{(s)}, \tag{28}$$

where f is friction coefficient. The criterion of the fuel rod coating abrasion can be expressed by the work of the friction forces during the period T of the first harmonic component of pressure pulsations

$$W_{u,g}^{(s)} = \int_0^T |P_{u,g}^{(s)}| \, \mathrm{d}t \text{ or } W_{v,g}^{(s)} = \int_0^T |P_{v,g}^{(s)}| \, \mathrm{d}t \tag{29}$$

from the moment of maximal dynamic force transmitted by extreme stressed spacer grid cell. The calculation of the dynamic forces $N_{q,g}^{(s)}$ defined in (25) transmitted by all couplings q inside and outside rod segments s on the all level of spacer grids g makes possible to identification of the maximal loaded spacer grid cell.

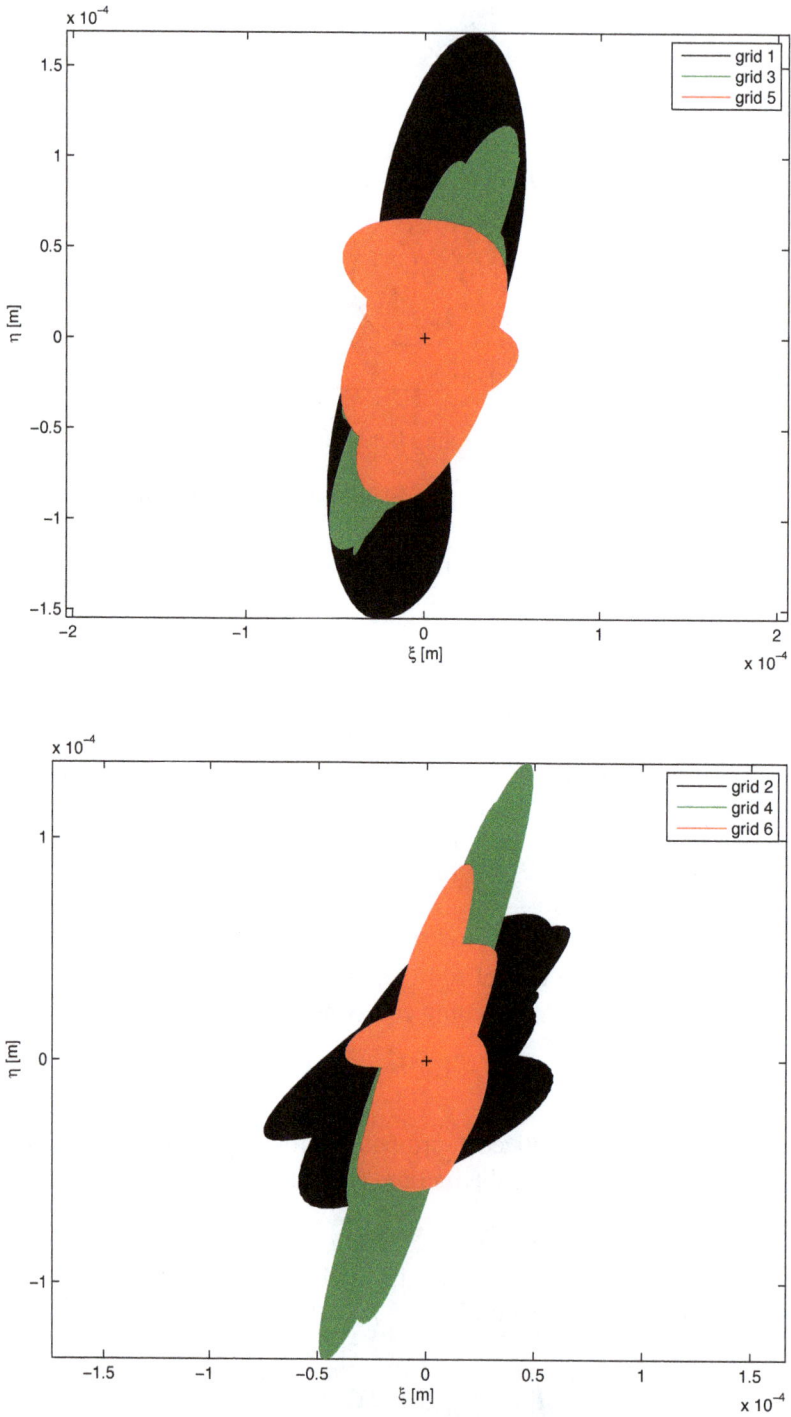

Fig. 5. Orbits of the fuel rod centre $r = 10$ in the first rod segment on the level spacer grids 1, 3, 5 (upper figure) and 2, 4, 6 (lower figure)

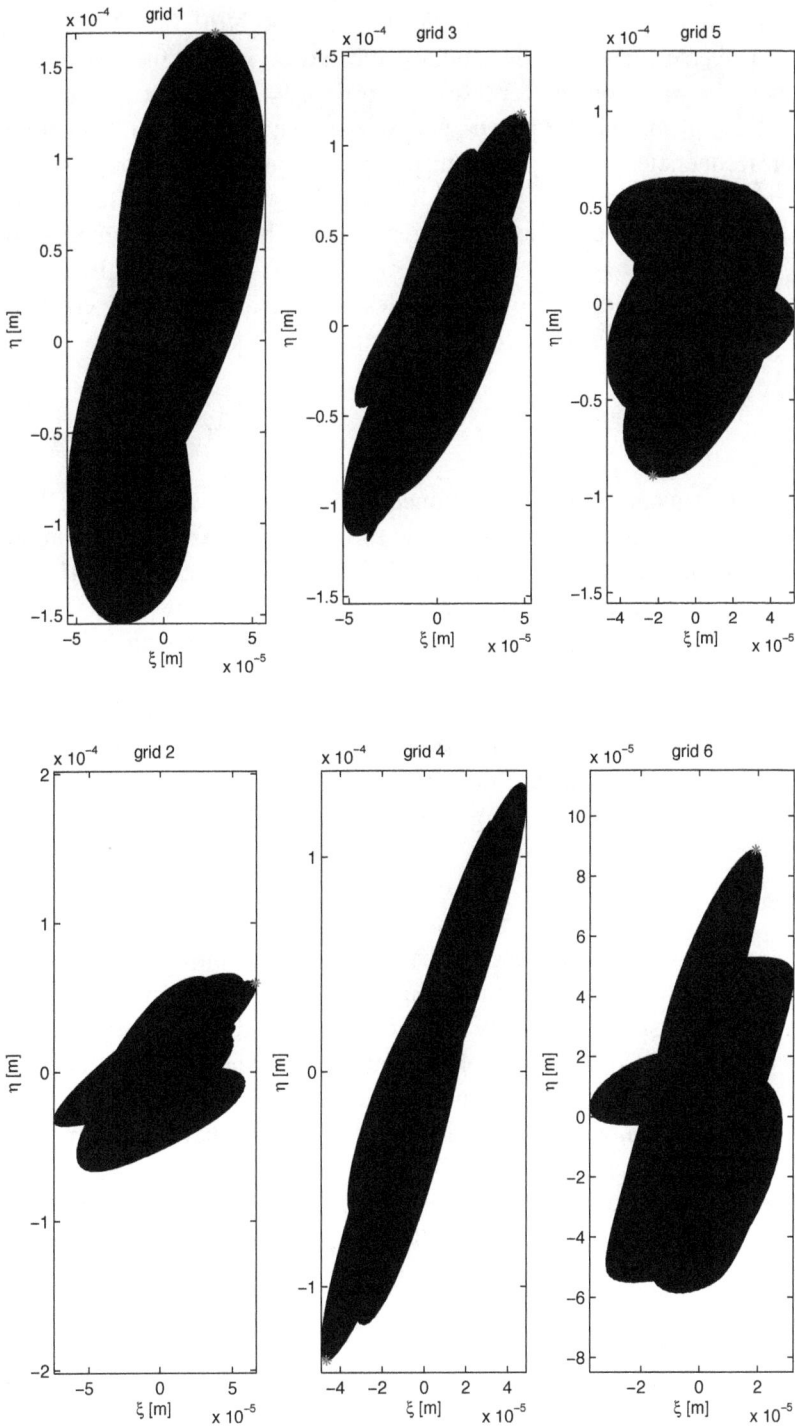

Fig. 6. Orbits from Fig. 5 depicted separately

4. Application

The presented methodology was applied for steady polyharmonic response of the Russian TVSA-T fuel assembly in the VVER 1000 reactor core in NPP Temelín. As an illustration, the orbits in transverse planes of the random selected ($r = 14$) fuel rod centre in the first fuel rod segment ($s = 1$) on the level spacer grids $g = 1$ to 6 caused by pressure pulsations generated by all circulation pumps [11] are shown in Fig. 5 and separately in Fig. 6. The revolution frequencies of the particular pumps in some coolant loops are slightly different $f_1 = f_2 = 16{,}635$ Hz and $f_3 = f_4 = 16{,}645$ Hz, whereas three harmonic components ($k = 1, 2, 3$) of the pressure pulsations were respected. The condensed model (20) with 3272 DOF ($m_s = 500, m_{CT} = n_{CT} = 32, m_{LS} = n_{LS} = 240$) was used for the calculation of the orbits. The accuracy of condensed model was tested in terms of relative errors of 125 lowest fuel assembly eigenfrequencies defined in the form

$$\varepsilon_\nu = \frac{|f_\nu(m_s) - f_\nu|}{f_\nu}, \quad \nu = 1, \ldots, 125, \tag{30}$$

where f_ν are eigenfrequencies of the full (noncondensed) model with 10832 DOF. The relative errors ε_ν for different condensation level of the rod segments expressed by number of the rod segment master eigenvectors $m_s = 100, 300, 500$ were investigated in [14]. Relative errors decrease with decreasing condensation level (m_s increases) in all FA eigenfrequencies. Upper limit of the relative error for $m_s = 500$ is in some higher eigenfrequencies 6 %. The orbits of these particular models distinguish only little. Time behaviour of the dynamic force transmitted by maximal loaded spacer grid cell (coupling $q = 147$ between fuel rod 6 in segment 3 and fuel rod 46 in segment 4) on the level of the first spacer grid ($g = 1$) is demonstrated in the Fig. 7. Time behaviour of the slip speed and friction power in the contact points of the mentioned full rods with spacer grid cell is shown in Fig. 8 (slip speed) and Fig. 9 (friction power).

5. Conclusion

The described method in direct sequence at the fuel assembly conservative mathematical model derived in [13] enables to investigate the flexural kinematic excited vibrations of all FA components. The vibrations are caused by spatial motion of the two horizontal support plates in the reactor core transformed into displacements of the kinematical excited nods of the FA components-fuel rods, guide thimbles, centre tube and skeleton angle pieces. The special coordinate system of radial and orthogonal lateral and flexural angular displacements around these directions of the fuel rods and guide thimbles enables to separate the hexagonal type FA into six identical revolved rod segments characterized in global FA mathematical model by identical mass, damping and stiffness matrices. These identical subsystems are linked each other and with centre tube and skeleton by spacer grids on the several level. All FA components are modelled as one dimensional continuum of beam type with nodal points in the gravity centres of their cross-sections on the level of the spacer grids.

The FA mathematical model has, in consequence of great number of fuel rods, to large DOF number for calculation of the dynamic response. Therefore is compiled condensed model based on reduction of the number rod segment eigenvectors conducive to FA dynamic response using modal synthesis method. The developed methodology was used for steady vibration analysis of the Russian type nuclear FA caused by motion of the support plates, excited by pressure pulsations generated by main circulation pumps in the coolant loops of the primary circuit. The

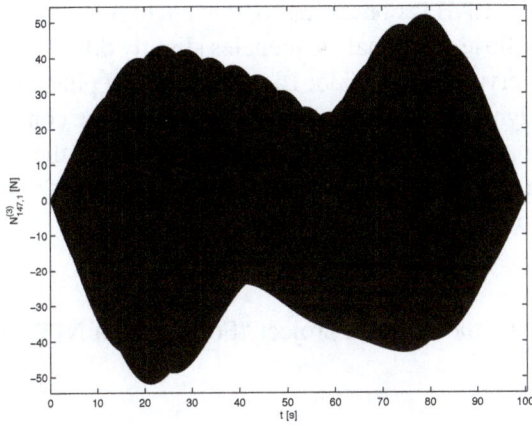

Fig. 7. Dynamic force transmitted by maximal loaded spacer grid cell (coupling 147, spacer grid 1)

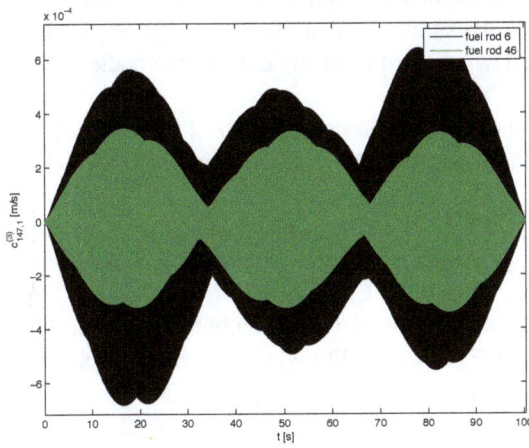

Fig. 8. Slip speed in the contact points specify in Fig. 7

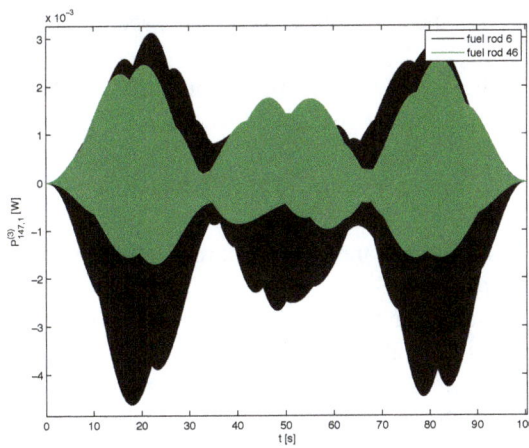

Fig. 9. Friction power in the contact points specify in Fig. 7

developed software in MATLAB is conceived so, that enables to choose an arbitrary configuration of operating pumps whose rotational frequencies slightly differentiate in the experimentally determined frequency interval $f \in \langle 16.635; 16.645 \rangle$ Hz. This phenomenon implicates beat vibrations, which amplify dynamic normal and friction forces in the contact of the fuel rod coating and spacer grid cells. The software enables an identification of the maximal dynamic loaded spacer grid cell and calculation of the maximal friction force work during defined time period. In this way the abrasion of fuel rod coating can be estimated.

Acknowledgements

This work was supported by the research project "Fuel cycle of NPP" of the NRI Řež plc.

References

[1] Fuel Assembly Mechanical Test Report, volume I a II, TEM-MC-04.RP (Rev 0), Property of JSC "TVEL" (inside information of NRI Řež, 2011).

[2] Hlaváč, Z., Zeman, V., The seismic response affection of the nuclear reactor WWER1000 by nuclear fuel assemblies, Engineering Mechanics 3/4 (17) (2010) 147–160.

[3] Hlaváč, Z., Zeman, V., Flexural vibration of the package of rods linked by lattices, Proceedings of the 8th conference Dynamics of rigid and deformable bodies 2010, Ústí nad Labem, 2010. (in Czech)

[4] Lavreňuk, P. I., Obosnovanije sovmestnosti TVSA-T PS CUZ i SVP s projektom AES Temelín, Statement from technical report TEM-GN-01, Sobstvennosť OAO TVEL (inside information of NRI Řež, 2009).

[5] Pečínka, L., Krupa, V., Klátil, J., Mathematical modelling of the propagation of the pressure pulsations in the piping systems of NPPs, Proceedings of the conference Computational Mechanics, UWB Plzeň, 1997, p. 203–210. (in Czech)

[6] Pečínka, L., Criterion assessment of fuel assemblies behaviour VV6 and TVSA-T at standard operating conditions of ETE V1000/320 type reactor, Research report DITI 300/406, NRI Řež, 2009. (in Czech)

[7] Smolík, J. and coll., Vvantage 6 Fuel Assembly Mechanical Test, Technical Report No. Ae 18018T, Škoda, Nuclear Machinery, Pilsen, Co. Ltd., 1995.

[8] Sýkora, M., Reactor TVSA-T fuel assembly insertion, part 4, Research report Pp BZ1, 2, ČEZ-ETE, 2009. (in Czech)

[9] Zeman, V., Hlaváč, Z., Pašek, M., Parametric identification of mechanical system based on measured eigenfrequencies and mode shapes. Zeszyty naukowe, nr. 8, Politechnika Slaska, Gliwice, 1998, p. 96–100.

[10] Zeman, V., Hlaváč, Z., Pečínka, L., Dynamic response of VVER1000/320 type reactor components excited by pressure pulsations generated by main circulation pumps, Report of the University of West Bohemia, No. 52120-02-07, Pilsen, 2007. (in Czech)

[11] Zeman, V., Hlaváč, Z., Dynamic response of VVER1000 type reactor excited by pressure pulsations, Engineering Mechanics 6 (15) (2008) 435–446.

[12] Zeman, V., Hlaváč, Z., Vibration of the package of rods linked by spacer grids, Vibration Problem ICOVP 2011, Springer, p. 227–233.

[13] Zeman, V., Hlaváč, Z., Modelling and modal properties of the nuclear fuel assembly, Applied and Computational Mechanics 5(2) (2011) 253–266.

[14] Zeman, V., Hlaváč, Z., Kinematical excited vibration of the nuclear fuel assembly, Proceedings of the 18th International Conference Engineering Mechanics 2012, Svratka, 2012, p. 1 597–1 602.

[15] Zeman, V., Hlaváč, Z., Dynamic response of nuclear fuel assembly in VVER 1000 type reactor caused by kinematical excitation, Report of the University of West Bohemia, No. 52120-04-12, Pilsen, 2012. (in Czech)

6

A computational method for determination of a frequency response characteristic of flexibly supported rigid rotors attenuated by short magnetorheological squeeze film dampers

J. Zapoměl[a,*], P. Ferfecki[a], L. Čermák[b]

[a] Centre of Smart Systems and Structures, Institute of Thermomechanics – Branch at VSB – Technical University of Ostrava, Czech Academy of Sciences, 17. listopadu 15, 708 33 Ostrava-Poruba, Czech Republic

[b] Institute of Mathematics, Brno University of Technology, Technická 2, 616 69 Brno, Czech Republic

Abstract

Lateral vibration of rotors can be significantly reduced by inserting the damping elements between the shaft and the casing. The theoretical analysis, confirmed by computational simulations, shows that to achieve the optimum compromise between attenuation of the oscillation amplitude and magnitude of the forces transmitted through the coupling elements between the rotor and the stationary part, the damping effect must be controllable. For this purpose, the squeeze film dampers lubricated by magnetorheological fluid can be applied. The damping effect is controlled by the change of intensity of the magnetic field in the lubricating film. This article presents a procedure developed for investigation of the steady state response of rigid rotors coupled with the casing by flexible elements and short magnetorheological dampers. Their lateral vibration is governed by nonlinear (due to the damping forces) equations of motion. The steady state solution is obtained by application of a collocation method, which arrives at solving a set of nonlinear algebraic equations. The pressure distribution in the oil film is described by a Reynolds equation modified for the case of short dampers and Bingham fluid. Components of the damping force are calculated by integration of the pressure distribution around the circumference and along the length of the damper. The developed procedure makes possible to determine the steady state response of rotors excited by their unbalance, to determine magnitude of the forces transmitted through the coupling elements in the supports into the stationary part and is intended for proposing the control of the damping effect to achieve optimum performance of the dampers.

Keywords: rotors, magnetorheological dampers, steady state response, collocation method

1. Introduction

The unbalance forces and moments of rotating parts are one of the main sources of lateral vibration of rotors working in industrial devices or means of transport. Their excessive vibration reduces the service life of all components of rotating machines, increases their noise and the forces transmitted through the coupling elements between the rotor and the stationary part. Excessive oscillations produce large deflection of the shaft, which may lead to exceeding the limit state of deformation and to occurrence of impacts between the discs and the rotor casing.

The damping devices inserted between the rotor and the stationary part can considerably reduce the vibration amplitude and magnitude of the transmitted forces. To achieve their efficient work, the damping effect must be controllable to be possible to adapt performance of the dampers to the current operating conditions.

*Corresponding author. e-mail: jaroslav.zapomel@vsb.cz.

2. The investigated rotor system

The investigated rotor (Fig. 1) consists of a shaft and of one disc. The rotor is mounted with rolling element bearings whose outer races are coupled with the casing by flexible elements. The system is symmetric relative to the middle plane of the disc. The rotor turns at constant angular speed and is loaded by its weight. In addition, it is excited by the centrifugal force produced by the disc unbalance.

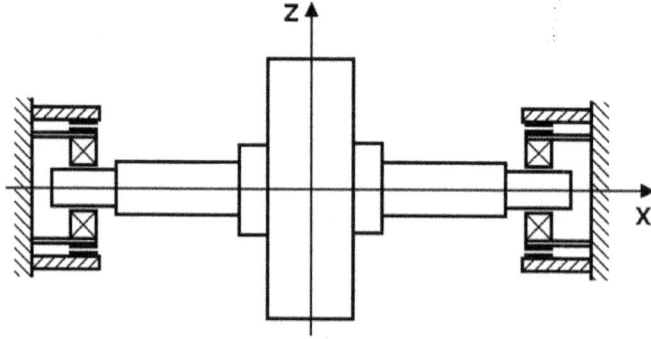

Fig. 1. Investigated rotor

To attenuate the rotor vibration, the damping devices should be inserted between the shaft journals and the casing. The task is to analyze their influence on the rotor steady state response. The attention should be focused on the dependence of amplitude of the vibration and magnitudes of the time varying forces transmitted through the coupling elements into the stationary part on the speed of the rotor rotation.

In the computational model, the rotor and the stationary part are considered as absolutely rigid and the spring elements supporting the rotor and the dampers as linear. Taking into account the system symmetry, lateral vibration of the rotor is described by two equations of motion

$$0.5m_R\ddot{y} + (b_D + 0.5b_P)\dot{y} + k_D y = 0.5m_R e_T \omega^2 \cos(\omega t + \psi_o), \tag{1}$$

$$0.5m_R\ddot{z} + (b_D + 0.5b_P)\dot{z} + k_D z = 0.5m_R e_T \omega^2 \sin(\omega t + \psi_o) - 0.5m_R g. \tag{2}$$

m_R is the mass of the rotor, b_D is the coefficient of linear damping of the damper, b_P is the damping coefficient of external damping (damping caused by the environment), k_D is stiffness of the supporting spring, e_T is eccentricity of the rotor centre of gravity, ψ_o denotes the phase lag of the unbalance force, y, z are the horizontal and vertical displacements of the rotor centre, ω is the angular speed of the rotor rotation, t is the time, g is the gravity acceleration and $(\dot{})$ and $(\ddot{})$ denote the first and second derivatives with respect to time.

On these conditions, the steady state trajectory of the rotor centre is a circle whose centre is slightly shifted in the vertical direction. Radius of the orbit and amplitude of the force transmitted via the spring and the damping elements in the rotor support depend on amount of the damping and on angular velocity of the rotor rotation

$$r = e_T \frac{\eta^2}{\sqrt{(1 - \eta^2)^2 + 4\xi^2\eta^2}}, \tag{3}$$

$$F_A = 0.5m_R e_T \omega^2 \sqrt{\frac{1 + 4\xi^2\eta^2}{(1 - \eta^2)^2 + 4\xi^2\eta^2}}, \tag{4}$$

where

$$\eta = \frac{\omega}{\Omega}, \qquad \Omega = \sqrt{\frac{2k_D}{m_R}}, \qquad \xi = \frac{b_D + 0.5b_P}{\sqrt{2k_D m_R}}. \tag{5}$$

r is the radius of the rotor centre trajectory, F_A is amplitude of the force transmitted through the coupling elements in each rotor support, Ω is the natural frequency of the rotor system, η is the frequency ratio and ξ is the damping ratio.

Analysis of relations (3) and (4) makes possible to draw several conclusions limiting application of passive and semiactive linear damping devices:

- rising damping always decreases amplitude of the rotor steady state vibration but for high revolutions the amplitude always approaches to eccentricity of the rotor unbalance and cannot be further reduced by the dampers,

- amplitude of the force transmitted via the coupling elements from the rotor into the stationary part with rising damping goes down if the frequency ratio η is lower than $\sqrt{2}$ and increases if η is greater than $\sqrt{2}$, but in this case its value is always less or equal to the centrifugal force caused by the rotor unbalance,

- for higher speeds of the rotor revolutions, amplitude of the rotor vibration is reduced only negligibly (is approximately equal to eccentricity of the rotor unbalance) but the forces transmitted through the coupling elements significantly rise.

It is evident that to achieve a compromise between attenuation of the amplitude of the rotor oscillation and magnitude of the force transmitted through the coupling elements the performance of the damper must be adaptable to the current operating conditions by means of the change of amount of damping in the supports.

3. Controllable magnetorheological squeeze film dampers

The control of the damping effect can be achieved by application of magnetorheological dampers. These damping devices are lubricated by magnetorheological liquids, which consist of the oil and of tiny ferromagnetic particles dispersed in it. If the magnetorheological liquid is not affected by magnetic field, it behaves as normal newtonian one. But if the magnetic field is applied, the flow begins only if the shear stress between two neighbouring layers exceeds the limit value (yield shear stress). In the areas where the limit value is not reached, the magnetorheological material forms a core in which the fluid behaves as a solid body.

In the mathematical models, the magnetorheological fluids are usually represented by Bingham or Bulkley-Herschel materials. Their properties, especially the relation between intensity of the magnetic field and the yielding shear stress, were studied e.g. by Kordonsky [5], Shulman et al. [6] and Si et al. [7].

The magnetorheological dampers consist of two rings, between which there is a thin film of magnetorheological liquid (Fig. 2). The rings are coupled with the casing of the rotating machine, the outer one directly, the inner ring by a squirrel spring. The shaft is supported by a rolling element bearing whose outer race is coupled with the inner ring of the damper. Vibration of the inner ring relative to the outer one squeezes the liquid in the lubricating layer, which produces the damping effect. In the stationary part of the damper, there are the coils, which are the source of magnetic field. Its intensity influences the resistance of the magnetorheological liquid against its flow and therefore the change of magnitude of the applied electric current can be used to control the damping effect.

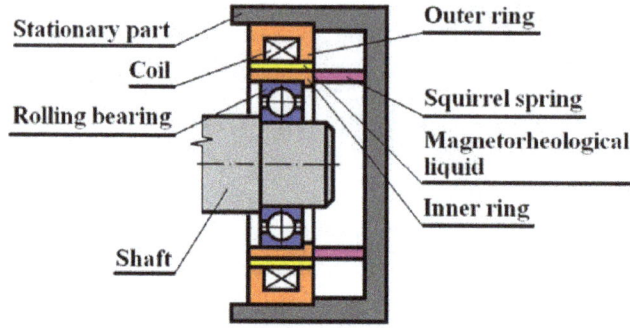

Fig. 2. Scheme of the squeeze film magnetorheological damper

The magnetorheological dampers have been a subject of intensive experimental and theoretical research since about the nineties of the 20th century. In [9], Wang et al. studied by means of experiments the vibration properties and the control method of a flexible rotor supported by a magnetorheological squeeze film damper. In [3, 4], Forte et al. presented results of the theoretical and experimental investigation of a long magnetorheological damper. In [8], Wang et al. developed a mathematical model of a long squeeze film magnetorheological damper based on modification of the Reynolds equation. The results of experiments carried out by Carmignani et al. with a squeeze film magnetorheological damper on a small test rotor rig were reported in [1, 2]. In [10] and [11], Zapomel and Ferfecki introduced the mathematical models of short and long squeeze film magnetorheological dampers. The developed model of a short damper was used for computational simulations of the transient response of a rigid rotor passing the critical speeds [12].

4. Mathematical modelling of a short magnetorheological squeeze film damper

In the developed mathematical model of a magnetorheological damper, it is assumed that (i) the inner and outer rings of the damper are absolutely rigid and smooth, (ii) the width of the damper gap is very small relative to the radii of both rings, (iii) ratio of the length of the damper to the diametre of its rings is small and the faces of the damper are not sealed (assumptions for a short damper), (iv) the lubricant behaves as Bingham liquid, (v) the yield shear stress depends on magnitude of the magnetic induction, (vi) the flow in the oil film (if occurs) is laminar and isothermal, (vii) the pressure of the lubricant in the radial direction is constant, (viii) the lubricant is considered to be massless, and (ix) the influence of the curvature of the oil film is negligible.

The thickness of the lubricating film depends on the position of the inner damper ring relative to the outer one

$$h = c - e_H \cos(\varphi - \gamma), \tag{6}$$

h is the thickness of the oil film, c is the width of the gap between the inner and outer rings of the damper, e_H is the journal eccentricity, φ is the circumferential coordinate, and γ is the position angle of the line of centres (Fig. 3).

The derivation, described in details in [11], arrives at relations for the pressure distribution in the lubricating layer

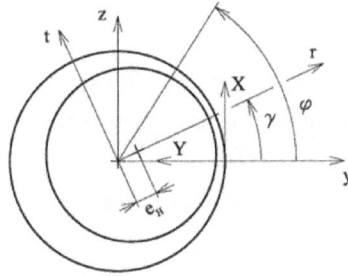

Fig. 3. The magnetorheological damper coordinate system

$$h^3 p'^3 + 3(h^2 \tau_y - 4\eta_B \dot{h} Z)p'^2 - 4\tau_y^3 = 0 \qquad \text{for} \quad p' < 0, \tag{7}$$

$$h^3 p'^3 - 3(h^2 \tau_y + 4\eta_B \dot{h} Z)p'^2 + 4\tau_y^3 = 0 \qquad \text{for} \quad p' > 0 \tag{8}$$

and for Y (radial) coordinate of the core boundary

$$h_1 = \frac{h}{2} + \frac{\tau_y}{p'} \qquad \text{for} \quad p' < 0, \tag{9}$$

$$h_1 = \frac{h}{2} - \frac{\tau_y}{p'} \qquad \text{for} \quad p' > 0. \tag{10}$$

p' denotes the pressure gradient in the axial direction, τ_y, η_B are the yield shear stress and viscosity of the Bingham liquid, h_1 is the radial coordinate of the core boundary on the side of the outer damper ring and Z is the axial coordinate.

The yielding shear stress τ_y depends on material properties and concentration of the ferromagnetic particles dispersed in the magnetorheological fluid, on intensity of the magnetic field in the damper gap and on several further parameters. Usually it is accepted

$$\tau_y = k_B H^{n_B}. \tag{11}$$

H denotes intensity of the magnetic field and k_B and n_B are the liquid material constants.

In the case of the simplest design of the damper, its inner and outer rings can be considered as a core of an electromagnet divided by two gaps and then the relation between the yield shear stress and the applied current in the coil can be expressed

$$\tau_y = k_d \left(\frac{I}{h}\right)^{n_B}, \tag{12}$$

where

$$k_d = k_B \left(\frac{N}{2}\right)^{n_B}. \tag{13}$$

N is the number of the coil turns, I is the current and k_d is a design parameter of the damper.

As evident from (12), the yield shear stress depends on the width of the damper gap and therefore, it changes around the circumference of the damper.

Determination of the pressure gradient for each value of the circumferential and axial coordinates requires solving cubic algebraic equations (7) and (8). Solution of each of them gives three roots. The one that has the physical meaning must satisfy the following three conditions

- it must be real (not complex),

- the conditions of validity of equations (7) and (8) must be satisfied, this means that the real roots obtained from (7) must be negative and the real ones obtained from (8) must be positive,

- $0 < h_1(p') < \frac{h}{2}$.

The pressure profile is calculated by integration of the pressure gradient

$$p = \int p' \, \mathrm{d}Z \tag{14}$$

with the boundary condition expressing that the pressure at the edge of the damper is equal to the atmospheric one

$$p = p_A \quad \text{for} \quad Z = \pm \frac{L}{2}. \tag{15}$$

p_A is the pressure in the surrounding space (atmospheric pressure) and L is the length of the damper.

If pressure at some location in the oil film drops to the critical level, a cavitation takes place. Further it is assumed that the cavitation occurs only in the area where the width of the damper gap increases with time and that pressure of the medium in cavitated areas is equal to the pressure in the ambient space. Then it holds with enough accuracy

$$p_d = p \qquad \text{for} \quad p \geq p_{CAV}, \tag{16}$$
$$p_d = p_{CAV} \quad \text{for} \quad p < p_{CAV}. \tag{17}$$

p_d is the pressure distribution in the layer of lubricant and p_{CAV} is the pressure in the cavitated area. Differentiation of (6) with respect to time gives the equation for calculation of the circumferential coordinates of the borders of the cavitated area

$$\dot{e}_H \cos(\varphi_{CAV} - \gamma) + e_H \dot{\gamma} \sin(\varphi_{CAV} - \gamma) = 0. \tag{18}$$

Its solution gives two roots that define the angular coordinates $(\phi_{CAV1}, \phi_{CAV2})$ of the beginning and end edges of the cavitated region.

Assuming that the damper is symmetric relative to its middle plane perpendicular to the shaft centre line, components of the damping force are obtained by integration of the pressure distribution around the circumference and along the length of the damper

$$F_{dy} = -2R \int_0^{2\pi} \int_0^{\frac{L}{2}} p_d \cos \varphi \, \mathrm{d}Z \, \mathrm{d}\varphi, \tag{19}$$

$$F_{dz} = -2R \int_0^{2\pi} \int_0^{\frac{L}{2}} p_d \sin \varphi \, \mathrm{d}Z \, \mathrm{d}\varphi. \tag{20}$$

F_{dy}, F_{dz} are the y and z components of the damping force respectively and R denotes the inner ring radius.

5. The equations of motion of the investigated rotor system

To control the damping effect, the magnetorheological dampers are inserted between the spring elements, which are mounted with the outer race of the rolling element bearings, and the casing. The springs are prestressed in the vertical direction to eliminate their deflection caused by the weight of the rotor.

Lateral vibration of the investigated rotor system is then described (taking into account the system symmetry) by two equations of motion

$$0.5m_R\ddot{y} + 0.5b_P\dot{y} + k_D y = F_{dy}(y, z, \dot{y}, \dot{z}) + 0.5m_{ReT}\omega^2\cos(\omega t + \psi_o), \tag{21}$$

$$0.5m_R\ddot{z} + 0.5b_P\dot{z} + k_D z = F_{dz}(y, z, \dot{y}, \dot{z}) + F_{PS} + 0.5m_{ReT}\omega^2\sin(\omega t + \psi_o) - 0.5m_R g \tag{22}$$

that are nonlinear and mutually coupled due to the hydraulic damping forces. F_{PS} denotes the prestress force

$$F_{PS} = 0.5m_R g. \tag{23}$$

Because of the prestress of the spring elements, the stiffness and damping properties in the supports are isotropic and the direction of the damping force in the damper depends only on the direction of excitation caused by the centrifugal force due to the disc unbalance and turns with the same angular speed as the rotor rotates. Therefore, trajectory of the rotor centre has a circular form. Nevertheless, its radius is not proportional to the loading magnitude because of nonlinear character of the damping force.

This enables to assume the steady state solution of the equations of motion (21) and (22) in the form

$$y = r\cos(\omega t + \psi_r), \tag{24}$$

$$z = r\sin(\omega t + \psi_r). \tag{25}$$

r is the radius of the rotor centre trajectory and ψ_r is the phase lag. Introducing the substitutions

$$r_C = r\cos\psi_r, \tag{26}$$

$$r_S = r\sin\psi_r, \tag{27}$$

the relationships (24) and (25) take the form

$$y = r_C\cos\omega t - r_S\sin\omega t, \tag{28}$$

$$z = r_C\sin\omega t + r_S\cos\omega t. \tag{29}$$

The unknown values of coefficients r_C and r_S can be calculated by application of a collocation method. This requires to substitute (28), (29) and their first and second derivatives with respect to time into (21) and (22) and to express the resulting equations at the collocation points of time.

As the number of unknown parameters and the number of equations is two, only one collocation point (collocation point of time) is needed. Carrying out the mentioned manipulations for the collocation time equal to 0 s arrives at a set of two nonlinear algebraic equations whose solution gives the values of the unknown parameters r_C and r_S

$$(k_D - 0.5m_R\omega^2)r_C - 0.5\omega b_P r_S - 0.5m_{ReT}\omega^2 - F_{dy}(r_C, r_S) = 0, \tag{30}$$

$$0.5\omega b_P r_C + (k_D - 0.5m_R\omega^2)r_S - F_{dz}(r_C, r_S) = 0. \tag{31}$$

6. Analysis of the investigated rotor with controllable magnetorheological dampers

The first task is to study amplitude of the rotor vibration (radius of the rotor centre trajectory) and of the force transmitted through the dampers and the flexible support elements into the stationary part during the rotor steady state running. The second task is to propose the dependence of magnitude of the electric current supplied into the coils on the speed of the rotor rotation so that the rotor could rotate at constant speed in the whole range of its working revolutions (including the resonances) and the time varying component of the force transmitted into the casing and its vibration amplitude could remain lower than 500 N and 0.2 mm.

Results of the computer simulations are evident from the figures. In Fig. 4 and 5, there are drawn the dependences of amplitude of the rotor vibration and amplitude of the force transmitted into the stationary part on angular velocity of the rotor rotation for four magnitudes of the applied electric current.

The results show that for rising rotor revolutions the increasing current contributes to reduction of the vibration amplitude only negligibly but the magnitude of the force transmitted into the rotor casing increases significantly. It is also evident that the increase of the current has a significant influence on the damping effect. If the damping is too strong, the supports behave as very stiff, the rotor almost does not oscillate and the force transmitted through the coupling

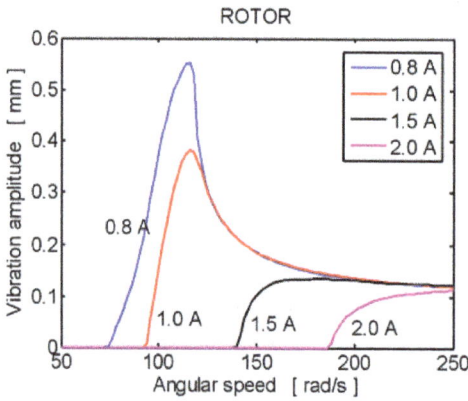

Fig. 4. Vibration amplitude — speed of rotation relation

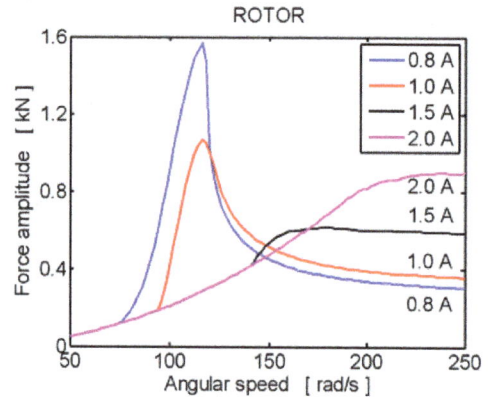

Fig. 5. Force amplitude — speed of rotation relation

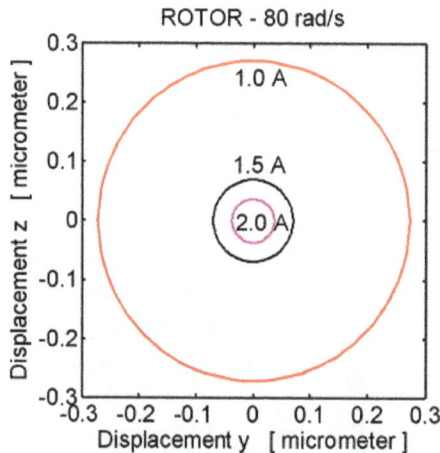

Fig. 6. Orbits of the rotor centre

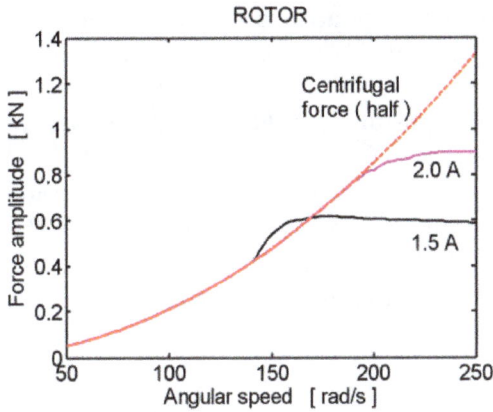

Fig. 7. Amplitude of the transmitted force

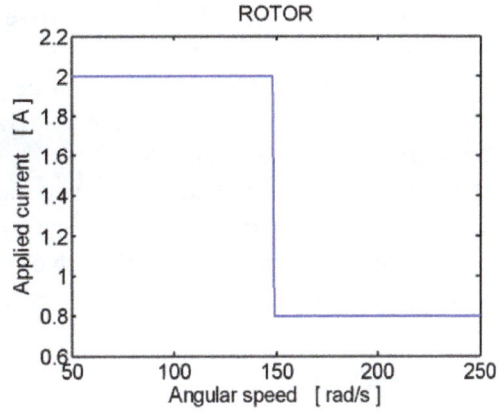

Fig. 8. Proposal of the current control

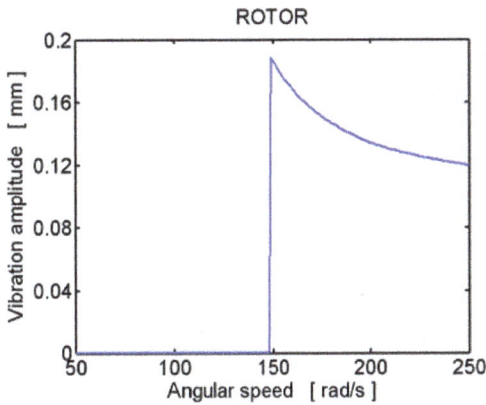

Fig. 9. Controlled vibration amplitude

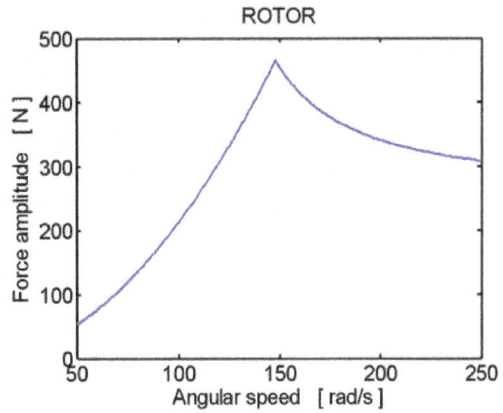

Fig. 10. Controlled force amplitude

elements is almost equal to the half of the whole centrifugal force produced by the rotor unbalance. This can be seen in Fig. 6 and 7. Fig. 8 shows the proposed dependence of the control current on the angular velocity of the rotor rotation to achieve the specified requirements. The corresponding amplitudes of the rotor vibration and of the transmitted forces are evident from Fig. 9 and 10.

7. Conclusions

The carried out analysis shows that to achieve efficient performance of the damping devices placed between the rotor and the stationary part and to reach the compromise between reduction of the rotor lateral vibration and magnitude of the forces transmitted via the rotor coupling elements, the damping effect must be controllable.

The approach described here represents a computational procedure for determination of the steady state response of a rigid symmetric rotor supported by flexible couplings combined with short magnetorheological squeeze film dampers. It is intended for determination of the steady state response of rotors excited by their unbalance and for determination of the magnitude of forces transmitted between the rotors and their casings through the coupling elements. The results can be used for judgement of the rotor limit state of deformation and for preparation of the input data for evaluation of the service life of the rotor components. From the mathematical point of view the presented procedure arrives at solving a set of nonlinear algebraic equations. For this purpose the Newton method was used. The pressure distribution in the lubricating film is described by a Reynolds equation modified for the case of short dampers and Bingham liquid. The damping forces are calculated by integration of the pressure distribution in the lubricating film utilizing the trapezoidal rule.

The carried out simulations show that there are some differences in behaviour of the rotors damped by classical linear and nonlinear magnetorheological squeeze film dampers and that the suitable change of the damping effect makes possible to satisfy the requirements put on maximum amplitude of the rotor vibration and of the forces transmitted to the stationary part.

Acknowledgements

This research work has been supported by the research grant projects P101/10/0209 and AVO Z20760514. The support is gratefully acknowledged.

References

[1] Carmignani, C., Forte, P., Rustighi, E., Design of a novel magneto-rheological squeeze-film damper, Smart Materials and Structures 15 (1) (2006) 164–170.

[2] Carmignani, C., Forte, P., Badalassi, P., Zini, G., Classical control of a magnetorheological squeeze-film damper, Proceedings of the conference Stability and Control Processes 2005, Saint-Petersburg, 2005, pp. 1 237–1 246.

[3] Forte, P., Paterno, M., Rustighi, E., A magnetorheological fluid damper for rotor applications, International Journal of Rotating Machinery 10 (3) (2004) 175–182.

[4] Forte, P., Paterno, M., Rustighi, E., A magnetorheological fluid damper for rotor applications, Proceedings of the IFToMM Sixth International Conference on Rotor Dynamics, Sydney, 2002, pp. 63–70.

[5] Kordonsky, W., Elements and devices based on magnetorheological effect, Journal of Intelligent Material Systems and Structures 4 (1) (1993) 65–69.

[6] Shulman, Z.-P., Kordonsky, V.-I., Zaltsgendler, E.-A., Prokhorov, I.-V., Khusid, B.-M., Demchk, S.-A., Structure, physical properties and dynamics of magnetorheological suspensions, International Journal of Multiphase Flow 12 (6) (1986) 935–955.

[7] Si, H., Peng, X., Li, X., A micromechanical model for magnetorheological fluids, Journal of Intelligent Material Systems and Structures 19 (1) (2008) 19–23.

[8] Wang, G.-J., Feng, N., Meng, G., Hahn, E.-J., Vibration control of a rotor by squeeze film damper with magnetorheological fluid, Journal of Intelligent Material Systems and Structures 17 (4) (2006) 353–357.

[9] Wang, J., Meng, G., Hahn, E.-J., Experimental study on vibration properties and control of squeeze mode MR fluid damper-flexible rotor system, Proceedings of the 2003 ASME Design Engineering Technical Conference & Computers and Information in Engineering Conference, Chicago, 2003, pp. 955–959.

[10] Zapoměl, J., Ferfecki, P., Mathematical modelling of a long squeeze film magnetorheological damper for rotor systems, Modelling and Optimization of Physical Systems, Wisła, 2010, pp. 97–102.

[11] Zapoměl, J., Ferfecki, P., Mathematical modelling of a short magnetorheological damper, Transactions of the VŠB – Technical University of Ostrava, Mechanical Series LV (1) (2009) 289–294.

[12] Zapoměl, J., Ferfecki, P., A computational investigation of vibration attenuation of a rigid rotor turning at a variable speed by means of short magnetorheological dampers, Applied and Computational Mechanics 3 (2) (2009) 411–422.

7

Semi-analytic solution to planar Helmholtz equation

M. Tukač[a,*], T. Vampola[a]

[a] *Faculty of Mechanical Engineering, Czech Technical University in Prague, Technická 4, 166 07 Praha 6, Czech Republic*

Abstract

Acoustic solution of interior domains is of great interest. Solving acoustic pressure fields faster with lower computational requirements is demanded. A novel solution technique based on the analytic solution to the Helmholtz equation in rectangular domain is presented. This semi-analytic solution is compared with the finite element method, which is taken as the reference. Results show that presented method is as precise as the finite element method. As the semi-analytic method doesn't require spatial discretization, it can be used for small and very large acoustic problems with the same computational costs.

Keywords: semi-analytic, acoustics, Helmholtz, Galerkin, weighted residuals method

1. Introduction

Numerical methods that are used to simulate acoustic properties inside or outside an acoustic domain are under constant development. Sound, as we hear it, is composed of many tones, that propagate at certain excitation frequencies. These frequencies must not be close to each other. Moreover it is probable that they span from lower over middle to high excitation frequencies. The wide range is a computational difficulty. Many computational methods exist. However limits of their usability restrict them to a limited range of acoustic frequencies. One of the most widely used methods for computing interior problems, the finite element method [14], is generally suitable only for lower excitation frequencies. As the frequency rises and the element size diminishes, pollution error deviates the correct solution. Boundary element method [4] is similar to the finite element method. This method is often used, when exterior or sound radiation problems are to be solved. Usability limits of these two method are given by the number and the size of used elements. The higher the excitation frequency is, the higher is the number of nodes and the smaller the elements become. At some element size numerical errors start to depreciate the solution. Then other methods, as e.g. the statistical energy analysis [13] has to be deployed. However there may be a range of excitation frequencies that is already too high for finite and boundary element method and too low for the statistical energy analysis. To address this "non-solvable" frequency range many extensions to before-mentioned methods were developed. Or new approaches based on some analytic properties were proposed. Approach based on the superposition theory and the integro-modal approach [1] is presented in [6]. Another possibility is to try to solve directly the Helmholtz equation (3). Either by an iterative procedure as in [10] or by the use of variational theory as in [8]. The paper [3] presents improved element free Galerkin method. Approaches that are based on the analytic solution to the Helmholtz equation are also under development. So called Trefftz methods [5] use as approximation functions harmonic

---

*Corresponding author. e-mail: martin.tukac@fs.cvut.cz.

and evanescent exponential functions that are the solution of the underlying partial differential equation in a rectangular or a block domain. Recently developed wave-based method [11] belongs to the group of Trefftz methods. However the methods based on analytic solution require simpler geometry.

In this text a novel approach based on analytic solution to Helmholtz equation in a rectangular domain (3) is described. Full derivation of the approximation functions and the application on a car-like interior cavity is described.

2. Mathematical description of the acoustic problem

Fluctuation of the acoustic pressure p in two-dimensional space is described in general by wave equation [12]

$$\frac{\partial^2 p}{\partial x^2} + \frac{\partial^2 p}{\partial y^2} = \frac{1}{c_0^2}\frac{\partial^2 p}{\partial t^2}. \tag{1}$$

In this text time-harmonic acoustic pressure behavior is considered

$$p(x,y,t) = p_0(x,y)\exp(j\omega t), \tag{2}$$

where $j = \sqrt{-1}$, c_0 [m/s] being the speed of sound in the acoustic fluid of the density ϱ [kg/m^3] and ω [s^{-1}] being the radial excitation frequency. The equation (1) is transformed into Helmholtz equation [7]

$$\frac{\partial^2 p_0}{\partial x^2} + \frac{\partial^2 p_0}{\partial y^2} + k^2 p_0 = 0, \tag{3}$$

where the wave number $k = \omega/c_0$. For the sake of notation simplicity, let's denote the acoustic pressure amplitudes $p_0(x,y)$ as $p(x,y)$.

The equation (3) is defined on a bounded region Ω. On the boundary Γ following types of boundary conditions may be applied.

- Normal acoustic velocity $\bar{v}$ on the boundary Γ_v is prescribed by Neumann boundary conditions

$$\frac{j}{\varrho\omega}\frac{\partial p(x,y)}{\partial\vec{n}} = \bar{v} \quad \text{on } \Gamma_v. \tag{4}$$

$\vec{n}$ being the outward pointing normal of the domain Ω.

- Acoustic pressure $\bar{p}$ on the boundary Γ_p is prescribed by Dirichlet boundary conditions

$$p(x,y) = \bar{p} \quad \text{on } \Gamma_p. \tag{5}$$

The solution of (3) in Cartesian coordinates in rectangular domain Ω is defined according to [7] as

$$p(x,y) = \left(A\cos\alpha x + B\sin\alpha x\right)\left(C\cos\beta y + D\sin\beta y\right). \tag{6}$$

The constants A, B, C, D and the wave numbers α and β are to be determined from the boundary conditions. Additionally for α and β holds

$$\alpha^2 + \beta^2 = k^2. \tag{7}$$

3. Finite element method

The finite element method, as proposed in [2, 15] operates with elements and nodes, that are defined in the domain Ω. Amplitudes of the acoustic pressure in an element p_f^e are described by polynomial functions (8). The superscript e denotes elements and the subscript f denotes the finite element method. In two dimensional problems the polynomials are usually bilinear functions $g(x^e, y^e)$. In that case p_f^e is defined as

$$p_f^e(x^e, y^e) = \sum_{i=1}^{m_n} p_i^e \cdot g_i(x^e, y^e), \tag{8}$$

where x^e, y^e are element local coordinates and p_i^e are the unknown acoustic pressures in m_n element's nodes. The vector $\boldsymbol{p}_f^e = [p_1^e, \ldots, p_{m_n}^e]^T \cdot e^{j\omega t}$ stores nodal pressures p_i^e of the element's nodes. As stated in [2], for every element mass matrix $\boldsymbol{M}_f^e$, damping matrix $\boldsymbol{B}_f^e$, stiffness matrix $\boldsymbol{K}_f^e$ and the vector of known nodal pressures $\boldsymbol{b}_f^e$ can be derived. Resulting damped element equations of motion are

$$\boldsymbol{M}_f^e \frac{\partial^2 \boldsymbol{p}_f^e}{\partial t^2} + \boldsymbol{B}_f^e \frac{\partial \boldsymbol{p}_f^e}{\partial t} + \boldsymbol{K}_f^e \boldsymbol{p}_f^e = \boldsymbol{b}_f^e. \tag{9}$$

The domain Ω consists of multiple elements. Total number of nodes in Ω is m. Mass, damping and stiffness matrices and right hand side vectors of individual elements are composed together to form global matrices of the whole system. Resulting steady-state acoustic problems are described by

$$\left(-\omega^2 \boldsymbol{M}_f + j\omega \boldsymbol{B}_f + \boldsymbol{K}_f\right) \boldsymbol{p}_f = \boldsymbol{b}_f. \tag{10}$$

The vector $\boldsymbol{p}_f = [p_1, \ldots, p_m]^T$ stores all unknown nodal acoustic pressures in Ω. Matrices in (10) are frequency independent and in case undamped system is being solved damping matrix $\boldsymbol{B}_f$ becomes zero.

The number of used elements and nodes is driven by the highest excitation component of ω. Eight elements per the shortest wavelength $\lambda_{min} = (2\pi c_0)/\omega_{max}$ are said to secure proper and precise solution.

4. Semi-analytic solution

Presented semi-analytic method was derived from the solution (6) and a specific set of boundary conditions applied to rectangular domain Ω.

4.1. Derivation of the basis functions

The approximation of the solution is based on the analytic solution (6). Derivation of the linear combination of the semi-analytic solution requires only Neumann boundary conditions. A specific set of boundary conditions is applied to the rectangular domain Ω with dimensions L_x and L_y, see Fig. 1. Unlike Dirichlet boundary conditions, that have to be continuous in the corners of the domain, Neumann boundary conditions may be discontinuous.

The course of the normal acoustic velocity excitation function $\bar{v}$ along the boundary Γ_v is in Fig. 1. Three of the four domain sides have zero normal acoustic velocity. On the last side a non-zero normal acoustic velocity function is prescribed. Substituted boundary conditions for

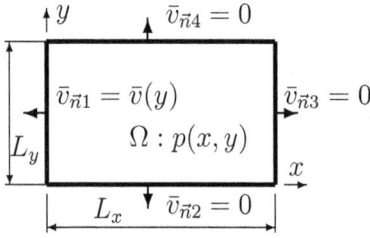

Fig. 1. Normal acoustic velocity boundary conditions for one domain problem

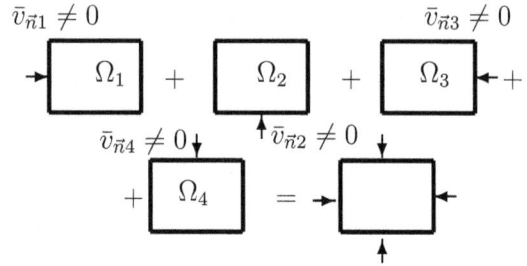

Fig. 2. Schematic procedure of computing rectangular cavities with non-zero boundary conditions on all sides

one domain are as follows:

$$v_{\bar{n}1}(0,y): \ -\bar{v}(y) = \frac{j\alpha}{\varrho\omega}(-A\sin\alpha 0 + B\cos\alpha 0)(C\cos\beta y + D\sin\beta y), \tag{11}$$

$$v_{\bar{n}2}(x,0): \ 0 = \frac{j\beta}{\varrho\omega}(A\cos\alpha x + B\sin\alpha x)(-C\sin\beta 0 + D\cos\beta 0), \tag{12}$$

$$v_{\bar{n}3}(L_x,y): \ 0 = \frac{j\alpha}{\varrho\omega}(-A\sin\alpha L_x + B\cos\alpha L_x)(C\cos\beta y + D\sin\beta y), \tag{13}$$

$$v_{\bar{n}4}(x,L_y): \ 0 = \frac{j\beta}{\varrho\omega}(A\cos\alpha x + B\sin\alpha x)(-C\sin\beta L_y + D\cos\beta L_y). \tag{14}$$

In the course of manipulation of equations (11)–(14) it can be assumed, that the space coordinates x, y, and the wavenumbers α and β can be non-zero values. Then the following relations can be determined:

$$D = 0 \quad \text{and} \quad \beta_n = n\frac{\pi}{L_y}, \quad n = 0,\dots,n_F. \tag{15}$$

Every wave number β_n is related to α_n by (7). Though α_n can be easily computed. From (13) the equation for A_n

$$A_n = B_n\frac{1}{\tan(\alpha_n L_x)} \tag{16}$$

is obtained. Substituting of (15), (16) and α_n into (6) and combining newly-emerged $B_n \cdot C_n = H_n$ also in (19) the final pressure approximation (17) in one rectangular domain is obtained as

$$\tilde{p}(x,y) = \sum_{n=0}^{n_F} H_n\left(\frac{1}{\tan(\alpha_n L_x)}\cos\alpha_n x + \sin\alpha_n x\right)\cos\beta_n y. \tag{17}$$

In matrix notation,

$$\tilde{p} = \mathbf{A}^\star \cdot \mathbf{h}^\star, \tag{18}$$

the basis functions are stored in a row vector $\mathbf{A}^\star = [A_0,\dots,A_{n_F}]$ and the unknown coefficients H_n are stored in a column vector $\mathbf{h}^\star = [H_0,\dots,H_{n_F}]^T$. Pressure approximation $\tilde{p}$ for the set of boundary conditions from Fig. 1 is a linear combination for unknown coefficients H_n.

From (11) a relation between the boundary value function $\bar{v}(y)$ and the unknown coefficients H_n is obtained as

$$H_n\alpha_n\cos\beta_n y_m = -\frac{\varrho\omega}{j}\bar{v}(y_m). \tag{19}$$

Unknown H_n from (19) can be obtained e.g. by using the least square method. In that case the boundary function $\bar{v}$ is evaluated at discrete positions $y = y_m$.

The number of basis functions n_F is derived from the requirement that the wavelength corresponding to the highest n is at least half the length of the wavelength in the acoustic fluid excited at the radial frequency ω.

4.2. Extension to more excited sides

Acoustic pressure approximation in (17) can only solve problems with one combination of boundary conditions. However, domain Ω can generally be excited on all sides by non-zero normal acoustic velocities. Acoustic problem is a linear problem of seeking a solution to a partial differential equation (1). For linear problems superposition theorem is valid.

Thus the solution of a rectangular domain excited on all sides is computed as the sum of four properly modified solutions (17). The scheme is in Fig. 2. The non-zero excitation functions can be discontinuous in the corners. As with only one excited side, the approximation for all excited sides can be written in matrix form. A row vector $\boldsymbol{A} = [\boldsymbol{A}_1^\star, \dots \boldsymbol{A}_4^\star]$ collects all basis functions of all four approximations and the column vector $\boldsymbol{h} = [\boldsymbol{h}_1^\star, \dots, \boldsymbol{h}_4^\star]^T$ stores all the unknown coefficients.

4.3. Application of the weighted residuals method

The unknown coefficients H_n can be obtained from (19) using least square method. This works for both the problem with single excited side and for the problem with all excited sides. In the latter case it requires computing the coefficients four times, individually for each side.

There may exist an acoustic domain, that is composed of two or more rectangular domains. In that case evaluating four individual solutions represents a problem for the coupling. Independently obtained solutions can not describe properly the continuity conditions on the common boundary between the domains.

Next, let assume an acoustic problem consisting of two rectangular domains. Between the domains there is interface boundary Γ_i. The outer boundary can be excited by both Dirichlet or Neumann boundary conditions.

System of linear equations is a convenient way of solving linear problems. As long as the approximation $\tilde{p}$ from (17) is a linear combination of basis function, Galerkin weighted residuals method described in [9] can be used. Using Galerkin weighted residuals method the unknown coefficient in $\tilde{p}$ are determined from a system of linear equations. In the Galerkin modification of weighted residuals method basis functions in $\tilde{p}$ are used as the weighting functions. The method minimizes the residuals in the domains Ω and on the boundary Γ. The residuals

$$R_\Omega = \frac{\partial^2 \tilde{p}}{\partial x^2} + \frac{\partial^2 \tilde{p}}{\partial y^2} + k^2 \tilde{p} \quad \text{on } \Omega, \tag{20}$$

$$R_{\Gamma_v} = \frac{j}{\varrho\omega}\frac{\partial \tilde{p}}{\partial \vec{n}} - \bar{v} \quad \text{on } \Gamma_v, \tag{21}$$

$$R_{\Gamma_p} = \tilde{p} - \bar{p} \quad \text{on } \Gamma_p \tag{22}$$

are the difference between the exact solution and the approximation. However R_Ω is in every domain identically equal to zero, because the basis functions are solution of the equation (3). Application of Galerkin weighted residuals method leads to the following equation in the case

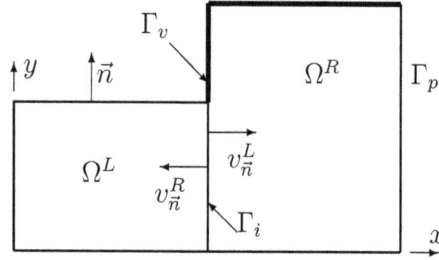

Fig. 3. Scheme of coupling of two domains when the Galerkin weighted residuals method is used

of one domain

$$
\underbrace{\int_{\Gamma_p} \frac{j}{\varrho\omega} \frac{\partial \boldsymbol{A}^T}{\partial \vec{n}} \boldsymbol{A} \, \mathrm{d}\Gamma_p \cdot \boldsymbol{h}}_{C_p} + \underbrace{\int_{\Gamma_v} \boldsymbol{A}^T \frac{j}{\varrho\omega} \frac{\partial \boldsymbol{A}}{\partial \vec{n}} \, \mathrm{d}\Gamma_v \cdot \boldsymbol{h}}_{C_v} =
$$

$$
\underbrace{\int_{\Gamma_p} \frac{j}{\varrho\omega} \frac{\partial \boldsymbol{A}^T}{\partial \vec{n}} \bar{p} \, \mathrm{d}\Gamma_p}_{b_p} + \underbrace{\int_{\Gamma_v} \boldsymbol{A}^T \bar{v} \, \mathrm{d}\Gamma_v}_{b_v} . \tag{23}
$$

On the boundary Γ_i both pressure and velocity continuity have to be enforced. Both requirements represent a boundary condition. The pressure continuity is applied to the domain on the "left" side of the interface and the velocity continuity condition to the "right" domain. Let's denote the domains around the boundary Ω^L and Ω^R. The residuals on the interface are

$$
R_{\Gamma_i}^L = \tilde{p}^L - \tilde{p}^R, \tag{24}
$$

$$
R_{\Gamma_i}^P = \frac{j}{\varrho\omega} \frac{\partial \tilde{p}^L}{\partial \vec{n}^L} + \frac{j}{\varrho\omega} \frac{\partial \tilde{p}^R}{\partial \vec{n}^R}. \tag{25}
$$

Equations (26) and (27) show the extension of the formulation (23) for a problem with two acoustic domains. The extension adds these members

$$
\ldots + \underbrace{\int_{\Gamma_i} \frac{j}{\varrho\omega} \frac{\partial \boldsymbol{A}^{L^T}}{\partial \vec{n}^L} \boldsymbol{A}^L \, \mathrm{d}\Gamma_i \cdot \boldsymbol{h}^L}_{C_{Bp}^L} + \underbrace{\int_{\Gamma_i} \frac{j}{\varrho\omega} \frac{\partial \boldsymbol{A}^{L^T}}{\partial \vec{n}^L} \boldsymbol{A}^R \, \mathrm{d}\Gamma_i \cdot \boldsymbol{h}^R}_{C_{LR}} = \ldots, \tag{26}
$$

$$
\ldots + \underbrace{\int_{\Gamma_i} \frac{j}{\varrho\omega} \frac{\partial \boldsymbol{A}^{R^T}}{\partial \vec{n}^R} \boldsymbol{A}^L \, \mathrm{d}\Gamma_i \cdot \boldsymbol{h}^L}_{C_{RL}} + \underbrace{\int_{\Gamma_i} \frac{j}{\varrho\omega} \frac{\partial \boldsymbol{A}^{R^T}}{\partial \vec{n}^R} \boldsymbol{A}^R \, \mathrm{d}\Gamma_i \cdot \boldsymbol{h}^R}_{C_{Bv}^R} = \ldots \tag{27}
$$

4.4. Resulting system of equations

Assembled matrices in (23) form a system of linear equations. Problem of one domain is solved with

$$
[\boldsymbol{C}_p + \boldsymbol{C}_v] \cdot \boldsymbol{h} = \boldsymbol{b}_v + \boldsymbol{b}_p. \tag{28}
$$

Fig. 4. Course of acoustic pressure boundary conditions. Points P_1 and P_2 are depicted as well as the interior boundaries of individual domains

In case more domains are coupled together, the system is extended with the matrices from (26) and (27). In this text only the situation of two coupled domains is shown

$$
\begin{bmatrix} C_p^L + C_v^L + C_{Bp}^L & C_{LR} \\ C_{RL} & C_p^R + C_v^R + C_{Bv}^R \end{bmatrix} \cdot \begin{pmatrix} h^L \\ h^R \end{pmatrix} = \begin{pmatrix} b_p^L + b_v^L \\ b_p^R + b_v^R \end{pmatrix} . \tag{29}
$$

Extension of the system (29) for more domains is straightforward. Unlike FEM, derived matrices are frequency dependent and the matrix elements have to be recomputed for every excitation frequency.

5. Application on a car-like acoustic cavity

Proposed acoustic solution was compared to the finite element solution on an example of a car-like interior.

5.1. Acoustic domain and the boundary conditions

The shape, depicted in Fig. 4, represents simplified car interior. Semi-analytic method is designed for the use on convex acoustic domains, that resemble rectangles. To fulfil this demand, the car-like cavity was divided into seventeen nearly rectangular domains, see Fig. 4. The whole system is solved using the Galerkin weighted residuals method applied to (17).

Only acoustic pressure boundary conditions are applied. The course and the location of applied boundary conditions is shown in the Fig. 4. Maximum amplitude of the exciting acoustic pressure is $\bar{p}_{max} = 2\,\mathrm{Pa}$. The course of the boundary condition function is continuous.

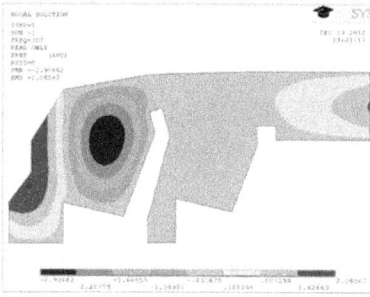

Fig. 5. Acoustic pressure from the finite element analysis. Excitation frequency of $f = 350$ Hz

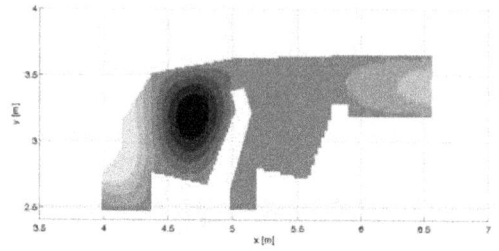

Fig. 6. Acoustic pressure from semi-analytic method. Excitation frequency of $f = 350$ Hz

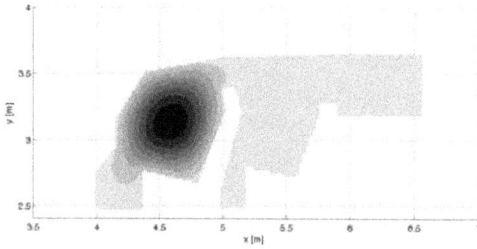

Fig. 7. Acoustic pressure from the semi-analytic method. Excitation frequency of $f = 296$ Hz, which is also one of computed system eigenfrequencies

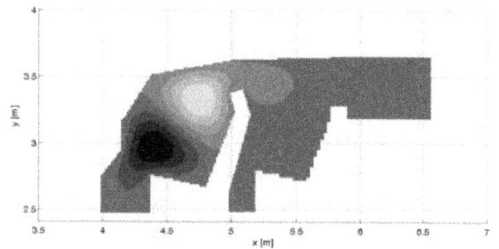

Fig. 8. Acoustic pressure from the semi-analytic method. Excitation frequency of $f = 438$ Hz, which is also one of computed system eigenfrequencies

5.2. Results

In the finite element analysis an convergence study was carried out. Model with 965 nodes was chosen as the optimal one. For finer meshes the improvement of accuracy was negligible. Matrices of the semi-analytic method are frequency dependent. Maximum number of unknowns H_n of the whole system was 748. This represents 22.5 % less unknowns in comparison with the finite element analysis. Computed transfer functions of both semi-analytic and the finite element methods in the points P_1 and P_2 are depicted in Fig. 9 and Fig. 10, respectively. Computation was done in the frequency range of

$$f = [144, \ldots, 474]\ \text{Hz.}$$

The solution of the system excited at $f = 350$ Hz computed by the finite element method is in Fig. 5 and the results of the semi-analytic method are in Fig. 6. Figs. 7 and 8 show the computed acoustic pressure at frequencies of $f = 296$ Hz and $f = 438$ Hz, respectively. These frequencies are the eigenfrequencies of the whole system.

The frequency characteristics show the sound pressure level (30). Sound pressure level was computed as follows

$$L_{sp} = 20 \log_{10} \frac{p}{|p_{ref}|}, \tag{30}$$

where $p_{ref} = 20 \cdot 10^{-6}$ Pa. As long as the excitation frequency is different from the eigenfrequency of the system, the results exhibit very good agreement with the finite element analysis results, that were taken as reference. Outside of the eigenfrequencies the relative error is lower

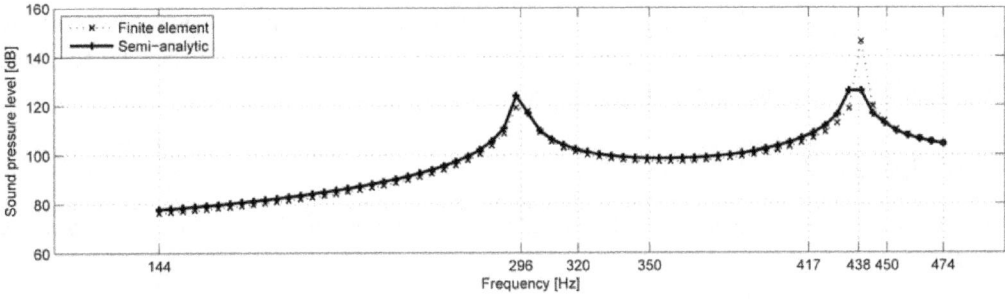

Fig. 9. Transfer functions of the acoustic pressure in the point P_1

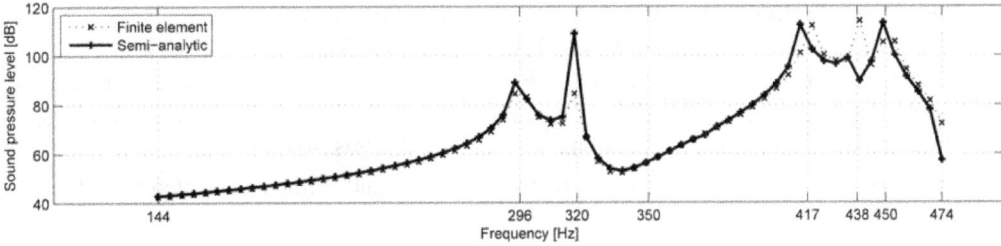

Fig. 10. Transfer functions of the acoustic pressure in the point P_2

than three percent. The semi-analytic method predicts the eigenfrequencies of the system very accurately. Pressure fields for excitation frequencies of $f = 296\,\text{Hz}$ in Fig. 7 and $f = 438\,\text{Hz}$ in Fig. 8 are correctly predicted. However these frequencies are the system eigenfrequencies and the amplitudes are deviated from the reference finite element solution. In case the excitation frequency equals any system eigenfrequency, the set of the independent basis functions A becomes linearly dependent. Thus the system of linear equations becomes badly conditioned. The precision of the results of such a system is lowered and the numerical results are deviated from the reference solution.

The bad conditioning could be avoided if damping was introduced. The example has only pressure boundary conditions and thus the system is undamped. When part of the boundary has impedance boundary condition, the system is damped. Then the maximum sound pressure level of a damped system is reduced and the precision improved. Basis functions of a damped semi-analytic method are supposed not to become linearly dependent, when the excitation frequency coincides with the eigenfrequency of the system. Implementation of impedance boundary conditions is the subject of further development.

6. Conclusion

A novel semi-analytic method for solving acoustic problems was suggested. It is based on the exact solution to the Helmholtz equation (3) in a rectangular domain. Acoustic pressure approximation (17) is in the form of linear combination. Derivation of basis functions A as well as the derivation of the system of linear equations was presented. Results of the semi-analytic method were compared with the finite element analysis. A study on a car-like cavity was carried out. The results of the undamped model show very good agreement. Acoustic pressure in points P_1 and P_2 for the excitation frequency that corresponds to the eigenfrequency of the system is a little deviated, however introduction of damping via impedance boundary conditions should restore good correspondence.

The semi-analytic method is based on the analytic solution. With its lower computational requirements it has the potential to solve problems, where other methods such finite element method would require too much computational force.

Acknowledgements

Supported by research project No. P101/12/1306 *Biomechanical modeling of human voice production*.

References

[1] Anyunzoghe, E., Cheng, L., Improved integro-modal approach with pressure distribution assessment and the use of overlapped cavities, Applied Acoustics 63 (11) (2002) 1 233-1 255.

[2] Bathe, K.-J., Finite element procedures in engineering analysis, 1st edition, Prentice-Hall, 1982.

[3] Bouillard, P., Lacroix, V., De Bel, E., A wave-oriented meshless formulation for acoustical and vibro-acoustical applications, Wave Motion 39 (4) (2004) 295-305.

[4] Kirkup, S., The boundary element method in acoustics, 2nd edition, Integrated Sound Software, 2007.

[5] Kita, E., Kamiya, N., Trefftz method: An overview, Advances in Engineering Software 24 (1-3) (1995) 3-12.

[6] Li, Y., Cheng, L., Vibro-acoustic analysis of a rectangular-like cavity with a tilted wall, Applied Acoustics 68 (7) (2007) 739-751.

[7] Merhaut, J., Theory of electro-acoustic devices, Czechoslovak Academy of Sciences, Praha, 1955. (in Czech)

[8] Momani, S., Abuasad, S., Application of He's variational iteration method to Helmholtz equation, Chaos, Solitons & Fractals 27 (5) (2006) 1 119-1 123.

[9] Nellis, G., Klein, S., The Galerkin weighted residual method, Cambridge, 2008, chap. 2.

[10] Otto, K., and Larsson, E., Iterative solution of the Helmholtz equation by a second-order method, UMIACS TR-96-95, (1996), 1-18.

[11] Pluymers, B., Wave based modelling methods for steady-state vibro-acoustics, Ph.D. thesis, Katholieke Universiteit Leuven, Department of Mechanical Engineering, Celestijnenlaan 300 B 3001, Heverlee, Belgium, 2006.

[12] Rienstra, S.-W., Hirschberg, A., An introduction to acoustics, Eindhoven University of Technology, 2004.

[13] Wijker, J., Statistical energy analysis, Random Vibrations in Spacecraft Structures Design, vol. 165 of Solid Mechanics and Its Applications. Springer Netherlands, 2009, pp. 51-320.

[14] Zienkiewics, O., Taylor, R., The finite element method: Fluid dynamics, 5th edition, vol. 3. Butterworth Heinemann, 2000.

[15] Zienkiewics, O., Taylor, R., The finite element method: The basis, 5th edition, vol. 1. Butterworth Heinemann, 2000.

8

Buckling analysis of thick isotropic plates by using exponential shear deformation theory

A. S. Sayyad[a,*], Y. M. Ghugal[b]

[a]Department of Civil Engineering, SRES's College of Engineering Kopargaon-423601, M.S., India
[b]Department of Applied Mechanics, Government Engineering College, Karad, Satara-415124, M.S., India

Abstract

In this paper, an exponential shear deformation theory is presented for the buckling analysis of thick isotropic plates subjected to uniaxial and biaxial in-plane forces. The theory accounts for a parabolic distribution of the transverse shear strains across the thickness, and satisfies the zero traction boundary conditions on the top and bottom surfaces of the plate without using shear correction factors. Governing equations and associated boundary conditions of the theory are obtained using the principle of virtual work. The simply supported thick isotropic square plates are considered for the detailed numerical studies. A closed form solutions for buckling analysis of square plates are obtained. Comparison studies are performed to verify the validity of the present results. The effects of aspect ratio on the critical buckling load of isotropic plates is investigated and discussed.

Keywords: shear deformation, isotropic plates, shear correction factor, buckling analysis, critical buckling load

1. Introduction

When plate is subjected to in-plane compressive forces, and if forces are sufficiently small the equilibrium of plate is stable. If the small additional disturbance result in a large response and the plate does not return to its original equilibrium configuration, the plate is said to be unstable. The onset of instability is called buckling. The magnitude of the in-plane compressive axial forces at which the plate becomes unstable is termed the critical buckling load. The magnitude of the critical buckling load depends on geometry, material properties, as well as on the buckling mode shape.

To predict the critical buckling load of plate, a number of plate theories have been proposed based on considering the transverse shear deformation effect. The well-known classical plate theory (CPT) which neglects the transverse shear deformation effect provides reasonably good results for thin plates and overpredicts the critical buckling loads for thick plates. The Reissner [7] and Mindlin [5] theories are known as the stress based and displacement based first-order shear deformation plate theory (FSDT) respectively, and account for the transverse shear effects by the way of linear variation of in-plane displacements through the thickness of plate. However, these theories do not satisfy the zero traction boundary conditions on the top and bottom surfaces of the plate, and need to use the shear correction factor to satisfy the constitutive relations for transverse shear stresses and shear strains. These shear correction factors are depends on the geometric parameters, boundary conditions and loading conditions.

---

*Corresponding author. e-mail: attu sayyad@yahoo.co.in. _

To overcome the drawbacks of the FSDT, a number of higher order shear deformation plate theories are developed. A recent reviews of such refined shear deformation theories are presented by Ghugal and Shimpi [1], Wanji and Zhen [13] and Kreja [4]. Recently Shimpi and Patel [9, 10] has developed two variable plate theory for the static and dynamic analysis of thick plate whereas Kim et al. [3] extended this theory for the buckling analysis of isotropic and orthotropic plates. Thai and Kim [12] employed Levy type solution for the buckling analysis of thick plate using two variable plate theory. Ghugal and Pawar [2] applied hyperbolic shear deformation theory for the buckling analysis of plates in which in-plane displacement field uses hyperbolic functions in terms of thickness coordinate to include the shear deformation effect.

In this paper, a displacement based an exponential shear deformation theory (ESDT) presented by Sayyad and Ghugal [8] is extended for the buckling analysis of thick isotropic square plates subjected to uniaxial and biaxial in-plane loads. Governing equations and associated boundary conditions are derived from the principle of virtual work. The Navier's solution is employed for solving the governing equations of square plates with all simply supported edges. The detail procedure of Navier's solution technique is given by Szilard [11]. Comparison studies are performed to verify the validity of the present results. The effects of aspect ratio on the critical buckling loads of isotropic plates are studied and discussed in detail.

2. Theoretical formulation

2.1. Isotropic plate under consideration

Consider a rectangular plate of sides 'a' and 'b' and a constant thickness of 'h' made up of isotropic material and subjected to in-plane compressice forces (N_{xx}^0, N_{yy}^0 and N_{xy}^0) as shown in Fig. 1. The co-ordinate system (x, y, z) chosen and the coordinate parameters are such a that, the plate occupies a region given by Eq. (1)

$$0 \leq x \leq a, \quad 0 \leq y \leq b, \quad -h/2 \leq z \leq h/2. \tag{1}$$

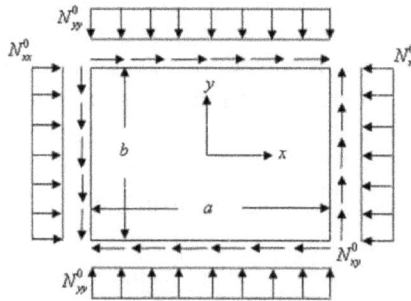

Fig. 1. Plate subjected to in-plane forces

2.2. The displacement field

The displacement field of the proposed plate theory is given by Sayyad and Ghugal [8]

$$u(x,y,z) = -z\frac{\partial w(x,y)}{\partial x} + z \exp\left[-2\left(\frac{z}{h}\right)^2\right]\phi(x,y),$$

$$v(x,y,z) = -z\frac{\partial w(x,y)}{\partial y} + z \exp\left[-2\left(\frac{z}{h}\right)^2\right]\psi(x,y), \tag{2}$$

$$w(x,y,z) = w(x,y).$$

Here u, v and w are the displacements in the x, y and z-directions respectively. The exponential function in terms of thickness coordinate in both the in-plane displacements u and v is associated with the transverse shear stress distribution through the thickness of plate. The functions ϕ and ψ are the unknown functions associated with the shear slopes. The strain field obtained by using strain-displacement relations can be given as

$$
\begin{aligned}
\varepsilon_x &= \frac{\partial u}{\partial x} = -z\frac{\partial^2 w}{\partial x^2} + f(z)\frac{\partial \phi}{\partial x}, \\
\varepsilon_y &= \frac{\partial v}{\partial y} = -z\frac{\partial^2 w}{\partial y^2} + f(z)\frac{\partial \psi}{\partial y}, \\
\gamma_{xy} &= \frac{\partial u}{\partial y} + \frac{\partial v}{\partial x} = -2z\frac{\partial^2 w}{\partial x \partial y} + f(z)\left(\frac{\partial \phi}{\partial y} + \frac{\partial \psi}{\partial x}\right), \\
\gamma_{zx} &= \frac{\partial u}{\partial z} + \frac{\partial w}{\partial x} = \frac{df(z)}{dz}\phi, \\
\gamma_{yz} &= \frac{\partial v}{\partial z} + \frac{\partial w}{\partial y} = \frac{df(z)}{dz}\psi,
\end{aligned}
\tag{3}
$$

where $f(z) = z\exp\left[-2\left(\frac{z}{h}\right)^2\right]$, $\frac{df(z)}{dz} = \exp\left[-2\left(\frac{z}{h}\right)^2\right]\left[1 - 4\left(\frac{z}{h}\right)^2\right]$.

2.3. Stress-strain relationships

The stress-strain relations of an isotropic plate can be written as:

$$
\begin{Bmatrix} \sigma_x \\ \sigma_y \\ \tau_{xy} \\ \tau_{zx} \\ \tau_{yz} \end{Bmatrix} = \frac{E}{1-\mu^2}\begin{bmatrix} 1 & \mu & 0 & 0 & 0 \\ \mu & 1 & 0 & 0 & 0 \\ 0 & 0 & \frac{1-\mu}{2} & 0 & 0 \\ 0 & 0 & 0 & \frac{1-\mu}{2} & 0 \\ 0 & 0 & 0 & 0 & \frac{1-\mu}{2} \end{bmatrix}\begin{Bmatrix} \varepsilon_x \\ \varepsilon_y \\ \gamma_{xy} \\ \gamma_{zx} \\ \gamma_{yz} \end{Bmatrix},
\tag{4}
$$

where E is the Young's modulus and μ is the Poisson's ratio of the material.

3. Governing equations and boundary conditions

The principle of virtural work of the plate can be written as

$$
\begin{aligned}
&\int_{z=-h/2}^{z=h/2}\int_{y=0}^{y=b}\int_{x=0}^{x=a}[\sigma_x\delta\varepsilon_x + \sigma_y\delta\varepsilon_y + \tau_{yz}\delta\gamma_{yz} + \tau_{zx}\delta\gamma_{zx} + \tau_{xy}\delta\gamma_{xy}]\,dx\,dy\,dz - \\
&\int_{y=0}^{y=b}\int_{x=0}^{x=a}q(x,y)\delta w\,dx\,dy - \\
&\int_{y=0}^{y=b}\int_{x=0}^{x=a}\left[N_{xx}^0\frac{\partial^2 w}{\partial x^2} + N_{yy}^0\frac{\partial^2 w}{\partial y^2} + 2N_{xy}^0\frac{\partial^2 w}{\partial x \partial y}\right]\delta w\,dx\,dy = 0,
\end{aligned}
\tag{5}
$$

where $q(x,y)$ is the transverse load acting in the downword z direction on surface $z = -h/2$. Substituting Eqs. (3)–(4) into Eq. (5) and integrating through the thickness of the plate, the governing differential equations and associated boundary conditions in-terms of stress resultants are as follows:

$$\frac{\partial^2 M_x}{\partial x^2} + 2\frac{\partial^2 M_{xy}}{\partial x \partial y} + \frac{\partial^2 M_y}{\partial y^2} + N_{xx}^0 \frac{\partial^2 w}{\partial x^2} + N_{yy}^0 \frac{\partial^2 w}{\partial y^2} + 2N_{xy}^0 \frac{\partial^2 w}{\partial x \partial y} + q(x, y) = 0,$$

$$\frac{\partial N_{sx}}{\partial x} + \frac{\partial N_{sxy}}{\partial y} - N_{Tcx} = 0, \qquad (6)$$

$$\frac{\partial N_{sy}}{\partial y} + \frac{\partial N_{sxy}}{\partial x} - N_{Tcy} = 0.$$

The boundary conditions at $x = 0$ and $x = a$ obtained are of the following form:

Either $V_x = 0$ or w is prescribed,

either $M_x = 0$ or $\dfrac{\partial w}{\partial x}$ is prescribed,   (7)

either $N_{sx} = 0$ or ϕ is prescribed,

either $N_{sxy} = 0$ or ψ is prescribed.

The boundary conditions at $y = 0$ and $y = b$ obtained are of the following form:

Either $V_y = 0$ or w is prescribed,

either $M_y = 0$ or $\dfrac{\partial w}{\partial y}$ is prescribed,   (8)

either $N_{sxy} = 0$ or ϕ is prescribed,

either $N_{sy} = 0$ or ψ is prescribed.

Reaction at the corners of the plate is of the following form:

Either $M_{xy} = 0$ or w is prescribed.   (9)

The stress resultants appear in the governing equations and boundary conditions are given as:

$$(M_x, M_y, M_{xy}) = \int_{-h/2}^{h/2} (\sigma_x, \sigma_y, \tau_{xy}) z \, \mathrm{d}z,$$

$$(N_{sx}, N_{sy}, N_{sxy}) = \int_{-h/2}^{h/2} (\sigma_x, \sigma_y, \tau_{xy}) f(z) \, \mathrm{d}z,$$

$$(N_{Tcx}, N_{Tcy}) = \int_{-h/2}^{h/2} (\tau_{zx}, \tau_{yz}) \frac{\mathrm{d}f(z)}{\mathrm{d}z} \, \mathrm{d}z, \qquad (10)$$

$$V_x = \frac{\partial M_x}{\partial x} + 2\frac{\partial M_{xy}}{\partial y},$$

$$V_y = \frac{\partial M_y}{\partial y} + 2\frac{\partial M_{xy}}{\partial x}.$$

The governing differential equations in-terms of unknown displacement variables used in the displacement field (w, ϕ and ψ) obtained are as follows:

$$D_1 \left(\frac{\partial^4 w}{\partial x^4} + 2\frac{\partial^4 w}{\partial x^2 \partial y^2} + \frac{\partial^4 w}{\partial y^4} \right) - D_2 \left(\frac{\partial^3 \phi}{\partial x^3} + \frac{\partial^3 \phi}{\partial x \partial y^2} + \frac{\partial^3 \psi}{\partial y^3} + \frac{\partial^3 \psi}{\partial x^2 \partial y} \right) =$$

$$q(x,y) + N_{xx}^0 \frac{\partial^2 w}{\partial x^2} + N_{yy}^0 \frac{\partial^2 w}{\partial y^2} + 2N_{xy}^0 \frac{\partial^2 w}{\partial x \partial y},$$

$$D_2 \left(\frac{\partial^3 w}{\partial x^3} + \frac{\partial^3 w}{\partial x \partial y^2} \right) - D_3 \left(\frac{\partial^2 \phi}{\partial x^2} + \frac{1-\mu}{2}\frac{\partial^2 \phi}{\partial y^2} \right) + D_4 \phi - D_3 \left(\frac{1+\mu}{2} \right) \frac{\partial^2 \psi}{\partial x \partial y} = 0, \quad (11)$$

$$D_2 \left(\frac{\partial^3 w}{\partial y^3} + \frac{\partial^3 w}{\partial x^2 \partial y} \right) - D_3 \left(\frac{1-\mu}{2}\frac{\partial^2 \psi}{\partial x^2} + \frac{\partial^2 \psi}{\partial y^2} \right) + D_4 \psi - D_3 \left(\frac{1+\mu}{2} \right) \frac{\partial^2 \phi}{\partial x \partial y} = 0.$$

The associated consistent boundary conditions in-terms of unknown displacement variables obtained along the edges $x = 0$ and $x = a$ are as below:

$$D_1 \left[\frac{\partial^3 w}{\partial x^3} + (2 - \mu)\frac{\partial^3 w}{\partial x \partial y^2} \right] -$$

$$D_2 \left[\frac{\partial^2 \phi}{\partial x^2} + (1-\mu)\frac{\partial^2 \phi}{\partial y^2} + \frac{\partial^2 \psi}{\partial x \partial y} \right] = 0 \qquad \text{or } w \text{ is prescribed,}$$

$$D_1 \left(\frac{\partial^2 w}{\partial x^2} + \mu\frac{\partial^2 w}{\partial y^2} \right) - D_2 \left(\frac{\partial \phi}{\partial x} + \mu\frac{\partial \psi}{\partial y} \right) = 0 \qquad \text{or } \frac{\partial w}{\partial x} \text{ is prescribed,}$$

$$D_2 \left(\frac{\partial^2 w}{\partial x^2} + \mu\frac{\partial^2 w}{\partial y^2} \right) - 2D_3 \left(\frac{\partial \phi}{\partial x} + \mu\frac{\partial \psi}{\partial y} \right) = 0 \qquad \text{or } \phi \text{ is prescribed,} \qquad (12)$$

$$D_3 \left(\frac{\partial \psi}{\partial x} + \frac{\partial \phi}{\partial y} \right) - D_2 \frac{\partial^2 w}{\partial x \partial y} = 0 \qquad \text{or } \psi \text{ is prescribed.}$$

The associated consistent boundary conditions in-terms of unknown displacement variables obtained along the edges $y = 0$ and $y = b$ are as below:

$$D_1 \left[\frac{\partial^3 w}{\partial y^3} + (2 - \mu)\frac{\partial^3 w}{\partial x^2 \partial y} \right] -$$

$$D_2 \left[\frac{\partial^2 \psi}{\partial y^2} + (1-\mu)\frac{\partial^2 \psi}{\partial x^2} + \frac{\partial^2 \phi}{\partial x \partial y} \right] = 0 \qquad \text{or } w \text{ is prescribed,}$$

$$D_1 \left(\mu\frac{\partial^2 w}{\partial x^2} + \frac{\partial^2 w}{\partial y^2} \right) - D_2 \left(\mu\frac{\partial \phi}{\partial x} + \frac{\partial \psi}{\partial y} \right) = 0 \qquad \text{or } \frac{\partial w}{\partial y} \text{ is prescribed,}$$

$$D_3 \left(\frac{\partial \psi}{\partial x} + \frac{\partial \phi}{\partial y} \right) - D_2 \frac{\partial^2 w}{\partial x \partial y} = 0 \qquad \text{or } \phi \text{ is prescribed,} \qquad (13)$$

$$D_2 \left(\mu\frac{\partial^2 w}{\partial x^2} + \frac{\partial^2 w}{\partial y^2} \right) - 2D_3 \left(\mu\frac{\partial \phi}{\partial x} + \frac{\partial \psi}{\partial y} \right) = 0 \qquad \text{or } \psi \text{ is prescribed.}$$

The boundary condition in-terms of unknown displacement variables (w, ϕ and ψ) obtained along the corners of plate is:

$$2D_1 \frac{\partial^2 w}{\partial x \partial y} - D_2 \left(\frac{\partial \phi}{\partial y} + \frac{\partial \psi}{\partial x} \right) = 0 \quad \text{or } w \text{ is prescribed,} \qquad (14)$$

where constants appeared in governing equations and boundary conditions are as follows:

$$D_1 = \frac{Eh^3}{12(1-\mu^2)}, \qquad D_2 = \frac{A_0 E}{(1-\mu^2)}, \qquad D_3 = \frac{B_0 E}{(1-\mu^2)}, \qquad D_4 = \frac{C_0 E}{2(1+\mu)} \qquad (15)$$

and

$$A_0 = \int_{-h/2}^{h/2} z f(z)\, dz, \qquad B_0 = \int_{-h/2}^{h/2} f^2(z)\, dz, \qquad C_0 = \int_{-h/2}^{h/2} \left[\frac{df(z)}{dz}\right]^2 dz. \qquad (16)$$

4. Buckling analysis of isotropic plates subjected to in-plane forces

The critical buckling loads of simply supported, isotropic, square plate will be determined in this paper by using the Navier solution. For the buckling analysis, we assume that the only applied loads are the in-plane forces and all other forces are zero, i.e. $q(x, y) = 0$. The governing equations of plate in case of static buckling when, $N_{xx}^0 = -N_0$, $N_{yy}^0 = kN_{xx}^0$ and $N_{xy}^0 = 0$ are given by

$$D_1\left(\frac{\partial^4 w}{\partial x^4} + 2\frac{\partial^4 w}{\partial x^2 \partial y^2} + \frac{\partial^4 w}{\partial y^4}\right) - D_2\left(\frac{\partial^3 \phi}{\partial x^3} + \frac{\partial^3 \phi}{\partial x \partial y^2} + \frac{\partial^3 \psi}{\partial y^3} + \frac{\partial^3 \psi}{\partial x^2 \partial y}\right) =$$
$$-N_0\left(\frac{\partial^2 w}{\partial x^2} + k\frac{\partial^2 w}{\partial y^2}\right),$$

$$D_2\left(\frac{\partial^3 w}{\partial x^3} + \frac{\partial^3 w}{\partial x \partial y^2}\right) - D_3\left(\frac{\partial^2 \phi}{\partial x^2} + \frac{1-\mu}{2}\frac{\partial^2 \phi}{\partial y^2}\right) + D_4\phi - D_3\left(\frac{1+\mu}{2}\right)\frac{\partial^2 \psi}{\partial x \partial y} = 0, \quad (17)$$

$$D_2\left(\frac{\partial^3 w}{\partial y^3} + \frac{\partial^3 w}{\partial x^2 \partial y}\right) - D_3\left(\frac{1-\mu}{2}\frac{\partial^2 \psi}{\partial x^2} + \frac{\partial^2 \psi}{\partial y^2}\right) + D_4\psi - D_3\left(\frac{1+\mu}{2}\right)\frac{\partial^2 \phi}{\partial x \partial y} = 0.$$

The following are the boundary conditions of the simply supported isotropic plate

$$w = \psi = M_x = N_{sx} = 0 \quad \text{at } x = 0 \text{ and } x = a, \qquad (18)$$
$$w = \phi = M_y = N_{sy} = 0 \quad \text{at } y = 0 \text{ and } y = b. \qquad (19)$$

4.1. The Navier's solution

The following displacement functions $w(x, y)$, $\phi(x, y)$, and $\psi(x, y)$ are chosen to automatically satisfy the boundary conditions in Eqs. (18)–(19)

$$w(x, y) = \sum_{m=1}^{\infty}\sum_{n=1}^{\infty} w_{mn} \sin \alpha x \sin \beta y,$$

$$\phi(x, y) = \sum_{m=1}^{\infty}\sum_{n=1}^{\infty} \phi_{mn} \cos \alpha x \sin \beta y, \qquad (20)$$

$$\psi(x, y) = \sum_{m=1}^{\infty}\sum_{n=1}^{\infty} \psi_{mn} \sin \alpha x \cos \beta y,$$

where $\alpha = m\pi/a$, $\beta = n\pi/b$ and w_{mn}, ϕ_{mn} and ψ_{mn} are unknown coefficients. Substitute Eq. (20) in the set of three governing differential Eq. (17) resulting the following matrix form

$$\left\{\begin{bmatrix} K_{11} & K_{12} & K_{13} \\ K_{21} & K_{22} & K_{23} \\ K_{31} & K_{32} & K_{33} \end{bmatrix} - N_0 \begin{bmatrix} N_{11} & 0 & 0 \\ 0 & 0 & 0 \\ 0 & 0 & 0 \end{bmatrix}\right\} \left\{\begin{matrix} w_{mn} \\ \phi_{mn} \\ \psi_{mn} \end{matrix}\right\} = 0, \qquad (21)$$

where

$$K_{11} = D_1(\alpha^4 + \beta^4 + 2\alpha^2\beta^2),$$
$$K_{12} = -D_2(\alpha^3 + \alpha\beta^2),$$
$$K_{13} = -D_2(\beta^3 + \alpha^2\beta),$$
$$K_{22} = D_3\left[\frac{1-\mu}{2}\beta^2 + \alpha^2\right] + D_4,$$
$$K_{23} = D_3\left(\frac{1+\mu}{2}\right)\alpha\beta,$$
$$K_{33} = D_3\left[\frac{1-\mu}{2}\alpha^2 + \beta^2\right] + D_4,$$
$$N_{11} = \alpha^2 + k\beta^2.$$

(22)

For nontrivial solution, the determinant of the coefficient matrix in Eq. (21) must be zero. This gives the following expression for buckling load:

$$N_0 = \frac{K_{11}\begin{vmatrix} K_{22} & K_{23} \\ K_{32} & K_{33} \end{vmatrix} - K_{12}\begin{vmatrix} K_{21} & K_{23} \\ K_{31} & K_{33} \end{vmatrix} + K_{13}\begin{vmatrix} K_{21} & K_{22} \\ K_{31} & K_{32} \end{vmatrix}}{N_{11}\begin{vmatrix} K_{22} & K_{23} \\ K_{32} & K_{33} \end{vmatrix}}.$$

(23)

For each choice of m and n, there is a corresponsive unique value of N_0. The critical buckling load is the smallest value of $N_0(m, n)$.

5. Numerical results and discussion

A simply supported square ($a = b$) plate subjected to the loading conditions, as shown in Fig. 2, is considered to illustrate the accuracy of the present theory in predicting the buckling behavior of the isotropic plate.

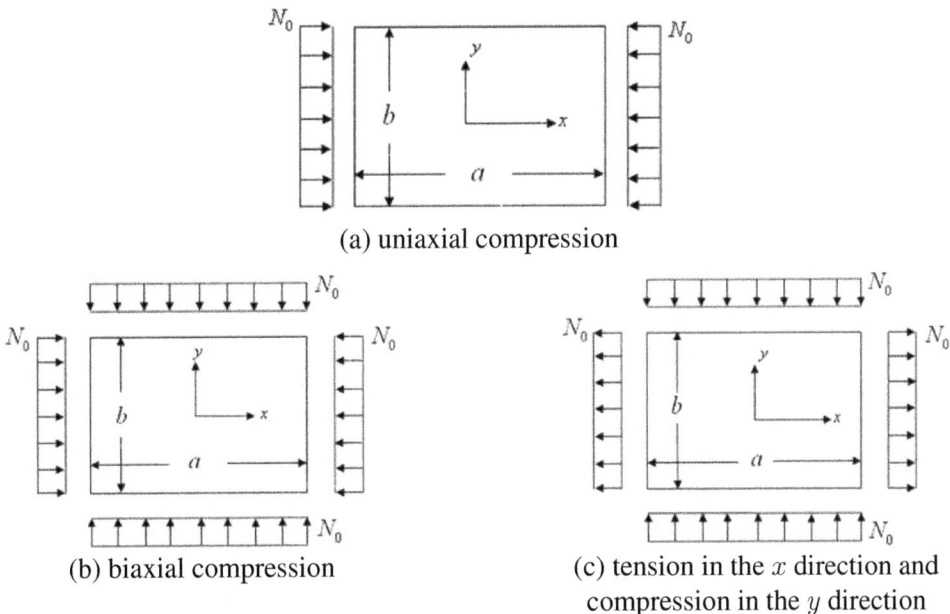

(a) uniaxial compression

(b) biaxial compression

(c) tension in the x direction and compression in the y direction

Fig. 2. The loading conditions of square plate for (a) uniaxial compression, (b) biaxial compression and (c) tension in the x direction and compression in the y direction

Results obtained for critical buckling load and the effects of aspect ratio on the critical buckling load of isotropic plates is investigated and discussed in detail. For verification purpose, corresponding results are also generated by higher order shear deformation theory (HSDT) of Reddy [6], classical plate theory (CPT) and first order shear deformation theory (FSDT) of Mindlin [5]. The exact elasticity solution for buckling analysis of plate is not available in the literature. Following material properties of isotropic plates are used

$$E = 210 \text{ GPa and } \mu = 0.3,$$
$$E = 70 \text{ GPa and } \mu = 0.33. \tag{24}$$

For convenience, the following nondimensional buckling load is used:

$$\bar{N}_{cr} = \frac{N_0 a^2}{E h^3}. \tag{25}$$

5.1. Discussion of results

The results of critical buckling load of simply supported square plates are presented in Tables 1–6 and Figs. 3–5. Tables 1–3 shows the comparison of critical buckling load for the steel plates whereas Tables 4–6 shows the comparison of critical buckling load for the aluminum plates subjected to in-plane forces. In case of plate subjected to uniaxial compression (Fig. 2a) and biaxial compression (Fig. 2b), buckling load is critical when mode for the plate is $(1,1)$ whereas in case of plate subjected to tension in x direction and compression in y direction (Fig. 2c), buckling load is critical when mode for the plate is $(1,2)$.

Table 1. Comparison of non-dimensional critical buckling load ($\bar{N}_{cr}$) of square plates subjected to uniaxial compression ($k = 0$, $E = 210$ GPa and $\mu = 0.3$)

Mode for the plate (m,n)	Theory	Aspect Ratio ($S = a/h$)				
		5	10	20	50	100
$(1,1)$	Present (ESDT)	2.9603	3.4242	3.5654	3.6072	3.6132
	Reddy (HSDT)	2.9512	3.4224	3.5649	3.6068	3.6130
	Mindlin (FSDT)	2.9498	3.4222	3.5649	3.6071	3.6130
	CPT	3.6152	3.6152	3.6152	3.6152	3.6152

Table 2. Comparison of non-dimensional buckling load ($\bar{N}_{cr}$) of square plates subjected to biaxial compression ($k = 1$, $E = 210$ GPa and $\mu = 0.3$)

Mode for the plate (m,n)	Theory	Aspect Ratio ($S = a/h$)				
		5	10	20	50	100
$(1,1)$	Present (ESDT)	1.4802	1.7121	1.7827	1.8038	1.8065
	Reddy (HSDT)	1.4756	1.7112	1.7825	1.8034	1.8065
	Mindlin (FSDT)	1.4749	1.7111	1.7825	1.8035	1.8065
	CPT	1.8076	1.8076	1.8076	1.8076	1.8076

Table 3. Comparison of non-dimensional critical buckling load ($\bar{N}_{cr}$) of square plates subjected tension in the x direction and compression in the y direction ($k = 1$, $E = 210$ GPa and $\mu = 0.3$)

Mode for the plate (m, n)	Theory	Aspect Ratio ($S = a/h$)				
		5	10	20	50	100
$(1, 2)$	Present (ESDT)	4.8798	6.6133	7.2777	7.4898	7.5212
	Reddy (HSDT)	4.8274	6.6024	7.2754	7.4893	7.5201
	Mindlin (FSDT)	4.8158	6.6010	7.2753	7.4895	7.5211
	CPT	7.5317	7.5317	7.5317	7.5317	7.5317

Table 4. Comparison of non-dimensional critical buckling load ($\bar{N}_{cr}$) of square plates subjected to uni-axial compression ($k = 0$, $E = 70$ GPa and $\mu = 0.33$)

Mode for the plate (m, n)	Theory	Aspect Ratio ($S = a/h$)				
		5	10	20	50	100
$(1, 1)$	Present (ESDT)	2.9991	3.4886	3.6388	3.6833	3.6898
	Reddy (HSDT)	2.9893	3.4866	3.6383	3.6833	3.6896
	Mindlin (FSDT)	2.9877	3.4865	3.6383	3.6832	3.6900
	CPT	3.6919	3.6919	3.6919	3.6919	3.6919

Table 5. Comparison of non-dimensional critical buckling load ($\bar{N}_{cr}$) of square plates subjected to biaxial compression ($k = 1$, $E = 70$ GPa and $\mu = 0.33$)

Mode for the plate (m, n)	Theory	Aspect Ratio ($S = a/h$)				
		5	10	20	50	100
$(1, 1)$	Present (ESDT)	1.4995	1.7443	1.8194	1.8416	1.8449
	Reddy (HSDT)	1.4947	1.7433	1.8192	1.8416	1.8448
	Mindlin (FSDT)	1.4939	1.7433	1.8192	1.8415	1.8450
	CPT	1.8459	1.8459	1.8459	1.8459	1.8459

Table 6. Comparison of non-dimensional critical buckling load ($\bar{N}_{cr}$) of square plates subjected tension in the x direction and compression in the y direction ($k = 1$, $E = 70$ GPa and $\mu = 0.33$)

Mode for the plate (m, n)	Theory	Aspect Ratio ($S = a/h$)				
		5	10	20	50	100
$(1, 2)$	Present (ESDT)	4.9083	6.7172	7.4208	7.6468	7.6803
	Reddy (HSDT)	4.8523	6.7055	7.4184	7.6465	7.6804
	Mindlin (FSDT)	4.8398	6.7040	7.4183	7.6465	7.6810
	CPT	7.6915	7.6915	7.6915	7.6915	7.6915

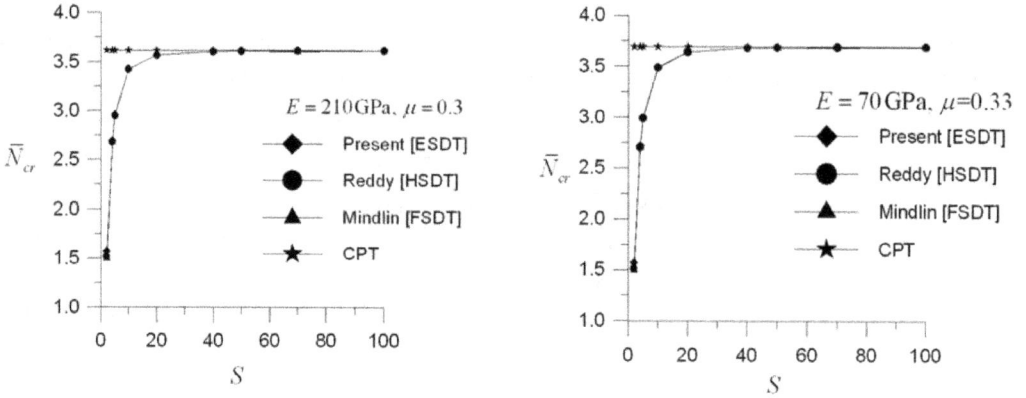

Fig. 3. The effect of aspect ratios on the critical buckling load of square plate subjected to uniaxial compression

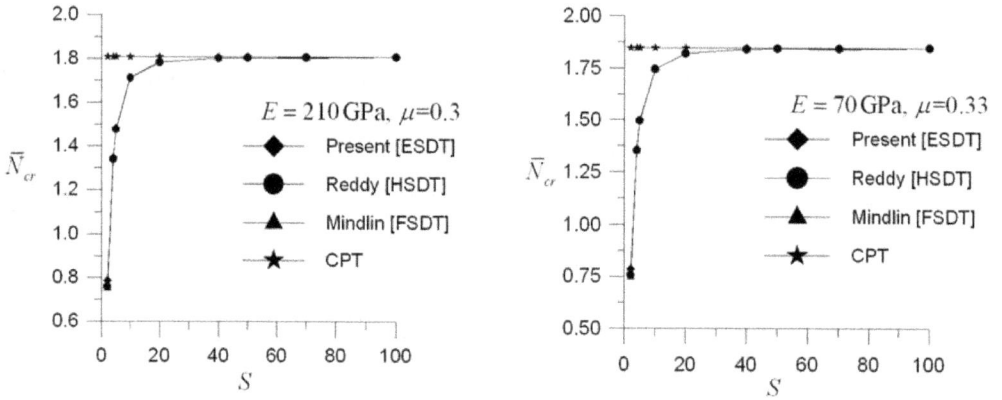

Fig. 4. The effect of aspect ratios on the critical buckling load of square plate subjected to biaxial compression

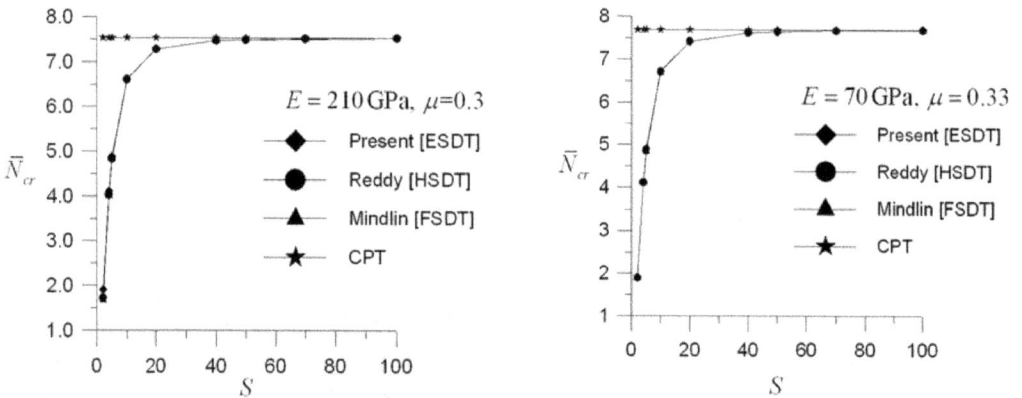

Fig. 5. The effect of aspect ratios on the critical buckling load of square plate subjected to tension in the x direction and compression in the y direction

From the examination of Tables 1–6, it is observed that, the critical buckling load obtained by present theory (ESDT) and Reddy's theory (HSDT) is in excellent agreement with each other even though the plate is very thick due to inclusion of effect of transverse shear deformation. It is also observed that, the value of critical buckling load is increased with increase in aspect (a/h) ratio. As compared to ESDT and HSDT, FSDT underestimates the values of critical buckling load in all cases due to use of shear correction factor (5/6) whereas CPT overestimates the same due to neglect of transverse shear deformation. In case of CPT, critical buckling load is independent of aspect ratio. Figs. 3–5 shows that, for the higher value of aspect (a/h) ratio, the results obtained by ESDT, HSDT, FSDT and CPT are more or less same.

6. Conclusions

An exponential shear deformation theory (ESDT) presented by Sayyad and Ghugal [8] has been applied in this paper for buckling behavior of thick isotropic plates. From the numerical results and discussions following conclusions are drawn.

1. The theory takes account of transverse shear effects and parabolic distribution of the transverse shear strains through the thickness of the plate, hence it is unnecessary to use shear correction factors.

2. It can be concluded that the presented theory can accurately predict the critical buckling loads of the isotropic plates.

3. Critical buckling load is increased with increase in aspect ratio.

References

[1] Ghugal, Y. M., Shimpi, R. P., A Review of Refined Shear Deformation Theories for Isotropic and Anisotropic Laminated Plates, Journal of Reinforced Plastics and Composites, 21 (2002) 775–813.

[2] Ghugal, Y. M., Pawar, M. D., Buckling and Vibration of Plates by Hyperbolic Shear Deformation Theory, Journal of Aerospace Engineering and Technology, 1(1) (2011) 1–12.

[3] Kim, S. E., Thai, H. T., Lee, J., Buckling Analysis of Plates Using the Two Variable Refined Plate Theory, Thin-Walled Structures 47 (2009) 455–462.

[4] Kreja, I., A Literature Review on Computational Models for Laminated Composite and Sandwich Panels, Central European Journal of Engineering, 1(1) (2011) 59–80.

[5] Mindlin, R. D., Influence of Rotatory Inertia and Shear on Flexural Motions of Isotropic, Elastic Plates, ASME Journal of Applied Mechanics, 18 (1951) 31–38.

[6] Reddy, J. N., A Simple Higher Order Theory for Laminated Composite Plates, ASME Journal of Applied Mechanics, 51 (1984) 745–752.

[7] Reissner, E., The Effect of Transverse Shear Deformation on the Bending of Elastic Plates, ASME Journal of Applied Mechanics, 12 (1945) 69–77.

[8] Sayyad, A. S., Ghugal, Y. M., Bending and Free Vibration Analysis of Thick Isotropic Plates by using Exponential Shear Deformation Theory, Applied and Computational Mechanics, 6(1) (2012) 65–82.

[9] Shimpi, R. P., Patel, H. G., A two Variable Refined Plate Theory for Orthotropic Plate Analysis, International Journal of Solids and Structures, 43 (2006) 6 783–6 799.

[10] Shimpi, R. P., Patel, H. G., Free Vibrations of Plate Using two Variable Refined Plate Theory, Journal of Sound and Vibration, 296 (2006) 979–999.

[11] Szilard, R., Theories and Applications of Plate Analysis: Classical, Numerical and Engineering Methods, John Wiley & Sons, Inc., Hoboken, New Jersey, (2004) 69–75.

[12] Thai, H. T., Kim, S. E., Levy-type Solution for Buckling Analysis of Orthotropic Plates Based on two Variable Refined Plate Theory, Composite Structures, 93 (2011) 1 738–1 746.

[13] Wanji, C., Zhen, W., A Selective Review on Recent Development of Displacement-Based Laminated Plate Theories, Recent Patents on Mechanical Engineering, 1 (2008) 29–44.

9

Modelling of flexi-coil springs with rubber-metal pads in a locomotive running gear

T. Michálek[a,*], J. Zelenka[a]

[a] University of Pardubice, Jan Perner Transport Faculty, Department of Transport Means and Diagnostics, Section of Rail Vehicles; Detached Branch of the Jan Perner Transport Faculty, Nádražní 547, 560 02 Česká Třebová, Czech Republic

Abstract

Nowadays, flexi-coil springs are commonly used in the secondary suspension stage of railway vehicles. Lateral stiffness of these springs is influenced by means of their design parameters (number of coils, height, mean diameter of coils, wire diameter etc.) and it is often suitable to modify this stiffness in such way, that the suspension shows various lateral stiffness in different directions (i.e., longitudinally vs. laterally in the vehicle-related coordinate system). Therefore, these springs are often supplemented with some kind of rubber-metal pads. This paper deals with modelling of the flexi-coil springs supplemented with tilting rubber-metal tilting pads applied in running gear of an electric locomotive as well as with consequences of application of that solution of the secondary suspension from the point of view of the vehicle running performance. This analysis is performed by means of multi-body simulations and the description of lateral stiffness characteristics of the springs is based on results of experimental measurements of these characteristics performed in heavy laboratories of the Jan Perner Transport Faculty of the University of Pardubice.

Keywords: flexi-coil spring, lateral stiffness, tilting rubber-metal pad, railway vehicle, multi-body simulations

1. Introduction

Flexi-coil springs are steel spiral springs which allow their lateral loading besides to the axial (vertical) loading. This property is often used in design of modern railway vehicles, especially in their secondary suspension stage (i.e., between the vehicle body and bogie frame). Sometimes we talk about a *bolsterless bogie concept* in such cases. Besides to the vertical suspension of the vehicle body (realized by means of the axial deformation of the springs), the secondary springs ensure by means of their lateral deformation also the lateral suspension of the vehicle body and yawing of the bogies (especially during the run of the vehicle through a curve). For the lateral suspension, the lateral deformation (in the vehicle-related coordinate system) is used; at the yawing of bogies, especially the longitudinal deformation (in the vehicle-related coordinate system) of the springs is significant.

The flexi-coil springs (and also the bolsterless bogie concept) have been used in the design of locomotive bogies for higher speeds approximately since 1960's; for example bogies of the legendary German six-axle electric locomotive Class 103 for speed 200 km/h or so-called Henschel bogies (see e.g. [1]) are well known in this context. At the development of modern railway bogies for high speed operation, a requirement on detailed knowledge of stiffness characteristics and credible modelling of these springs has arisen together with a significant progress in computer simulations in the branch of railway vehicle dynamics. For calculation

---

*Corresponding author. e-mail: Tomas.Michalek@upce.cz.

of the lateral stiffness of the flexi-coil springs, several theories were formulated and many experiments were realized by different authors (see e.g., report [7] or paper [8]); also the FEM calculations can be used for these purposes (see [8]). Nowadays, some models of flexi-coil springs, which allow their description on basis of the knowledge of their parameters, are implemented into the commonly used multi-body simulation tools as SIMPACK, ADAMS/VI-Rail etc. A state-of-the-art of modelling of suspension elements in multi-body vehicle dynamics is summarized e.g. in paper [2].

Sometimes it is suitable to modify the lateral stiffness of the flexi-coil springs in a desirable way. That modification is usually related with the requirement on a minimization of the resistance against bogie rotation (i.e., on a reduction of the lateral stiffness of the springs in longitudinal direction) at a preservation of the relatively high lateral stiffness of the springs in the lateral direction because of a needful value of the stiffness of lateral suspension of the vehicle body. This directionally dependent modification of the lateral stiffness of the flexi-coil springs can be reached for example by means of an application of tilting rubber-metal pads.

2. Influence of lateral stiffness of flexi-coil springs on parameters of running gear

The lateral stiffness of a flexi-coil spring is influenced by means of its design parameters, i.e., especially height, number of coils, mean diameter of coils and wire diameter. Values of these parameters follow from a proposal of the spring which must reflect the required vertical (axial) stiffness as well as conditions of state of stress in the spring. Therefore, the lateral stiffness of the spring, which is intended for a particular application, cannot be simply modified by means of a change of the design parameters of this spring. As it was mentioned in the introduction, the lateral stiffness of secondary flexi-coil springs determines in case of vehicles with the bolsterless bogie concept these two important basic parameters of their running gear:

- *lateral stiffness of the secondary suspension* which is determined by means of the total lateral stiffness of the secondary springs in lateral direction (in the vehicle-related coordinate system) and possibly also by means of the additional lateral stiffness of a mechanism for transmission of forces between the vehicle body and bogie frame;

- *resistance against bogie rotation* which is determined especially by means of the lateral stiffness of the secondary springs in longitudinal direction (in the vehicle-related coordinate system) because of their dominant component of deformation at the rotation of bogie around its vertical axis (see the scheme in Fig. 1).

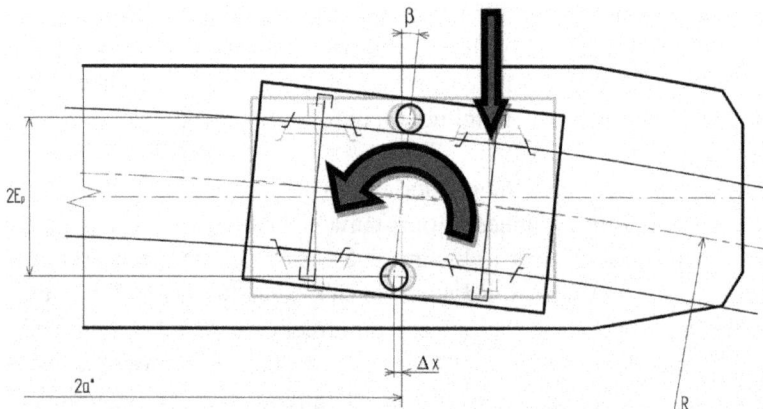

Fig. 1. Principle of acting of the moment against bogie rotation on the front bogie of railway vehicle during its run through a curve

Especially the resistance against bogie rotation is a very important quantity from the point of view of the vehicle running performance because of its influence on lateral force interaction between the vehicle and the track during the run through curves as well as on stability of vehicle run at higher speeds in a straight track. During the run of the vehicle through a curve with a radius R, the bogie turns around its vertical axis about the angle β. If the lateral distance of centres of the springs of secondary suspension on the bogie is marked as $2E_p$, a moment against bogie rotation (which is outlined in Fig. 1) can be simply estimated as:

$$M_z = 2 \cdot k_x \cdot E_p^2 \cdot \beta, \tag{1}$$

where k_x is a lateral stiffness of the flexi-coil springs situated on one side of the bogie in longitudinal direction. Then, the resistance against bogie rotation can be expressed as:

$$\gamma = \frac{M_z}{\beta} = 2 \cdot k_x \cdot E_p^2. \tag{2}$$

2.1. General requirements on lateral stiffness of secondary flexi-coil springs

A negative consequence of higher values of the resistance against bogie rotation is a higher value of the *quasistatic guiding force* which acts on the outer wheel of the first wheelset of the vehicle during its run through a curve (and which is also outlined in Fig. 1). This force is related with wear of wheels and rails in curves and its value is assessed in framework of the approval process of new railway vehicles according to relevant standards (e.g. the Technical Specifications for Interoperability which refer to the European Standard EN 14363 [3]). Because it is usually problematic to meet the requirements of the standards on a limit value of the quasistatic guiding force in small-radius curves, the resistance against bogie rotation should be as low as possible. On the other hand, higher values of the resistance against bogie rotation contribute to a better *riding stability* of the vehicle at higher speeds during the run in a straight track. Therefore, a proposal of the secondary suspension from the point of view of the resistance against bogie rotation — i.e., from the point of view of the lateral stiffness of the secondary springs, as well — is always a certain compromise.

In the design stage, the lateral stiffness of a flexi-coil spring can be estimated with using of some of existing empirical formulae, e.g., according to Gross, Wahl, Sparing, Timoshenko-Ponomarev or British Standard (see e.g. the overview in paper [8]). The lateral stiffness of the spring itself is practically directionally independent. A certain influence of orientation of the end coils of the spring can be observed; however, this effect (in relation with the absolute magnitude of the lateral stiffness of the springs used in design of running gears of railway vehicles) is not too significant. However, it is desired to reach a low lateral stiffness of the flexi-coil spring in longitudinal direction (because of the minimization of the resistance against bogie rotation) and simultaneously preserve a higher value of the lateral stiffness of the flexi-coil spring in lateral direction (because of its function as an element of the lateral suspension of the vehicle body). Therefore, the application of special rubber-metal pads is practically the only effective way how to reach the directionally dependent modification of the lateral stiffness of the flexi-coil springs.

2.2. Influence of tilting rubber-metal pads on lateral stiffness of flexi-coil springs

A frequently used type of the pads, which is used for purpose of the directionally dependent modification of the lateral stiffness of flexi-coil springs, is a tilting rubber-metal pad. This pad (see e.g. the visualisation in Fig. 2) can be placed on the upper or bottom end of the spring. At the lateral deformation of the assembly spring/pad, these pads allow tilting of the adjoining end

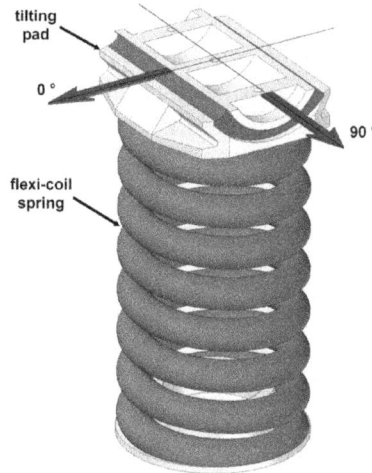

Fig. 2. Flexi-coil spring with tilting rubber-metal pad GMT

coil of the spring in relevant direction. This tilting effect is related with a significant change of the lateral stiffness of the assembly spring/pad as well as a reduction of stress in the spring. In the perpendicular direction, the tilting of the end coil of the spring is practically impossible. Therefore, the influence of the pad on the resulting lateral stiffness of the assembly is marginal in this direction.

Because the above mentioned formulae for calculation of the lateral stiffness of flexi-coil springs assume fixed end coils, it is not possible to use them for estimation of the lateral stiffness of a spring with the tilting pad. Therefore, a new analytical formula considering an angular stiffness of the pad was derived on the basis of the Gross' formula in following form:

$$
k_x^{sp} = \left[\frac{\frac{1}{k} \cdot \tan kH - H - \gamma_p \cdot \left(1 - \frac{1}{\cos kH}\right) \cdot \frac{1 - \cos kH - k \cdot H \cdot \sin kH}{F_z + \gamma_p \cdot k \cdot \sin kH}}{F_z \cdot \left(1 + \gamma_p \cdot k \cdot (1 - \cos kH) \cdot \frac{\tan kH}{F_z + \gamma_p \cdot k \cdot \sin kH}\right)} + \frac{H}{S} \right]^{-1}, \tag{3}
$$

where F_z is a vertical load of the assembly spring/pad, H is a length of the loaded spring, S is a shear stiffness of the spring according to Gross, γ_p is the angular stiffness of the tilting pad and k is a constant, which is defined as:

$$
k = \sqrt{\frac{F_z}{B \cdot \left(1 - \frac{F_z}{S}\right)}}, \tag{4}
$$

where B is a bending stiffness of the spring according to Gross. A detailed description of derivation of the formula (3) can be found in paper [4]. An analysis of this formula shows that decreasing angular stiffness and increasing vertical load of the assembly spring/pad lead to decreasing lateral stiffness of the assembly. Lower values of the angular stiffness of the pad in combination with a higher vertical load can even lead to a negative value of the resulting lateral stiffness. However, the description of the lateral stiffness of the assembly by means of the formula (3) is applicable only for the relevant direction of the lateral loading and also only in case of pads with linear angular characteristics.

In order to verify the results of the formula (3) and describe the lateral stiffness characteristic of the assembly spring/pad (see Fig. 2) completely, a measurement of this lateral characteristic was performed on the dynamic test stand in heavy laboratories of the Jan Perner Transport Faculty of the University of Pardubice in 2013. Results of the measurement, which was carried out with using of a secondary flexi-coil spring from a bogie of a modern electric locomotive supplemented with a tilting rubber-metal pad GMT, show following facts:

- the direction of loading has a dominant influence on the resulting lateral stiffness;
- in the direction, in which the pad allows tilting of the end coil of the spring (i.e., the direction, which is defined as $0°$ in Fig. 2), the reduction of the lateral stiffness of the assembly is most significant;
- in the direction, in which the pad allows tilting of the end coil of the spring, the measurement results confirm the analytically calculated trend that increasing vertical load of the assembly leads to decreasing resulting lateral stiffness;
- because of a non-linear angular characteristic of the real tilting pad, the lateral stiffness of the assembly spring/pad depends on its lateral deformation, as well.

The performed measurement allows to describe the lateral characteristic of the investigated assembly spring/pad completely. Especially in the directions which are not parallel or perpendicular to the direction, in which the pad allows tilting of the end coil of the spring, an analytical description of the resulting lateral stiffness would be too complicated. On basis of the measurement results, the lateral characteristic of the assembly was approximated by means of mathematical functions — in a general form in dependency on the vertical load F_z, direction of lateral loading β_p (relative to the orientation of the tilting pad — see Fig. 2) and resulting lateral deformation of the assembly Δ as:

$$k_\Delta = f\left(F_z, \beta_p, \Delta\right). \tag{5}$$

A graphical representation of this characteristic is depicted for a chosen value of the vertical load F_z (which corresponds to the static load of the spring on the locomotive) in Fig. 3. This characteristic was used as an input to the multi-body model of an electric locomotive.

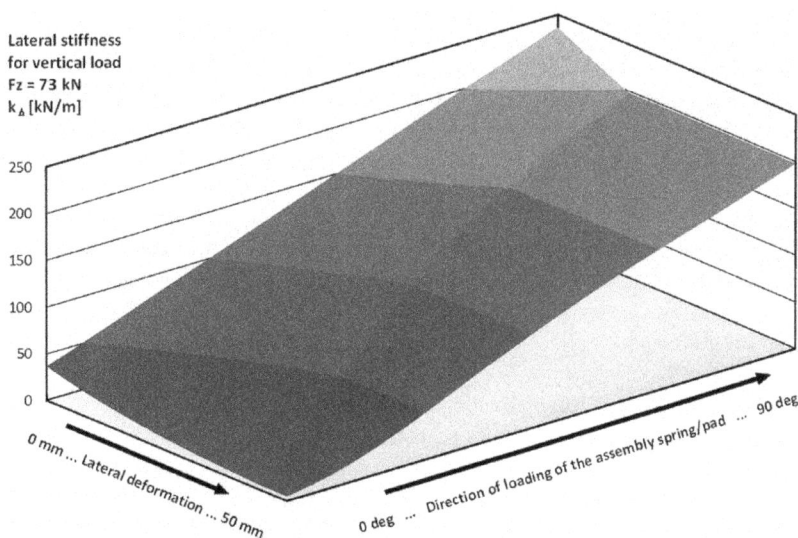

Fig. 3. Approximation of measured lateral characteristic of flexi-coil spring with tilting rubber-metal pad for the static vertical load 73 kN; the basic directions of loading ($0°$ and $90°$) are defined in Fig. 2

3. Modelling of flexi-coil springs with tilting rubber-metal pads in multi-body models for simulations of railway vehicle dynamics

The measured characteristic of lateral stiffness of the flexi-coil spring supplemented with the tilting rubber-metal pad (see Fig. 3) was implemented into a multi-body simulation model of an electric locomotive. The aim of the subsequently performed simulations was an assessment of influence of the newly arranged secondary suspension on dynamic properties of the vehicle. The simulation model was created in the multi-body simulation tool "SJKV" which is being developed at the Detached Branch of the Jan Perner Transport Faculty in Česká Třebová. A more detailed description of this original simulation tool is presented e.g. in the paper [5].

For purposes of modelling of the flexi-coil springs with tilting pads, several modifications of the simulation software had to be carried out. The first change is related with a level of detail of the dynamic model. Newly, the model of the locomotive consists of 15 rigid bodies and has 58 degrees of freedom. This requirement on the more detailed computational model is connected especially with the new principle which was applied at the modelling of joints between individual bodies. Originally, some joints were characterized with reduced parameters. It means that e.g. lateral stiffness of wheelset guiding was described by means of only one characteristic which covered relevant stiffness of all elements forming the wheelset guiding (primary springs, longitudinal rods, bump stops etc.). Similarly, the resistance against bogie rotation was usually characterized with a constant value, at least in case of the bolsterless bogies without friction elements. This approach to modelling of joints is very useful especially in such cases, in which the needed characteristics of these joints can be measured (e.g. on a prototype of the railway vehicle on a special test stands).

In the new version of the program system "SJKV", each joint element is modelled separately. The reason for that way of modelling is the knowledge of the characteristics of these individual joint elements. Especially in case of the flexi-coil springs supplemented with the tilting rubber-metal pads, the lateral stiffness of these elements shows a strong dependency on the direction of loading and therefore, the new way of joint modelling seems to be more correct than the original one. The process of computing of the forces acting in the springs with tilting pads in horizontal plane was algorithmized into following steps:

- a couple of coordinates is assigned to each of 4 spring/pad assemblies on each bogie (see the scheme in Fig. 4) — these coordinates determine the position of the upper and bottom end of these assemblies, i.e., their position on the bogie frame $[x_{fi}, y_{fi}]$ and on the vehicle body $[x_{bi}, y_{bi}]$; these coordinates are coupled with the vehicle body-related reference frame with the origin in the centre of the bogie pivot;

- for a general position of the bogie relative to the vehicle body, i.e., for the longitudinal displacement R_{bfxi}, lateral displacement R_{bfyi} and angle of bogie rotation β_i (see Fig. 4), actual coordinates can be calculated for individual spring/pad assemblies as:

$$[x_{bi}, y_{bi}] = [\pm V_p, \pm E_p], \tag{6}$$

$$x_{fi} = R_{bfxi} \pm \sqrt{\frac{E_p^2 + V_p^2}{1 + \tan^2\left[\arctan\left(\frac{E_p}{V_p}\right) \pm \beta_i\right]}}, \tag{7}$$

$$y_{fi} = R_{bfyi} \pm \tan\left[\arctan\left(\frac{E_p}{V_p}\right) \pm \beta_i\right] \cdot (x_{fi} - R_{bfxi}), \tag{8}$$

where V_p is a longitudinal distance of a centre of the spring/pad assembly from the lateral bogie axis and E_p is a lateral distance of the centre of the assembly from the longitudinal

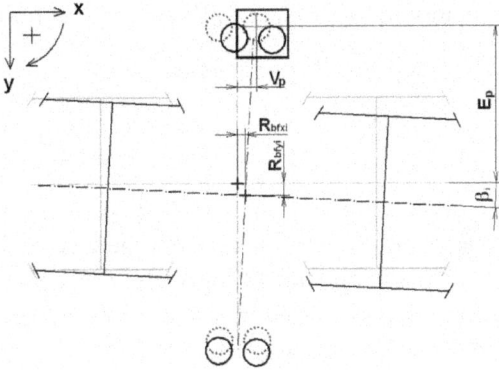

Fig. 4. General position of a bogie under a vehicle body

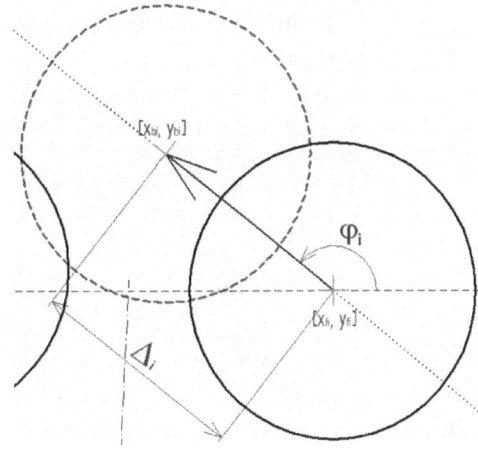

Fig. 5. Detail view on a part of Fig. 4

bogie axis (see Fig. 4); the expressions (7) and (8) can be derived by means of analytical geometry with using of the condition that the actual position of the observed points on the bogie frame is given as intersection points of a circle, defined with a shifted centre $[R_{bfxi}, R_{bfyi}]$ and a radius corresponding to the distance of the spring centres from the bogie centre, and straight lines, connecting these spring centres in the turned position, i.e., with respect to the angle of bogie rotation β_i;

- on basis of the actual position of the upper and bottom end of the spring/pad assemblies, the resulting horizontal deformation Δ_i and the direction of this deformation φ_i can be calculated for individual assemblies according to the scheme in Fig. 5:

$$\Delta_i = \sqrt{(x_{bi} - x_{fi})^2 + (y_{bi} - y_{fi})^2}, \tag{9}$$

$$\varphi_i = \arctan \frac{y_{bi} - y_{fi}}{x_{bi} - x_{fi}}; \tag{10}$$

- the direction of deformation φ_i of individual spring/pad assemblies is transformed (with respect to the orientation of the tilting pads) to the direction of lateral loading of these assemblies β_{pi} (i.e., relative to the orientation of the tilting pad);

- on basis of knowledge of the vertical load F_{zi} of the spring/pad assemblies, the horizontal forces acting in the assemblies in relevant direction, which is given by means of the angle φ_i, can be calculated with using of the mathematical description of the lateral stiffness characteristic of the assembly (5), which is depicted in Fig. 3, as:

$$F_{\Delta i} = \int_{(\Delta_i)} k_{\Delta i}(F_{zi}, \beta_{pi}, \Delta_i) \, d\Delta_i. \tag{11}$$

4. Simulations of dynamic behaviour of locomotive with modified secondary suspension

The new model of an electric locomotive with a total weight of 90 t was subjected to investigation of running and guiding behaviour. The simulation results show that the tilting pads influence the resistance against bogie rotation of the vehicle in a very specific way with all consequences on the vehicle running performance. Because of the new approach to modelling of joints, this resistance is no more an input quantity into the simulation. Newly, the *resistance against bogie rotation* is an output quantity which depends on specific situation and which can be approximately calculated for the actual value of angle of bogie rotation β with using of the horizontal forces $F_{\Delta j}$ acting in the spring/pad assemblies on each bogie as:

$$\gamma \doteq \frac{1}{\beta} \cdot E_p \cdot \sum_{(j)} F_{\Delta j} \cdot \cos \varphi_j. \tag{12}$$

The influence of implementation of the tilting pads into the secondary suspension on the *stability of vehicle run* is negative. Under given conditions, the critical speed of the locomotive is (significantly) lower than in case of a locomotive without the tilting pads in secondary suspension. This effect is related with a significant reduction of the resistance against bogie rotation and corresponds to general conclusions presented in the paper [6].

The influence of the tilting pads on the *guiding behaviour in small-radius curves* is more complicated because of its dependency on the actual cant deficiency. The cant deficiency is an important quantity which characterizes the unbalanced lateral force acting on the vehicle in curves and which can be calculated on the basis of actual speed and constructional-technical parameters of relevant curve (see e.g. [9]). Similarly to the run of the locomotive in a straight track, the resistance against bogie rotation has a very low value in case of run of the vehicle through a curve with low values of the cant deficiency. A consequence of this fact is a similar behaviour of both bogies in such situations. It can be demonstrated on the example of distributions of the quasistatic guiding forces on individual wheels of the locomotive at the simulations of its run through a 300 m curve which are presented in Fig. 6. However, in case of run of the vehicle through curves with higher (absolute) values of the cant deficiency, the resistance against bogie rotation has higher values as a consequence of the strong dependence of lateral stiffness of the spring/pad assemblies on the direction of their loading (see Fig. 3) and the fact, that the increasing unbalanced lateral force acting on the vehicle body causes higher lateral deformations of the secondary suspension leading to a significant change of the direction of horizontal loading of the individual spring/pad assemblies in comparison with the cases corresponding to the small cant deficiency or to the run in a straight track. From the graph in Fig. 6, it is evident that especially the difference of the guiding forces acting on the leading wheels of both bogies (i.e., on wheels 12 and 32) increases with increasing value of the cant deficiency.

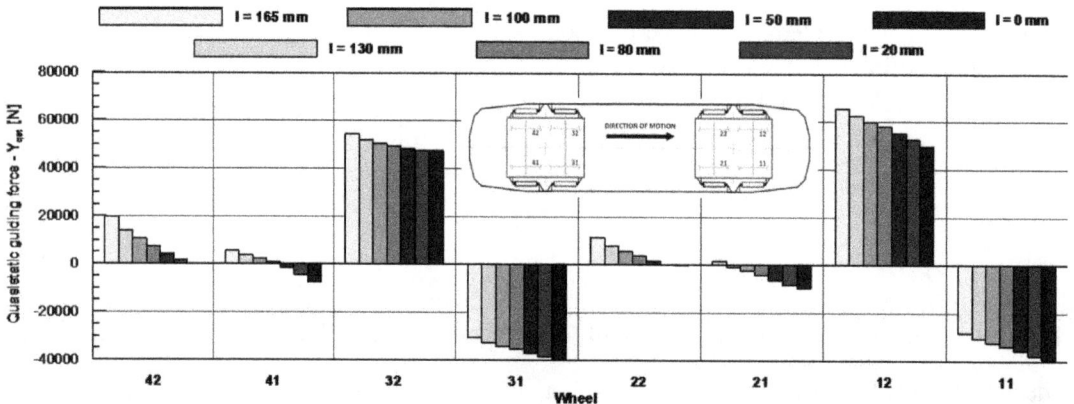

Fig. 6. Quasistatic guiding forces acting on individual wheels of a 90 t locomotive with the secondary suspension supplemented with tilting pads during the run through a 300 m curve at various values of cant deficiency

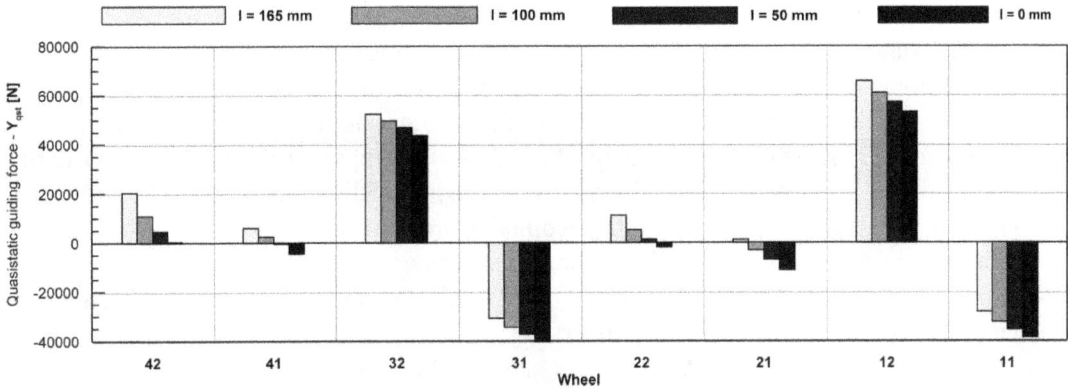

Fig. 7. Quasistatic guiding forces acting on individual wheels of a 90 t locomotive with the secondary suspension without the tilting pads during the run through a 300 m curve at various values of cant deficiency

For comparison, simulation results of a similar locomotive with the same parameters, which is not equipped with the tilting pads in the secondary suspension, can be shown for analogous situations defined with chosen values of cant deficiency at the run through the 300 m curve. The obtained distributions of quasistatic guiding forces on individual wheels of the locomotive are demonstrated in Fig. 7 for that case. Identically to the case of locomotive equipped with the tilting pads, these simulations were also performed for conditions of dry rails (i.e., for a value of friction coefficient in wheel/rail contact $f = 0.40$) and characteristics of wheel/rail contact geometry corresponding to theoretical wheel profiles S1002 and rail profiles 60E1 at nominal values of wheelset gauge (1 425 mm) and track gauge (1 435 mm).

From the graph in Fig. 7, it is evident that the quasistatic guiding force on the leading wheel of the front bogie (i.e., on the wheel 12) has a higher value in case of the locomotive without tilting pads in all cases of the simulated cant deficiency. However, this difference between the locomotive with the tilting pads and without them decreases with increasing cant deficiency. At the zero cant deficiency, the locomotive without tilting pads shows by 3.3 kN higher force Y_{qst12} than the locomotive with pads; at the cant deficiency of $I = 165$ mm, this difference is only 0.3 kN. The second well observable effect is the dependency of difference of the guiding forces acting on the leading wheels of both bogies (i.e., on wheels 12 and 32) on the cant deficiency. In case of the locomotive without tilting pads, a change of this difference with increasing cant deficiency is not so significant. At zero cant deficiency, the difference $Y_{qst12} - Y_{qst32}$ has a value of 6.5 kN (in case of the locomotive with tilting pads, it is only 2.3 kN); for the cant deficiency of $I = 165$ mm, the difference is 13.3 kN (and 11.1 kN for the locomotive with tilting pads). These results confirm the dependency of the resistance against bogie rotation of the locomotive with secondary suspension supplemented with the tilting pads on the cant deficiency. While the results of simulations of guiding behaviour of both observed computational models are similar at higher values of the cant deficiency; at lower values, the above mentioned differences can be observed.

5. Conclusion

This paper deals with a possibility of application of tilting rubber-metal pads into the secondary suspension of an electric locomotive where these pads are able to modify the lateral stiffness of the currently used flexi-coil springs in dependency on their horizontal loading. Section 2 deals with general requirements on the secondary suspension of railway vehicles from the point of view of their running performance; a theoretical calculation of influence of the tilting pad on the lateral stiffness of the spring/pad assembly and results of measurement of complete lateral characteristics of that assembly are presented in this section, as well. Section 3 describes the implementation of model of joint representing the flexi-coil springs with tilting pad into the multi-body model of the locomotive created in the simulation tool "SJKV". Results of the performed simulations of running performance of the locomotive are presented in section 4. These results point out very specific properties of the secondary suspension consisting of the flexi-coil springs with tilting pads. Especially the resistance against bogie rotation, which depends on the cant deficiency, shows following important properties:

- at the run with a low cant deficiency and in a straight track, the resistance is lowest which has a negative influence on the stability of the vehicle at higher speeds;
- at the run through curves with higher values of cant deficiency, the desired softening of the resistance is not too significant, and therefore its contribution to the decrease of the guiding force acting on the leading wheel of the first wheelset is limited;
- the dependency of the resistance against bogie rotation on the lateral deformation of the secondary suspension (i.e., on the cant deficiency and on the speed in curves) complicates the possibility of using of bogie couplings as well as the assessment of safety against derailment; to this point, an attention will be paid in the next research.

Acknowledgements

This work has been supported by the project No. TE01020038 "Competence Centre of Railway Vehicles" of the Technology Agency of the Czech Republic.

Reference

[1] Baur, K. G., Drehgestelle – Bogies, EK-Verlag, Freiburg, 2006.
[2] Bruni, S., Vinolas, J., Berg, M., Polách, O., Stichel, S., Modelling of suspension components in a rail vehicle dynamics context, Vehicle System Dynamics 49 (7) (2011) 1 021–1 072.
[3] EN 14363:2005. Railway applications — Testing for the acceptance of running characteristics of railway vehicles — Testing of running behaviour and stationary tests, CEN, Brussels, 2005.
[4] Michálek, T., Vágner, J., Kohout, M., Zelenka, J., Analytical calculation and experimental verification of lateral stiffness of a flexi-coil spring with tilting rubber-metal pad, Scientific Papers of the University of Pardubice, Series B — The Jan Perner Transport Faculty 19 (2014) 167–179.
[5] Michálek, T., Zelenka, J., Reduction of lateral forces between the railway vehicle and the track in small-radius curves by means of active elements, Applied and Computational Mechanics 5 (2) (2011) 187–196.
[6] Michálek, T., Zelenka, J., Sensitivity analysis of running gear parameters on dynamic behaviour of an electric locomotive, Proceedings of the 21st Conference with international participation "Current Problems in Rail Vehicles", Česká Třebová, University of Pardubice, 2013, pp. 85–92 (in Czech).
[7] Mohyla, M., Transversal stiffness of helical springs — Theory and measurement results, Technical Reports of the Research Centre of Rail Vehicles 2, 1971 (in Czech).
[8] Vágner, J., Hába, A., Possibilities of assessing lateral stiffness of flexi-coil springs, The Scientific and Technological Anthology 30, 2010 (in Czech).
[9] Zelenka, J., Michálek, T., Theory of Vehicles — Study Material, Jan Perner Transport Faculty of the University of Pardubice, Pardubice, 2014.

10

Numerical analysis of bypass model geometrical parameters influence on pulsatile blood flow

A. Jonášová^{a,*}, J. Vimmr^a, O. Bublík^a

^{a}*Faculty of Applied Sciences, University of West Bohemia, Univerzitní 22, 306 14 Plzeň, Czech Republic*

Abstract

The present study is focused on the analysis of pulsatile blood flow in complete idealized 3D bypass models in dependence on three main geometrical parameters (stenosis degree, junction angle and diameter ratio). Assuming the blood to be an incompressible Newtonian fluid, the non-linear system of Navier-Stokes equations is integrated in time by a fully implicit second-order accurate fractional-step method. The space discretization is performed with the help of the cell-centred finite volume method formulated for unstructured tetrahedral grids. In order to model a realistic coronary blood flow, a time-dependent flow rate taken from corresponding literature is considered. For the analysis of obtained numerical results, special emphasis is placed on their comparison in the form of velocity isolines at several selected cross-sections during systolic and diastolic phases. The remainder of this paper is devoted to discussion of walls shear stress distribution and its oscillatory character described by the oscillatory shear index with regard to areas prone to development of intimal hyperplasia or to thrombus formation.

Keywords: bypass model, pulsatile blood flow, stenosis degree, junction angle, diameter ratio, fractional-step method, FVM

1. Introduction

One of the main problems of implanted bypass grafts is related to intimal hyperplasia, [2], an abnormal healing process occurring at the distal anastomosis. It is typical for thickening of tunica intima (the innermost layer of an artery) and after several years also for graft patency loss and bypass failure. Many clinical studies indicate the importance of hemodynamics during healing processes after serious vessel surgeries. Local hemodynamics is also hypothesized to be a trigger of several serious vascular diseases such as atherosclerosis and intimal hyperplasia, [10]. Several hemodynamical factors are currently assumed to be responsible for the activation of endothelial cells in tunica intima, [11], resulting in metabolical and morphological changes within the vessel wall. In the case of intimal hyperplasia, low and oscillating wall shear stress τ_W is often mentioned, [10], since the thickening of tunica intima is similar to the development of atherosclerotic lesions in early stages. In practice, blood's influence on the endothelial cells is often analyzed in the form of several selected hemodynamical factors such as wall shear stress (WSS) and oscillatory shear index (OSI). For more details on the theme of hemodynamical factors and their significance in connection with cardiovascular diseases and disorders see [8].

Although the understanding of hemodynamics and its influence on arterial wall morphology can be crucial for patient's survival, the actual connection between blood flow and the

---

*Corresponding author. e-mail: jonasova@kme.zcu.cz.

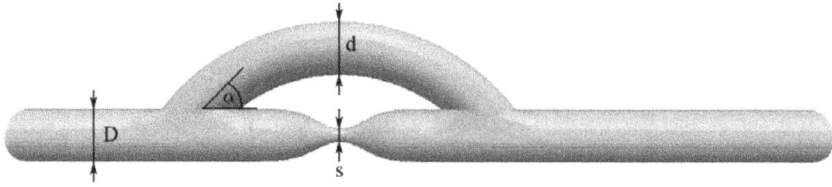

Fig. 1. Selected geometrical parameters for an idealized end-to-side bypass model with stenosed artery

occurrence of pathological changes still remains unknown. In this case, the investigation of hemodynamics in the form of numerical simulations represents one of possible contributions to the understanding of this problem. In the last decade, quite many studies dealt with the analysis of blood flow and geometry optimization, e.g., [3, 5, 7]. Considering all the surgery techniques usually used for bypass grafting of end-to-side anastomoses, it is possible to name several important geometrical parameters crucial for the resulting flow field. Beside the so-called distance of grafting (distance between the distal anastomosis and the narrowed or occluded artery part) analyzed for example in [1], following three parameters are usually mentioned.

- **Stenosis degree** characterizes the narrowing of the native artery and is stated as

$$\text{stenosis degree} = \frac{D - s}{D} \cdot 100\,\%\,,$$

  where D is the native artery diameter and s is the narrowed artery diameter, see Fig. 1. In clinical practice, it is an important indicator of arterial damage and its severity. In the case of bypass graft surgery, the degree of artery narrowing determines the need for a surgical intervention, which is usually above 70 % depending on artery damage and surgeon's experience. From published numerical studies dealing with various stenosis degrees in relation to bypass hemodynamics, it is possible to mention [5]. The authors observed hemodynamical changes for steady boundary conditions in five different bypass models (0 %, 70 %, 80 %, 90 % and 100 %) with the outcome that stenosis severity has a significant impact on the final distribution of velocity and wall shear stress within the bypass model.

- **Junction angle** (α) also known as anastomosis or bifurcation angle is the angle between native artery and implanted bypass graft, Fig. 1. The choice of the angle may be decided by the surgeon during a bypass surgery depending on the actual condition of the native artery (small angles require larger artery incision and longer suture lines). One of the first studies that investigated this geometrical parameter in a complete idealized bypass model was [7], which dealt with angles 30°, 45°, 60°, 75° and 90° for steady flow conditions. The authors concluded that the angle values have a significant impact on the occurrence of secondary flows and recirculation zones in the bypass model.

- **Diameter ratio** ($D : d$) represents the ratio between the diameter of native artery D and that of bypass graft d, Fig. 1. Its importance lies in the application of grafts with adequate inner diameters so that the amount of restored blood supply to the distal part of the native artery is sufficient for downstream tissue perfusion. Although the graft diameter may seem to be one of the most optional parameters used in bypass surgery, in reality the surgeon's choice in selecting an appropriate vascular graft is limited by the performed surgery type. For example, bypass grafting in the case of coronary arteries is currently done only with autologous arterial and venous grafts with given inner diameters. On the other hand, peripheral vascular bypass grafting is mostly carried out with synthetic grafts whose inner diameter is guaranteed by their manufacturer. Despite these limitations it is very advisable for the surgeons to be aware what the selection of inadequate graft may cause. In this regard, the authors of this paper have not be able to find a numerical study dealing with this kind of problem. A similar problem may be seen in the application of the so-called Miller cuff bypass graft mentioned, e.g., in [3], whose distal anastomosis is artificially enlarged in comparison to the native artery diameter.

Since the authors of this paper were unable to find a study that would be solely focused on the analysis of geometrical parameters influence on *pulsatile* bypass hemodynamics, the main objective of the study presented here is to perform a more in-depth comparison of geometry influence on physiological blood flow through a complete idealized end-to-side bypass model. The flow changes will be observed for three different values of stenosis degree, junction angle and diameter ratio. For results evaluation, special emphasis will be placed on the analysis of velocity distribution during the two phases of cardiac cycle (systole and diastole). Distribution of wall shear stress (WSS) and its oscillatory tendencies in the form of oscillatory shear index (OSI), [6], will be taken into account as well.

2. Problem formulation

In comparison to the majority of published studies, which modelled blood flow mainly in distal bypasses, e.g., [1, 3], this paper considers an idealized stenosed 3D bypass model with both proximal and distal parts, Fig. 1. This step is made in order to develop an adequate flow field at the distal anastomosis and also to enable us to observe flow changes inside the bypass graft in dependence on the studied geometrical parameter. For the purpose of results comparison, the bypass model proportions are set uniform regardless of the chosen geometrical parameter value, i.e., the native artery length is set to $L = 50$ mm and artery diameter is equal to $D = 3$ mm corresponding to an average right coronary artery. Taking into account the shape variability of all atherosclerotic plaques, the arterial narrowing is approximated with the help of the Gaussian function providing a smooth and realistic transition between the damaged and non-damaged native artery parts. In accordance to similar studies, e.g., [5, 7], we assume all bypass walls to be rigid and impermeable. This assumption is based on the knowledge that the majority of implanted bypass grafts is either of venous origin or is made of special synthetic materials, whose elastic properties are rather negligible, [10].

An overview of bypass model configurations studied in this paper is shown in Fig. 2. The left column represents models with various stenosis degrees (50 %, 75 % , 100 % — fully occluded native artery), the middle column stands for models with various junction angles ($\alpha = 30°$, $45°$, $70°$) and the right column shows models with various graft diameters ($D : d = 1 : 0.5; 1 : 1; 1 : 1.5$). At this point, note that in the case of configurations with 75 % stenosis, junction angle $\alpha = 45°$ and diameter ratio $D : d = 1 : 1$, the bypasses are all identical and will be later refereed to as the reference bypass model. The unstructured computational mesh for each considered bypass model configuration is generated with the help of the commercial software package Altair Hypermesh, see Fig. 3. In order to obtain grid independent results, numerical experiments were carried out for computational grids with around 150 000, 300 000 and 500 000 tetrahedral cells. In the end, it was stated that results computed for grids with 300 000 and

stenosis degree	junction angle	diameter ratio
(50 %; 75 %; 100 %)	$(\alpha = 30°; 45°; 70°)$	$(D\!:\!d = 1 : 0.5; 1 : 1; 1 : 1.5)$

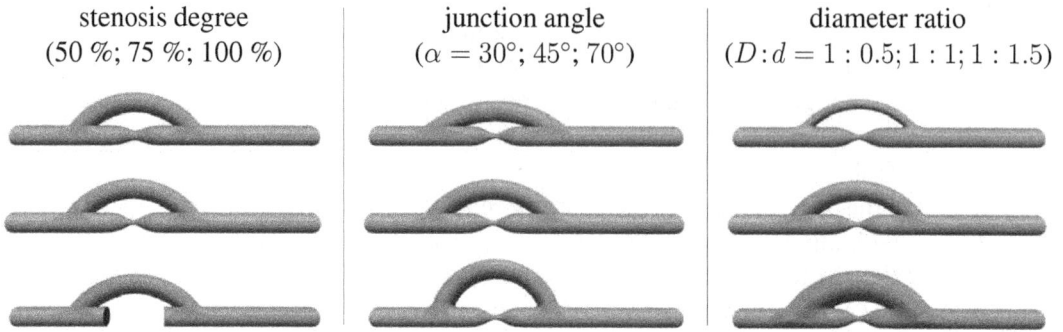

Fig. 2. Bypass model configurations with considered geometrical parameters

500 000 cells were comparable leading to the application of 300 000 cells mesh in all considered bypass configurations.

Since the aim of this paper is to investigate the problem of geometry influence on physiological bypass hemodynamics, several assumptions are made. Firstly, blood flow is modelled as an isothermal laminar pulsatile flow of incompressible Newtonian fluid with density $\varrho = 1\,060\,\mathrm{kg} \cdot \mathrm{m}^{-3}$ and dynamic viscosity $\eta = 3.45 \cdot 10^{-3}\,\mathrm{Pa} \cdot \mathrm{s}$. Regarding the supposed blood's Newtonian behaviour, the authors of this paper have taken into account conclusions resulting from their previous numerical simulations of non-Newtonian blood flow, [13, 14]. For steady flow conditions, it was stated that the occurrence of non-Newtonian effects in coronary bypass models is minimal. Hence, blood may be treated as Newtonian fluid in the case of coronary bypasses. The second assumption made in this paper is related to the prescription of realistic blood flow conditions, since blood flow in human arteries is always pulsatile. The problem is approached by considering a time-dependent inlet flow rate $Q(t)$, Fig. 4. The data are taken from [1] and should correspond to flow rate values measured in the right coronary artery during rest. For the purpose of numerical simulations, the flow rate is prescribed in the form of following Fourier series

$$Q(t) = A_0 + \sum_{k=1}^{5} A_k \cos\left(k\omega t - \varphi_k\right), \tag{1}$$

where $\omega = 2\pi/T$ is the angular frequency determined from cardiac period $T = 1.68\,\mathrm{s}$, $A_0 = 65.07\,\mathrm{ml} \cdot \mathrm{min}^{-1}$ represents the steady flow rate component and A_k and φ_k, $k = 1, \ldots, 5$ are the amplitude and the phase angle, respectively.

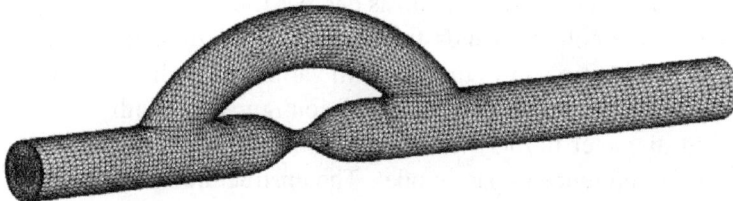

Fig. 3. Unstructured tetrahedral computational mesh — reference bypass model

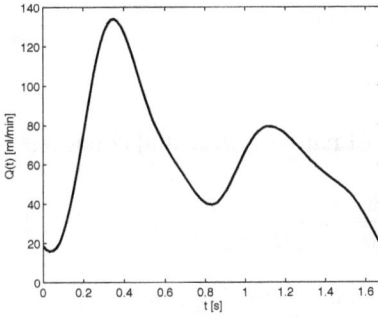

k	$A_k \,[\mathrm{ml}\cdot\mathrm{min}^{-1}]$	$\varphi_k \,[\mathrm{rad}]$
1	18.149	1.944
2	34.828	2.836
3	12.329	-2.124
4	9.107	-1.875
5	2.944	-0.447

Fig. 4. Prescribed time-dependent inlet flow rate $Q(t)$ corresponding to the right coronary artery during rest, [1], and overview of amplitude and phase values used in the Fourier series (1)

3. Mathematical model and numerical method

Let us consider a time interval $(0, \mathcal{T})$, $\mathcal{T} > 0$ and a bounded computational domain $\Omega \subset \mathbf{R}^3$ with boundary $\partial\Omega = \partial\Omega_I \cup \partial\Omega_O \cup \partial\Omega_W$, where $\partial\Omega_I$, $\partial\Omega_O$ and $\partial\Omega_W$ denote the inlet, the outlet and the rigid walls, respectively. The governing equations describing the pulsatile motion of blood in the given computational domain Ω constitute the non-linear system of the incompressible Navier-Stokes (NS) equations written in the non-dimensional form

$$\frac{\partial v_j}{\partial y_j} = 0, \tag{2}$$

$$\frac{\partial v_i}{\partial t} + \frac{\partial}{\partial y_j}(v_i v_j) + \frac{\partial p}{\partial y_i} = \frac{1}{Re}\frac{\partial^2 v_i}{\partial y_j \partial y_j} \quad \text{for } i,j = 1,2,3 \tag{3}$$

in the space-time cylinder $\Omega_T = \Omega \times (0, \mathcal{T})$, where $t \in (0, \mathcal{T})$ is the time, v_i is the i-component of the velocity vector $\boldsymbol{v} = [v_1, v_2, v_3]^T$ corresponding to the Cartesian coordinate $y_i \in \Omega$, p is the pressure, ϱ and η are the density and the molecular viscosity of blood, respectively. The reference Reynolds number is given as $Re = U_{ref} D_{ref} \varrho / \eta = 141.4$, where $U_{ref} = 0.153\,\mathrm{m\cdot s^{-1}}$ is the reference velocity corresponding to the average inlet velocity and D_{ref} is the reference length equal the native artery diameter $D = 3\,\mathrm{mm}$.

Considering a discretized 3D computational domain in the form of an unstructured tetrahedral grid with control volumes Ω_k, a hybrid grid system is applied for all calculations, [4]. Beside the values of pressure and velocity vector components computed in cell-centers, the hybrid grid system defines also a normal velocity V_m at the cell-faces. For the m-th face Γ_k^m of the control volume Ω_k, the normal velocity is calculated as $V_m = v_{im} \cdot {}^i n_k^m$, where $\boldsymbol{n}_k^m = [{}^1 n_k^m, {}^2 n_k^m, {}^3 n_k^m]^T$ is the outward unit vector normal to the cell face Γ_k^m, v_{im} are the velocity components at the m-th face Γ_k^m of the control volume Ω_k.

For time discretization of the non-linear system of the Navier-Stokes equations (2)–(3), a fully implicit second order accurate fractional step algorithm is used, [4]. The application of the implicit second order Crank-Nicolson scheme to the convective and viscous terms and the implicit Euler scheme to the pressure term in (3) leads to

$$\frac{v_i^{n+1} - v_i^n}{\Delta t} + \frac{1}{2}\frac{\partial}{\partial y_j}\left(v_i^{n+1}v_j^{n+1} + v_i^n v_j^n\right) = -\frac{\partial p^{n+1}}{\partial y_i} + \frac{1}{2Re}\frac{\partial^2}{\partial y_j \partial y_j}\left(v_i^{n+1} + v_i^n\right), \tag{4}$$

where $\Delta t = t_{n+1} - t_n$ is the time step between time levels t_n and t_{n+1}. Next, the convective terms in (4) are linearized with second order accuracy

$$v_i^{n+1} v_j^{n+1} = v_i^{n+1} v_j^n + v_i^n v_j^{n+1} - v_i^n v_j^n + \mathcal{O}(\Delta t^2) \,. \tag{5}$$

The subtitution of (5) into (4) gives a linearized system of partial differential equations

$$\frac{v_i^{n+1} - v_i^n}{\Delta t} + \frac{1}{2} \frac{\partial}{\partial y_j} \left(v_i^{n+1} v_j^n + v_i^n v_j^{n+1} \right) = -\frac{\partial p^{n+1}}{\partial y_i} + \frac{1}{2Re} \frac{\partial^2}{\partial y_j \partial y_j} \left(v_i^{n+1} + v_i^n \right) \,. \tag{6}$$

Finally, the application of the fractional step algorithm yields following equations

$$\frac{\hat{v}_i - v_i^n}{\Delta t} + \frac{1}{2} \frac{\partial}{\partial y_j} \left(\hat{v}_i v_j^n + v_i^n \hat{v}_j \right) = -\frac{\partial p^n}{\partial y_i} + \frac{1}{2Re} \frac{\partial^2}{\partial y_j \partial y_j} \left(\hat{v}_i + v_i^n \right) \,, \tag{7}$$

$$\frac{\partial^2 \Psi}{\partial y_i \partial y_i} = \frac{\partial \hat{v}_i}{\partial y_i} \,, \tag{8}$$

$$p^{n+1} = p^n + \frac{1}{\Delta t} \Psi \,, \tag{9}$$

$$v_i^{n+1} = \hat{v}_i - \frac{\partial \Psi}{\partial y_i} \,, \tag{10}$$

where Ψ is the pressure correction function and $\hat{v}_i$ are the velocity vector components introduced by the numerical method.

For spatial discretization of the system of equations (7)–(10), the cell-centered finite volume method is used. After several mathematical operations, following system of algebraic equations can be obtained

$$(\delta \hat{v}_i)_k + \frac{\Delta t}{2|\Omega_k|} \sum_{m=1}^{4} \left(\delta \hat{v}_{i\,m} V_m^n + v_{i\,m}^n \cdot {}^j n_k^m \, \delta \hat{v}_{j\,m} + 2 v_{i\,m}^n V_m^n \right) |\Gamma_k^m| =$$

$$-\frac{\Delta t}{|\Omega_k|} \sum_{m=1}^{4} p_m^n \cdot {}^i n_k^m |\Gamma_k^m| + \frac{\Delta t}{2Re|\Omega_k|} \sum_{m=1}^{4} \frac{\partial}{\partial n_k^m} \left(\delta \hat{v}_{i\,m} + 2 v_{i\,m}^n \right) |\Gamma_k^m|, \tag{11}$$

$$\sum_{m=1}^{4} \frac{\partial \Psi}{\partial n_k^m} |\Gamma_k^m| = \sum_{m=1}^{4} \hat{v}_{i\,m} \cdot {}^i n_k^m |\Gamma_k^m| \equiv \sum_{m=1}^{4} \hat{V}_m |\Gamma_k^m| \,, \tag{12}$$

$$\left(p^{n+1} \right)_k = (p^n)_k + \frac{1}{\Delta t} (\Psi)_k \,, \tag{13}$$

$$(v_i^{n+1})_k = (\hat{v}_i)_k - \frac{1}{|\Omega_k|} \sum_{m=1}^{4} \Psi_m \cdot {}^i n_k^m |\Gamma_k^m| \,, \tag{14}$$

$$V_m^{n+1} = \hat{V}_m - \frac{\partial \Psi}{\partial n_k^m} \,, \tag{15}$$

where $\delta \hat{v}_i = \hat{v}_i - v_i^n$, $\hat{V}_m$ is the hat velocity normal to the m-th cell face, $|\Gamma_k^m|$ is the area of the m-th face Γ_k^m of the control volume Ω_k, $|\Omega_k|$ is the volume of the control volume Ω_k, ${}^i n_k^m$ are the components of the outward unit vector normal to the cell face Γ_k^m. The adding of equation (15) to the system of equations (11)–(14) helps to suppress unwanted oscillations in the pressure field and to satisfy the continuity equation. Namely, it can be shown that the calculated velocity components $\left(v_i^{n+1} \right)_k$ do not generally satisfy the continuity equation at the $(n+1)$-th time level.

On the other hand, the normal velocities V_m^{n+1} give a divergence-free velocity field. Therefore, normal velocity values V_m^{n+1} determined in (15) are used in (11) and (12) at the next time level instead of V_m^n.

For evaluation of quantities and their derivatives at the m-th cell face Γ_k^m of the control volume Ω_k in (11)–(15), a second order interpolation method is used, [12]. Let us consider an arbitrary quantity Φ in the computed flow field. Its value Φ_m at the m-th face Γ_k^m can be determined using the second order interpolation method as

$$\Phi_m = (\Phi)_k + \frac{(\Phi)_{l_m} - (\Phi)_k}{\gamma_k + \gamma_{l_m}} \cdot \gamma_k = \frac{\gamma_{l_m}(\Phi)_k + \gamma_k(\Phi)_{l_m}}{\gamma_k + \gamma_{l_m}}, \tag{16}$$

where γ_k and γ_{l_m} are the minimal distances to the cell-face Γ_k^m from cell-centers S_k and S_{l_m} of the neighbouring control volumes Ω_k and Ω_{l_m}, respectively. The derivative of the quantity Φ in the direction of the unit outer vector $\boldsymbol{n}_k^m$ normal to the m-th face Γ_k^m of the control volume Ω_k can be approximated as

$$\left. \frac{\partial \Phi}{\partial n_k^m} \right|_{\Gamma_k^m} \approx \frac{(\Phi)_{l_m} - (\Phi)_k}{\gamma_k + \gamma_{l_m}}. \tag{17}$$

Regarding the application of boundary conditions, following approach has to be adopted. At the inlet, velocity vector components v_{iI} and zero pressure derivative $\frac{\partial p}{\partial n} = 0$ are prescribed. At the outlet, a known pressure p_O is given leading to the condition

$$-p\boldsymbol{n} + \frac{1}{Re} \cdot \frac{\partial \boldsymbol{v}}{\partial n} = -p_O \boldsymbol{n}. \tag{18}$$

At the solid walls, non-slip boundary condition $\boldsymbol{v} = \boldsymbol{0}$ and zero pressure derivative $\frac{\partial p}{\partial n} = 0$ is considered. For the numerical solution of the Poisson equation (12), zero pressure correction derivative $\frac{\partial \Psi}{\partial n} = 0$ at the inlet and at the solid walls are applied. In order to keep the required pressure value $p = p_O$ at the outlet, Dirichlet boundary condition $\Psi = 0$ is prescribed. The correction of the normal velocity (15) guarantees the satisfaction of the continuity equation for all control volumes except the ones with a face lying at the outlet boundary $\partial \Gamma_O$. In this case, normal velocity values have to be controlled so that the computed velocity field satisfies the continuity equation within these control volumes.

Taking into account boundary conditions mentioned above, equation (11) leads to the system of linear algebraic equations $\boldsymbol{A}\boldsymbol{x} = \boldsymbol{b}$ with sparse matrix $\boldsymbol{A}$ of order $3N_{CV}$, where N_{CV} is the total number of control volumes. Similarly, the Poisson equation for the pressure correction (12) with implemented boundary conditions leads to the system of linear algebraic equations $\tilde{\boldsymbol{A}}\tilde{\boldsymbol{x}} = \tilde{\boldsymbol{b}}$ with sparse matrix $\tilde{\boldsymbol{A}}$ of order N_{CV}. For the numerical solution of both these systems, the BICGSTAB iterative method with the incomplete LU preconditioner implemented in the UMFPACK library is used.

4. Numerical results and discussion

Since the objects of interest in this paper are three different geometrical parameters considered in similar bypass model configurations, only one version of unsteady boundary conditions is prescribed at corresponding computational domain boundaries:

- *inlet* $\partial \Omega_I$ — fully developed time-dependent velocity profile

$$v_{1I}(r,t) = \frac{8Q(t)}{\pi D^2}\left[1 - \left(\frac{2r}{D}\right)^2\right], \qquad v_{2I} = v_{3I} = 0 \text{ m} \cdot \text{s}^{-1},$$

where $r = \sqrt{y_2^2 + y_3^2}$ denotes the distance from artery center-line, D is the diameter of the coronary native artery and $Q(t)$ is the time-dependent flow rate, Fig. 4;

- *outlet* $\partial\Omega_O$ — constant pressure $p_O = 12\,\text{kPa}$ corresponding to average arterial pressure;

- *impermeable and rigid walls* $\partial\Omega_W$ — non-slip boundary condition, i.e., $v = 0$.

The application of fully developed velocity profile instead of a commonly used Womersley velocity profile results from prescribed inlet values. According to [9], the oscillatory Womersley velocity profile may be satisfactorily approximated by a time-varying parabolic velocity profile, also often denoted as 'quasi-steady', when the Womersley number of the pulsatile flow is sufficiently low ($Wo < 3$). In our case, the Womersley number is stated as $Wo = 0.5D\sqrt{\omega\varrho/\eta} = 1.61$ and enables us to prescribe the velocity profile mentioned above.

4.1. Velocity distribution

For a better comparison between each considered bypass model, velocity isolines at five different cross-sections are presented for various stenosis degrees, junction angles and diameter ratios in Table 1, Table 2 and Table 3, respectively. The upper part of each table shows results from systolic phase ($t_1 = 0.34\,\text{s}$) and the lower part from diastolic phase ($t_2 = 1.12\,\text{s}$) of the cardiac cycle, Fig. 5. The selection of all displayed cross-sections was made in order to take the flow character of each bypass configuration and their geometrical differences into account. In this regard, not every cross-section is located at the same distance from the inlet. On the other hand, all cross-sections should be approximately situated at following positions. The first cross-section denoted as A should show the flow field before the arterial narrowing and together with B, which is located near the graft inlet, should present the proximal bypass hemodynamics. On the other hand, the downstream hemodynamics near the distal anastomosis, which is of main interest in this paper, should be given by cross-sections C, D and E.

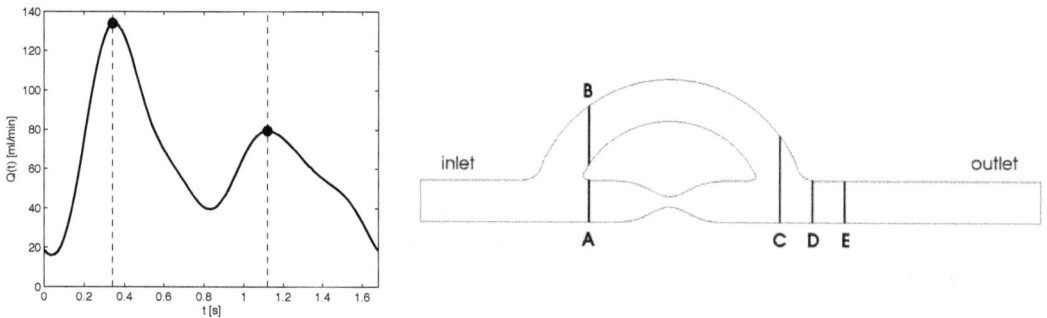

Fig. 5. Time instants $t_1 = 0.34\,\text{s}$ and $t_2 = 1.12\,\text{s}$ used for visualization of velocity isolines at selected cross-sections A–E along the idealized 3D bypass model

Despite the different maximum velocities during systolic and diastolic phases, the flow fields shown in Table 1 for the three stenosis degrees do not appear to be too dissimilar in dependence on the current inlet velocity. On the other hand, the comparison between each bypass configuration indicates several hemodynamical changes regardless of the current time instant. For example, the region around the arterial narrowing (cross-section A) in the case of the 100 % stenosis (occlusion) is filled with a large and relatively high-velocity recirculation zone in comparison to other bypasses. Another more distinct difference is the skewed shape of velocity profiles at

Table 1. Velocity isolines at cross sections A–E for various stenosis degrees, Fig. 5

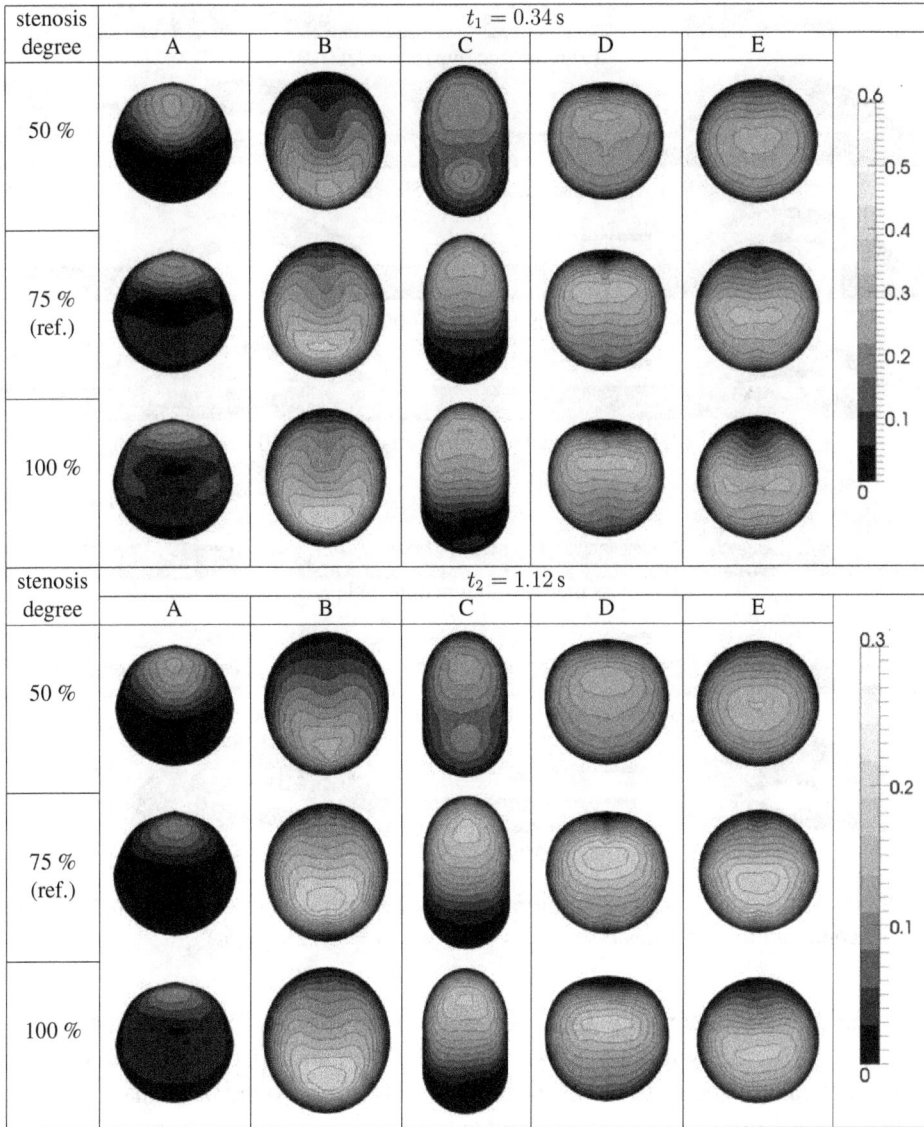

the graft inlet (cross-section B) and downstream from the distal anastomosis (cross-section C). In order to better understand the connection between stenosis degree and hemodynamics, the authors of this paper deem it necessary to present corresponding longitudinal sections of the three considered bypass models, Fig. 6a-c. At first sight, it is apparent that the overall blood flow in the 50 % stenosis configuration is mostly directed towards the stenosis, leading to a relatively low-velocity flow in the bypass graft and to a development of recirculation zones near the graft inlet, Fig. 6b. Some improvements can be found in the reference model with typical recirculation zones around the stenosis, Fig. 6a. With regard to the complete stream redirection in the case of the occluded bypass model, Fig. 6c, several small flow disturbances are visible in the vicinity of both anastomosis regions.

As to the junction angle influence on the resulting flow field, Table 2, once again no distinct differences between both cardiac phases are observed. More important in this regard seems to

(a) reference bypass model

(b) bypass model with 50% stenosis

(c) bypass model with 100% stenosis

(d) bypass model with junction angle $\alpha = 30°$

(e) bypass model with junction angle $\alpha = 70°$

(f) bypass model with diameter ratio $D : d = 1:0.5$

(g) bypass model with diameter ratio $D : d = 1:1.5$

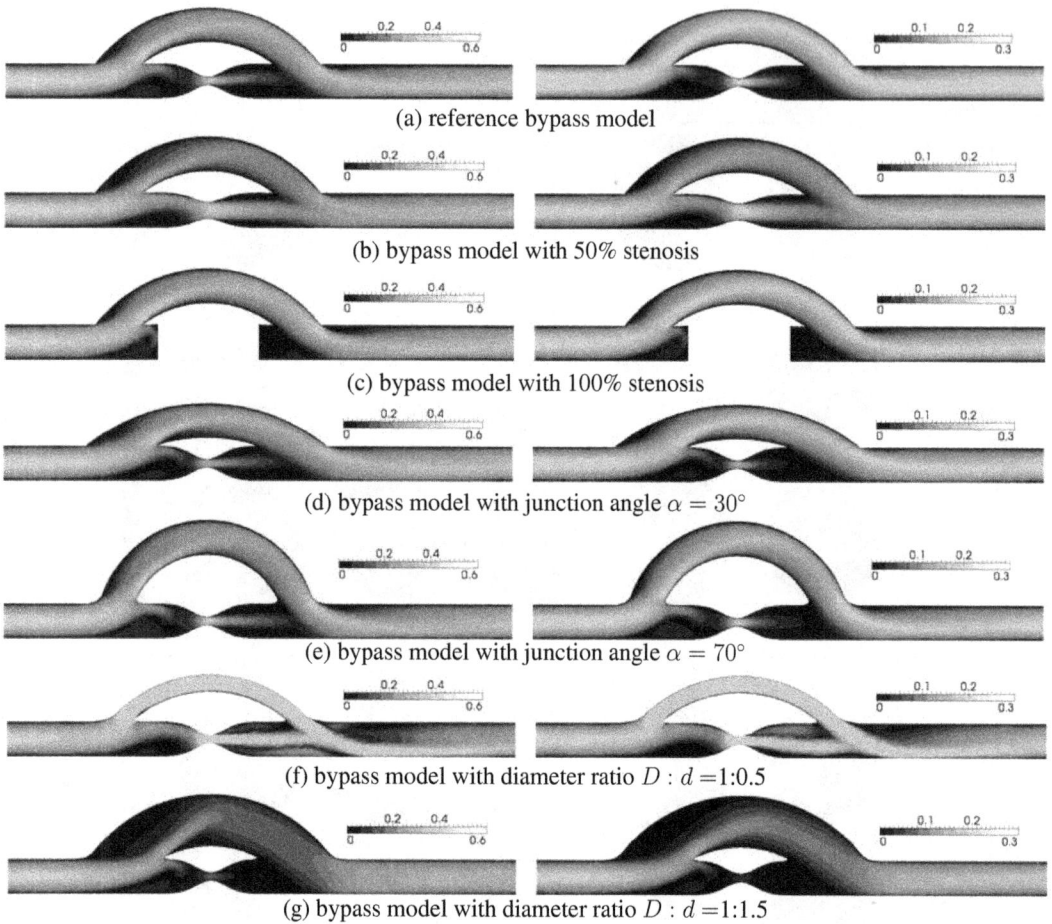

Fig. 6. Distribution of velocity magnitude $|v|$ $[m \cdot s^{-1}]$ in the longitudinal section through all considered bypass model configurations at times $t_1 = 0.34\,s$ (*left*) and $t_2 = 1.12\,s$ (*right*)

be the occurrence of secondary flows and 'M'-shaped velocity profiles inside the bypass graft, especially in the $70°$ configuration (cross-section B). In comparison to previous results, the junction angle influence seems to be mostly limited to regions around both anastomoses (cross-sections B and C). In this case, the hemodynamics changes have the form of sudden stream redirections and occurrence of several recirculation zones as is apparent from longitudinal sections shown in Fig. 6d–e. Referring to the $30°$ and $45°$ configurations, the small angle model indicates minimal improvement of the overall bypass flow field in comparison to the reference model, e.g., cross-sections C and D in Table 2 or Fig. 6a and 6d. On the other hand, the $70°$ model leads to the worst possible velocity distribution within the three considered angle configurations, Fig. 6e. The presence of relatively high-velocity recirculation zones (cross-section A) and the occurrence of strong secondary flows (B and E) is rather undesirable.

From all geometrical parameters considered in this paper, the diameter ratio is probably the most interesting one in relation to the resulting velocity distribution, Table 3. The main reason for the non-uniform colorbars at both time instants lies in the maximum velocity observed in the case of the bypass model, whose graft has the diameter equal to the native artery radius ($D : d = 1 : 0.5$). Since the decrease of cross-sectional area within this model entails a considerate increase in the velocity magnitude (up to $1.8\,m \cdot s^{-1}$ during systole), both colorbars

Table 2. Velocity isolines at cross sections A–E for various junction angles α, Fig. 5

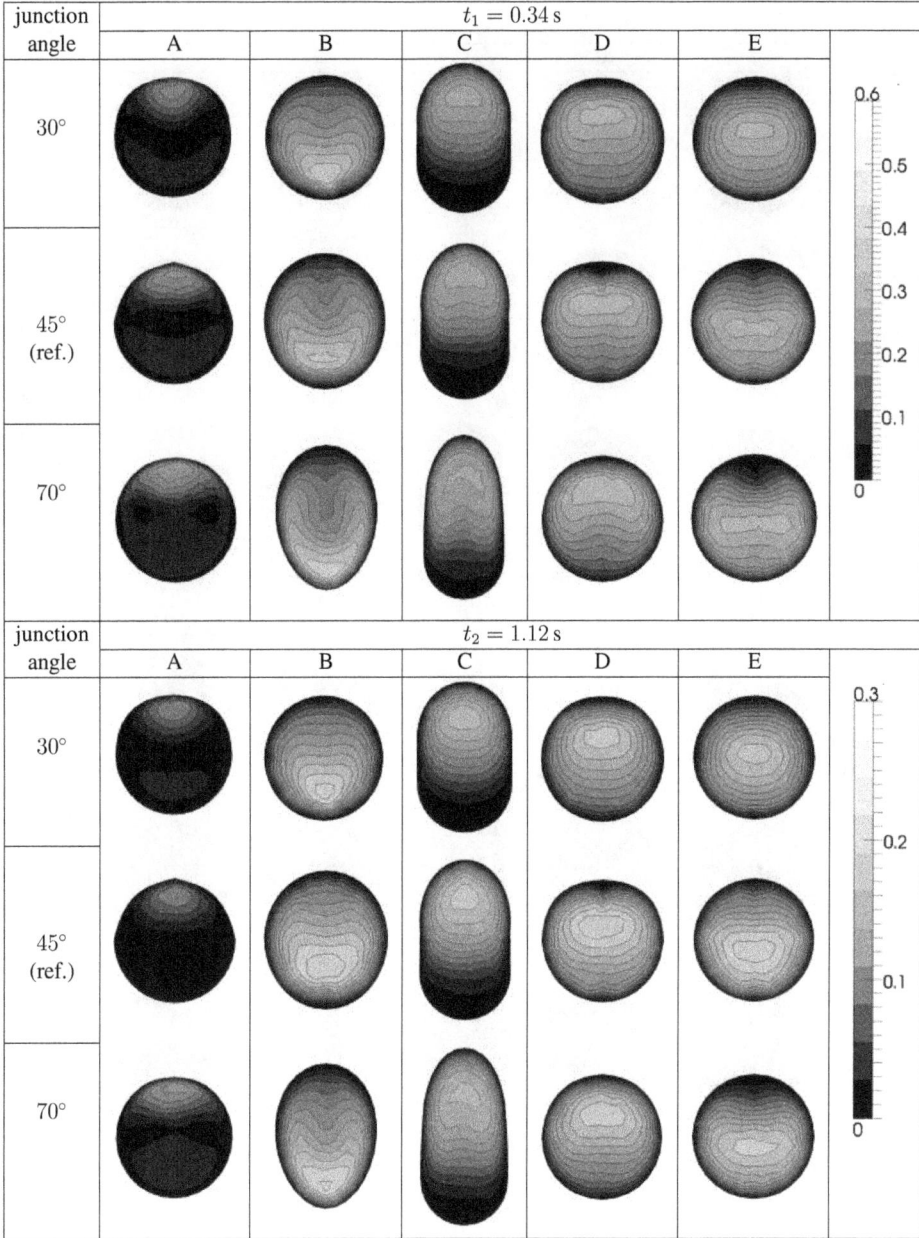

were chosen as non-uniform in order to capture the much lower velocities in the two remaining bypass configurations. For a better understanding, the authors of this paper present beside the cross-sections also the longitudinal sections through both small and large graft diameter models in Fig. 6f–g. Comparing the flow fields obtained for the three graft diameters, it is possible to deduce that the small diameter model gives the worst results presented for the distal anastomosis yet. The strongly disturbed and non-uniform hemodynamics in this bypass part is very well apparent from cross-sections C–E in Table 3 or from Fig. 6f. On the other hand, the large graft diameter configuration seems to be typical for relatively uniform flow field consisting of blunt velocity profiles (cross-sections C–E). Considerable disadvantage of this bypass geometry is

Table 3. Velocity isolines at cross sections A–E for various diameter ratios $D : d$, Fig. 5

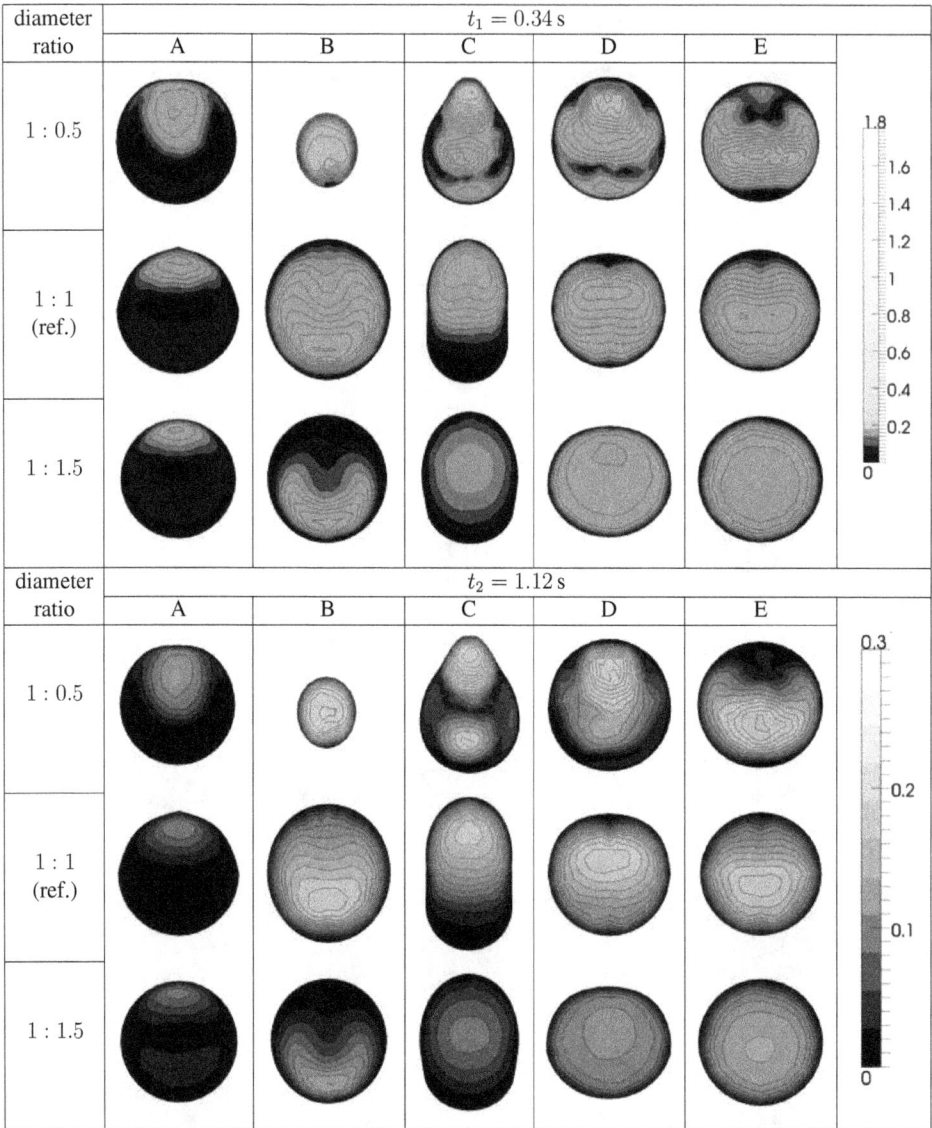

also the large recirculation zone that prevails during the whole cardiac cycle in the first third of the bypass graft, Fig. 6g, and is also apparent from cross-section B in Table 3.

4.2. WSS and OSI distributions

In this section, only the most interesting numerical results of WSS and OSI are going to be shown and discussed. For the WSS distribution at two time instances, the selected bypass model configurations are the reference model, the model with junction angle $\alpha = 70°$ and lastly the model with diameter ratio $D : d = 1 : 1.5$, Fig. 7a–c. For the OSI distribution, the reference model, the model with 50 % stenosis and the models with diameter ratios $D : d = 1 : 0.5$ and $D : d = 1 : 1.5$ are chosen, Fig. 8a–d. However, before the results discussion, it is appropriate

(a) reference bypass model (75 % stenosis, $\alpha = 45°$, $D : d = 1 : 1$)

(b) bypass model with junction angle $\alpha = 70°$

(c) bypass model with diameter ratio $D : d = 1 : 1.5$

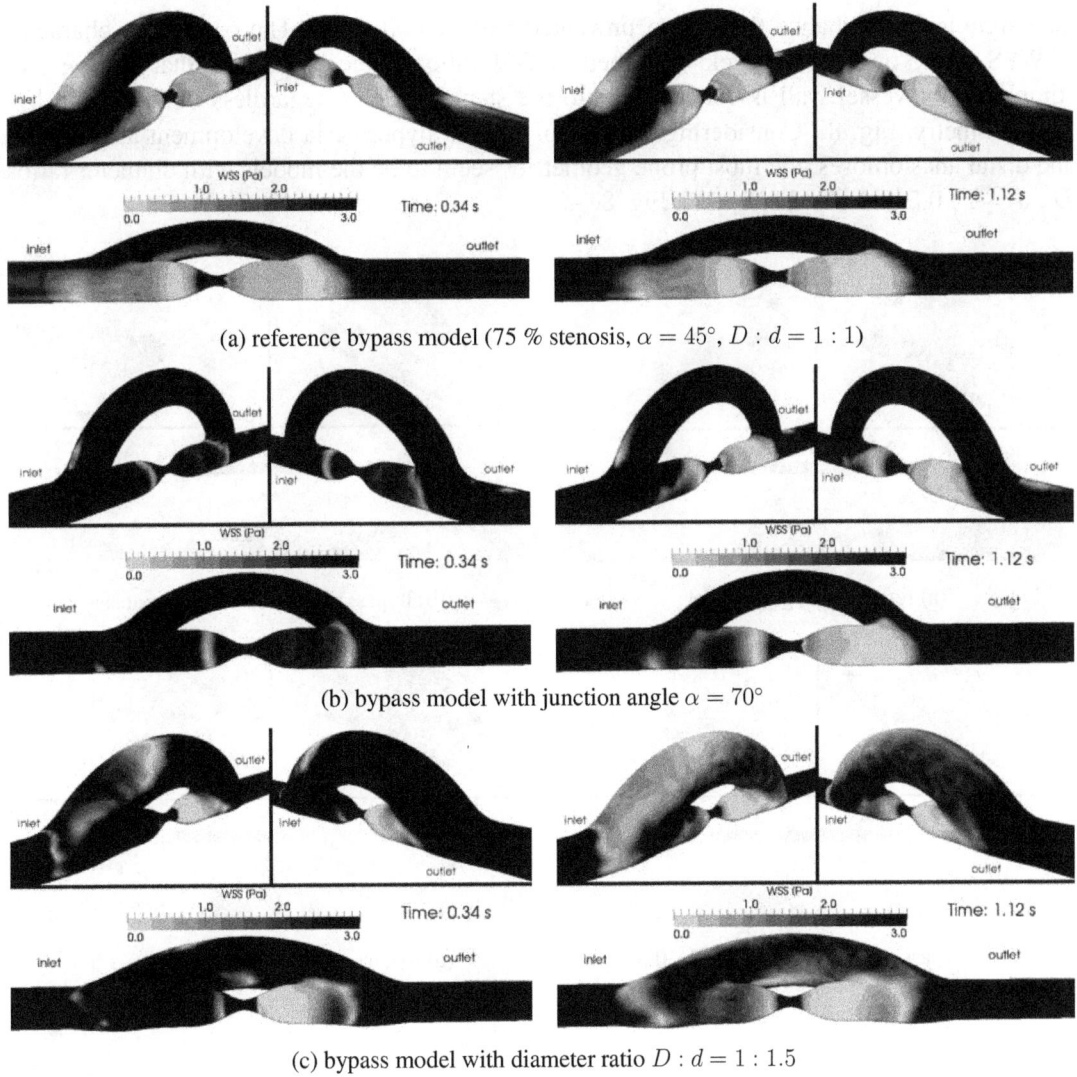

Fig. 7. WSS distribution for three different bypass model configurations at times $t_1 = 0.34\,\text{s}$ (*left*) and $t = 1.12\,\text{s}$ (*right*)

to mention the definition of OSI used in this paper, which is consistent with its introduction by Ku et al. in [6],

$$OSI = \frac{1}{2} \left(1 - \frac{\left| \int_0^T \tau_W dt \right|}{\int_0^T |\tau_W| \, dt} \right). \tag{19}$$

Since low WSS values are known to negatively influence the vessel wall morphology and are, therefore, of interest here, all ranges in Fig. 7 are lowered to values between $0 \div 3$ Pa. This way, several areas within the bypass model may be denoted as to be prone to the development of atherosclerotic lesion (stenosis region) or intimal hyperplasia (proximal and distal anastomoses). The most interesting WSS distribution is shown for the large graft diameter model, Fig. 7c, whose graft walls are exposed to low WSS most of the cardiac cycle and, therefore, may significantly enhance the risk of unwanted wall remodelling. The oscillatory character of WSS within one cardiac cycle described by OSI validates the assumption that the negative stimulation of vessel wall is concentrated to the stenosis region regardless of the actual by-pass geometry, Fig. 8. Considering the risk of intimal hyperplasia development at proximal and distal anastomoses, the most prone geometries seem to be the models with diameter ratios $D : d = 1 : 0.5$ and $D : d = 1 : 1.5$, Fig. 8c–d.

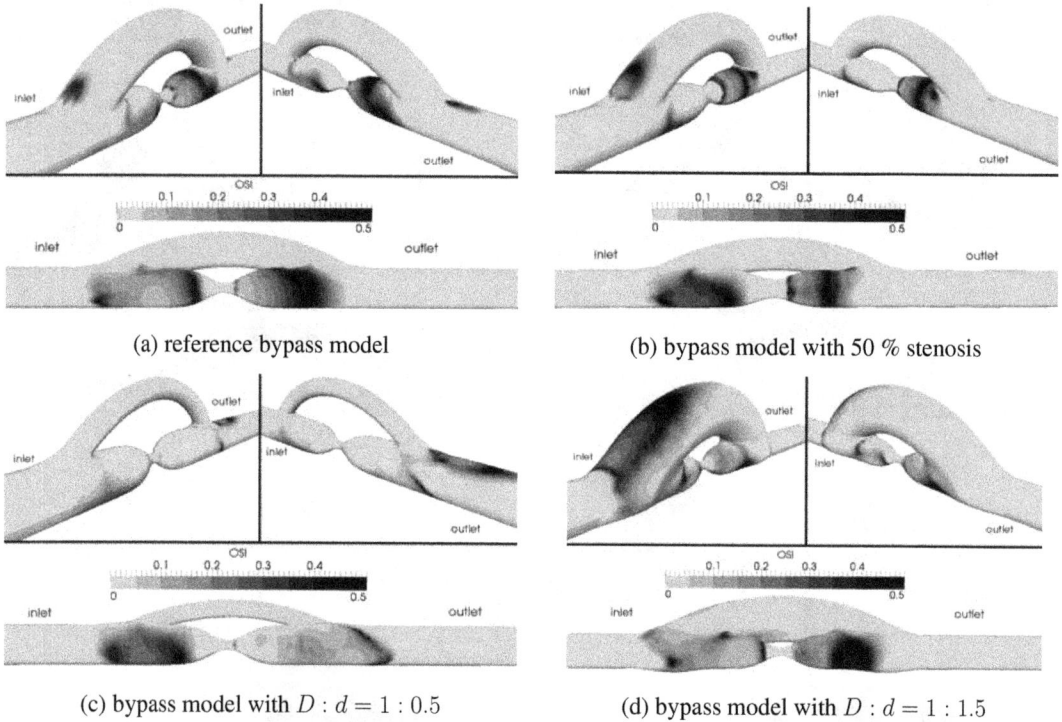

(a) reference bypass model

(b) bypass model with 50 % stenosis

(c) bypass model with $D : d = 1 : 0.5$

(d) bypass model with $D : d = 1 : 1.5$

Fig. 8. OSI distribution for four different bypass model configurations

5. Conclusion

The main objective of this paper was focused on the analysis of end-to-side bypass hemody-namics in dependence on three significant geometrical parameters (stenosis degree, junction angle and diameter ratio) by prescribing unsteady boundary conditions corresponding to physi-ological pulsatile blood flow. Special emphasis was placed on the application of own developed incompressible NS solver based on the fully implicit fractional-step method and the cell-centred finite volume method for unstructured tetrahedral grids. In comparison to their previous studies dealing with steady non-Newtonian blood flow, [13, 14], the authors dealt with the analysis of unsteady Newtonian blood flow.

The numerical results obtained for three different stenosis degrees showed the importance of residual blood flow through the arterial stenosis and its influence on the overall bypass hemo-dynamics. For example, the blood flow observed in the graft of the 50 % stenosis bypass configuration during the cardiac cycle showed a hemodynamics prone to thrombus formation, which after several months would lead to complete graft closure. Therefore, an graft implan-tation at this stage would in most cases result in an unsuccessful surgery. On the other hand, a bypass with occluded native artery showed a relatively uniform flow field, whose only disadvantage lies in a possible red blood cells damage due to stream impact on the proximal part of the atherosclerotic lesion during systole. In the case of the junction angle $30°$, small flow field improvements were observed in comparison to the reference bypass model with $\alpha = 45°$. However, the main disadvantage of all small angle bypasses can be seen in the need for a longer suture line resulting in a serious artery damage during surgery. The presence of several recirculation zones near both anastomoses for the $70°$ configuration may enhance the risk of blood cell accumulation and possibly lead to thrombus formation in corresponding bypass parts. Probably the most interesting results were obtained by varying the graft diameter d in comparison to the constant native artery diameter D. Considering the graft cross-sectional area in the bypass model with diameter ratio $D : d = 1 : 0.5$, the considerable velocity increase, especially during systole, produced a strongly disturbed flow field with several high-velocity recirculation zones near the distal anastomosis. Therefore, it is possible to declare this bypass configuration as one with very high probability of damage and accumulation of red blood cells with very high risk of thrombus formation. On the other hand, the bypass model with large graft diameter ($D : d = 1 : 1.5$) showed relatively undisturbed flow field. The only and probably very significant exception could be found in the first third of the graft, where a large low-velocity recirculation zone prevailed during the whole cardiac cycle. From the biomechanical point of view, the corresponding WSS and OSI distribution for this bypass configuration led us to the assumption that the inadequate velocity distribution near the proximal anastomosis may very well trigger unwanted changes in the vessel wall morphology and so enhance the risk of intimal hyperplasia development. The risk of blood cell accumulation inside the bypass graft is also not negligible.

Taking all the observations into account, each described bypass configuration seems to be favourable in some way — be it adequate WSS and OSI values or uniform velocity distribution during the cardiac cycle. In general, the most optimal flow field was achieved for the reference bypass model with average values of stenosis degree (75 %), junction angle ($45°$) and diameter ratio ($D = d$). The authors of this paper are, however, convinced that a more favourable flow field could be achieved for the complete bypass model by considering a small modification in the graft geometry. Namely, a graft with gradually increasing diameter would combine the advantages of small diameter graft at the proximal anastomosis and the advantages of the large diameter one at the distal anastomosis.

By presenting hemodynamics analysis in a complete idealized bypass model for various geometrical parameters, the authors of this paper would like to finish their previous research in the field of hemodynamics for this type of bypass models. In the future, their approach should be mainly focused on modelling of blood flow in patient-specific bypass models reconstructed from CT data.

Acknowledgements

This investigation was supported by the research project MSM 4977751303 of the Ministry of Education, Youth and Sports of the Czech Republic and by the internal student grant project SGS-2010-077 of the University of West Bohemia.

References

[1] Bertolotti, C., Deplano, V., Fuseri, J., Dupouy, P., Numerical and experimental model of post-operative realistic flows in stenosed coronary bypasses, Journal of Biomechanics 34 (8) (2001) 1 049–1 064.

[2] Haruguchi, H., Teraoka, S., Intimal hyperplasia and hemodynamic factors in arterial bypass and arteriovenous grafts: a review, Journal of Artificial Organs 6 (4) (2003) 227–235.

[3] Henry, F. S., Küpper, C., Lewington, N. P., Simulation of flow through a Miller cuff bypass graft, Computer methods in Biomechanics and Biomedial Engineering 5 (3) (2002) 207–217.

[4] Kim, D., Choi, H., A second-order time-accurate finite volume method for unsteady incompressible flow on hybrid unstructured grids, Journal of Computational Physics 162 (2) (2000) 411–428.

[5] Ko, T. H., Ting, K., Yeh, H. C., Numerical investigation on flow fields in partially stenosed artery with complete bypass graft: An in vitro study, International Communications in Heat and Mass Transfer 34 (6) (2007) 713–727.

[6] Ku, D. N., Giddens, D. P., Zarins, C. K., Glagov, S., Pulsatile flow and atherosclerosis in the human carotid bifurcation – Positive correlation between plaque location and low oscillating shear stress, Arteriosclerosis, Thrombosis, and Vascular Biology 5 (3) (1985) 293–302.

[7] Lee, D., Su, J. M., Liang, H. Y., A numerical simulation of steady flow fields in a bypass tube, Journal of Biomechanics 34 (11) (2001) 1 407–1 416.

[8] Kleinstreuer, C., Buchanan, J. R., Lei, M., Truskey, G. A., Computational analysis of particle-hemodynamics and prediction of the onset of arterial diseases, In Biomechanical Systems: Techniques and Applications, Volume II: Cardiovascular Techniques, Cornelius T. Leondes (editor), CRC Press, New York, 2000.

[9] Loudon, C., Tordesillas, A., The use of the dimensionless Womersley number to characterize the unsteady nature of internal flow, Journal of Theoretical Biology 191 (1) (1998) 63–78.

[10] Loth, F., Fischer, P. F., Bassiouny, H. S., Blood flow in end-to-side anastomoses, Annual Review of Fluid Mechanics 40 (2008) 367–393.

[11] Malek, A. M., Izumo, S., Mechanism of endothelial cell shape change and cytoskeletal remodeling in response to fluid shear stress, Journal of Cell Science 109 (1996) 713–726.

[12] Rhie, C. M., Chow, W. L., Numerical study of the turbulent flow past an airfoil with trailing edge separation, AIAA Journal 21 (11) (1983) 1 525–1 532.

[13] Vimmr, J., Jonášová, A., On the modelling of steady generalized Newtonian flows in a 3D coronary bypass, Engineering Mechanics 15 (3) (2008) 193–203.

[14] Vimmr, J., Jonášová, A., Non-Newtonian effects of blood flow in complete coronary and femoral bypasses, Mathematics and Computers in Simulation 80 (6) (2010) 1 324–1 336.

11

Bending of a nonlinear beam reposing on an unilateral foundation

J. Machalová[a,*], H. Netuka[a]

[a] *Faculty of Science, Palacký University in Olomouc, 17. listopadu 1192/12, 771 46 Olomouc, Czech Republic*

Abstract

This article is going to deal with bending of a nonlinear beam whose mathematical model was proposed by D. Y. Gao in (Gao, D. Y., Nonlinear elastic beam theory with application in contact problems and variational approaches, Mech. Research Communication, 23 (1) 1996). The model is based on the Euler-Bernoulli hypothesis and under assumption of nonzero lateral stress component enables moderately large deflections but with small strains. This is here extended by the unilateral Winkler foundation. The attribution unilateral means that the foundation is not connected with the beam. For this problem we demonstrate a mathematical formulation resulting from its natural decomposition which leads to a saddle-point problem with a proper Lagrangian. Next we are concerned with methods of solution for our problem by means of the finite element method as the paper (Gao, D. Y., Nonlinear elastic beam theory with application in contact problems and variational approaches, Mech. Research Communication, 23 (1) 1996) has no mention of it. The main alternatives are here the solution of a system of nonlinear nondifferentiable equations or finding of a saddle point through the use of the augmented Lagrangian method. This is illustrated by an example in the final part of the article.

Keywords: nonlinear beam, unilateral Winkler foundation, saddle-point formulation, finite element method, augmented Lagrangians

1. Introduction

It is well known that the classical beam theory is based on the Euler-Bernoulli hypothesis. It states that plane sections perpendicular to the longitudinal axis of the beam before deformation remain plane, undeformed and perpendicular to the axis after deformation. The standard mathematical model for large deflection can be derived using the displacement field

$$u_x(x, y) = u(x) - y\theta(x), \quad u_y(x, y) = w(x), \quad u_z(x, y) = 0, \tag{1}$$

where u_x and u_y are axial and transverse displacement components of an arbitrary beam material point, w and u denotes transverse and horizontal displacements of the middle axis $y = 0$. θ is the bending angle and it holds $\theta = \tan^{-1}(w') \approx w'$. The motion in the z direction is of no interest. Under the assumption concerning the stress components $\sigma_x \neq 0, \sigma_y = 0$ one can derive (for details see e.g. [11, 13]) the following governing equations

$$\left(EA \left[u' + \frac{1}{2}(w')^2 \right] \right)' = \tilde{f}, \tag{2}$$

$$(EI\, w'')'' - \left(EA\, w' \left[u' + \frac{1}{2}(w')^2 \right] \right)' = \tilde{q}, \tag{3}$$

where E is the Young's modulus, A is the cross-section area, I is the moment of inertia, $\tilde{f}(x)$ is the distributed axial load (per unit length) and $\tilde{q}(x)$ is the distributed transverse load (per unit

---

*Corresponding author. e-mail: jitka.machalova@upol.cz.

length). We can consider (2)–(3) as the 1D von Kármán equations. $\tilde{f} \equiv 0$ is a common case and it implies that after some rearrangements we obtain

$$(EI\,w'')'' - \left(EA \left[u' + \frac{1}{2}(w')^2 \right] \right) w'' = \tilde{q}, \tag{4}$$

where the coefficient by w'' has a constant value and consequently (4) is only a linear equation.

This inadequacy was revised by D. Y. Gao in his paper [3] by the change of the assumption about the stress components to $\sigma_x \neq 0, \sigma_y \neq 0$. After a short recapitulation of the Gao's model we want to propose a suitable finite element (hereafter we will use abbr. FE) solution for this beam because the paper [3] has no remark about it. Then we are going to concern ourself with the system of this nonlinear beam plus unilateral Winkler foundation. First we have to establish a suitable formulation for the bending problem. Next we want to analyze the system in order to solve such problem and finally we intend to obtain a computational model for the considered problem.

As for the aforementioned foundation, the classical works were concerned with beams in fixed connection with a foundation. Such problems are *linear* provided the beam model is linear too. However, some applications have the different matter because the beam is not firmly connected with the given foundation. These are *nonlinear* problems regardless using beam model and in such cases we can speak about the *unilateral* foundation. Some works on this field have been done for the classical Euler-Bernoulli beam model, see e.g. [6, 8] and [12], but there are no papers concerning nonlinear beams with unilateral foundation.

2. The nonlinear beam by D. Y. Gao

Here we want to present only a brief introduction of the nonlinear beam model from the paper [3]. Let us consider an elastic beam whose cross section in the x–y plane is a rectangle $[0, L] \times [-h, h]$ and in the y–z plane a rectangle $[-h, h] \times [0, b]$, i.e. the beam's length is L, its thickness 2h and its width b.

Displacements of such a beam are described by means of components (1). The Green-St Venant strain tensor for $x_1 = x$, $x_2 = y$, $x_3 = z$ has following components

$$\begin{pmatrix} \varepsilon_{11} & \varepsilon_{12} \\ \varepsilon_{12} & \varepsilon_{22} \end{pmatrix} = \begin{pmatrix} u' - y\theta' + \frac{1}{2}(u' - y\theta')^2 + \frac{1}{2}(w')^2 & \frac{1}{2}(w' - \theta) - \frac{1}{2}(u' - y\theta')\theta \\ \frac{1}{2}(w' - \theta) - \frac{1}{2}(u' - y\theta')\theta & \frac{1}{2}\theta^2 \end{pmatrix}. \tag{5}$$

This gives us after neglecting small terms $(u' - y\theta')^2$, $(u' - y\theta')\theta$ and substituting $\theta = w'$

$$\varepsilon_{11} \equiv \epsilon_x = u' - yw'' + \frac{1}{2}(w')^2, \tag{6}$$

$$\varepsilon_{22} \equiv \epsilon_y = \frac{1}{2}(w')^2, \tag{7}$$

$$\varepsilon_{12} = 0. \tag{8}$$

More details can be found in [3]. The nonzero stress components now can be obtained by the following constitutive relation

$$\begin{pmatrix} \sigma_x \\ \sigma_y \end{pmatrix} = \frac{E}{1 - \nu^2} \begin{pmatrix} 1 & \nu \\ \nu & 1 \end{pmatrix} \begin{pmatrix} \epsilon_x \\ \epsilon_y \end{pmatrix} \tag{9}$$

with ν denoting the Poisson's ratio.

Next we will suppose the beam is subject of a transversal load $\tilde{q}(x)$. The potential energy of a beam represented by a domain Ω is then as follows (see e.g. [13])

$$\Pi(u,w) = \frac{1}{2}\int_\Omega (\sigma_x \epsilon_x + \sigma_y \epsilon_y)\, d\Omega - \int_0^L \tilde{q}w\, dx = \tag{10}$$

$$\frac{E}{2(1-\nu^2)}\int_\Omega (\epsilon_x^2(x,y) + 2\nu\epsilon_x(x,y)\epsilon_y(x) + \epsilon_y^2(x))\, d\Omega - \int_0^L \tilde{q}w\, dx. \tag{11}$$

Using the Gâteaux derivatives (or first variations technique) for this functional we can get after some computation (see [3]) the system of two nonlinear equations for $x \in (0,L)$

$$u'' + (1+\nu)w'w'' = 0, \tag{12}$$

$$EI\, w^{IV} - 2hbE\left[(1+\nu)(2(w')^2 + u')w'' + \nu w'u''\right] = f, \tag{13}$$

assuming E is a constant, $I = \frac{2}{3}h^3b$ and denoting $f = (1-\nu^2)\tilde{q}$. The system can be reduced by integrating its first equation (12). We obtain

$$u' = -\frac{1}{2}(1+\nu)(w')^2 \tag{14}$$

and substituting this result into (13) we finally get

$$EI\, w^{IV} - E\alpha\, (w')^2 w'' = f \qquad \forall x \in (0,L), \tag{15}$$

where

$$\alpha = 3hb(1-\nu^2) \tag{16}$$

is a positive constant. The beam model described by the equation (15) is known as the *Gao beam* and it can be extended into a time-dependent model (see e.g. [4]).

3. Finite element model for the Gao beam

As the paper [3] contains only the beam theory, we are going to present here the FE approximation of the Gao beam. First we need a variational formulation of our problem. Let V be the space of kinematically admissible deflections v such that

$$H_0^2((0,L)) \subseteq V \subseteq H^2((0,L)). \tag{17}$$

Let us remember that the *Sobolev space* $H^2((0,L))$ consists of those functions $v \in L^2((0,L))$ for which derivatives v' and v'' (in the distribution sense) belong to the space $L^2((0,L))$. The *Lebesgue space* $L^2((0,L))$ is defined as the space of all measurable functions on $(0,L)$ which squares have a finite Lebesgue integral. And

$$H_0^2((0,L)) = \{v \in H^2((0,L)) : v(0) = v'(0) = 0 = v(L) = v'(L)\} \tag{18}$$

(more information can be found e.g. in [1]). It is well known that the finite element method distinguishes between *natural* and *essential boundary conditions*. The first ones are contained in the space V, the second ones are built into the variational formulation. Without a loss of generality we will assume for definiteness the clamped boundary conditions, i.e. $V = H_0^2((0,L))$, since another boundary conditions will not change in principle our approach.

From (15) after using integration by parts we can now immediately deduce

$$EI \int_0^L w''v'' \, \mathrm{d}x + \frac{1}{3}E\alpha \int_0^L (w')^3 v' \, \mathrm{d}x = \int_0^L fv \, \mathrm{d}x \qquad \forall v \in V. \tag{19}$$

This is in fact the equation for a stationary point of the potential energy of the Gao beam, which can be formally written as

$$\Pi_B'(w; v) = 0 \qquad \forall v \in V. \tag{20}$$

$\Pi_B'(w; v)$ denotes the Gâteaux derivative of Π_B at the point w in the direction v (see e.g. [1]). (19) with (20) imply that the functional of potential energy has the form

$$\Pi_B(w) = \frac{1}{2}EI \int_0^L (w'')^2 \, \mathrm{d}x + \frac{1}{12}E\alpha \int_0^L (w')^4 \, \mathrm{d}x - \int_0^L fw \, \mathrm{d}x. \tag{21}$$

It is easy to prove that this functional is strictly convex. Then the equation (20) can be consequently rewritten as

$$\Pi_B(w) = \min_{v \in V} \Pi_B(v). \tag{22}$$

The problem of finding a function $w \in V$ such that (22) holds we will call the *variational formulation* of the Gao beam bending. The convexity implies the unique solution of the minimization problem (22) and also the fact that (22) can be equivalently represented by the equation (19).

Now we proceed to a FE discretization of our problem. For this purpose we have to construct some dividing of the interval $[0, L]$ into subintervals $K_i = [x_{i-1}, x_i]$, where we have generated *nodes* $0 = x_0 < x_1 < \ldots < x_n = L$. Formally, the *discrete problem* reads as follows:

Find $w_h \in V_h$ such that

$$EI \int_0^L w_h'' v_h'' \, \mathrm{d}x + \frac{1}{3}E\alpha \int_0^L (w_h')^3 v_h' \, \mathrm{d}x = \int_0^L fv_h \, \mathrm{d}x \qquad \forall v_h \in V_h. \tag{23}$$

V_h is a finite-dimensional subspace of the given space V. In our case it has the form

$$V_h = \{v_h \in V : v_h|_{K_i} \in P_3(K_i) \quad \forall i = 1, \ldots, n\} \tag{24}$$

and contains piecewise polynomial functions from $C^1([0, L])$, i.e. continuous on $[0, L]$ together with its first derivatives. $P_3(K_i)$ denotes the set of cubic polynomials defined on K_i.

Now we can continue as it is usual for the standard FE beam model. We define the *Hermite basis functions* for our space (24) (see e.g. [8]) and afterwards the *shape functions* on a single element, which are beneficial from the practical computation point of view (see e.g. [8, 10]). But contrary to the standard FE solution process the second term in (23) prevents us to obtain a system of linear equations, as it is a rule in the classical beam theory. In matrix form we get formally

$$[\boldsymbol{K}_1 + \boldsymbol{K}_2(\boldsymbol{w})] \, \boldsymbol{w} = \boldsymbol{f}. \tag{25}$$

Into the vector $\boldsymbol{w}$ we assembled all the unknowns. This system contains the matrix $\boldsymbol{K}_1$ from the first integral in (23), which is well known from the linear FE model, and the matrix $\boldsymbol{K}_2$ from the second integral in (23), which depends on the vector of unknowns $\boldsymbol{w}$ and therefore cannot be evaluated explicitly (similar cases are described e.g. in [11]). Formulas concerning this matrix are quite cumbersome and we omit them here. Traditional method for solution of (25) is the Newton method (see e.g. [9, 11]). Of course, the infamous property of the Newton method is its sensitivity to a good initial guess, which can occasionally cause divergence of our computational process.

A fair alternative is return to the minimization problem (22) instead of the nonlinear system solution. First we formulate the discrete optimization problem to (22) as follows:

Find $w_h \in V_h$ such that

$$\Pi_B(w_h) = \min_{v_h \in V_h} \Pi_B(v_h). \tag{26}$$

The same discretization process as above leads here not to a system of equations but to the minimization of the strictly convex function of N unknowns

$$F_B(\boldsymbol{w}) = \min_{\boldsymbol{v} \in \mathbb{R}^N} F_B(\boldsymbol{v}), \tag{27}$$

which gradient is formally done by the expression

$$\nabla F_B(\boldsymbol{v}) = [\boldsymbol{K}_1 + \boldsymbol{K}_2(\boldsymbol{v})]\, \boldsymbol{v} - \boldsymbol{f}. \tag{28}$$

The methods such as the conjugate gradient method or the BFGS method require only computation of the gradient and some inexact line-search algorithm to determine a step size. For the details we encourage the gentle reader to look through some book concerning optimization methods, e.g. [9].

4. Problem with an unilateral foundation

In this section we are going to present a new extension of Gao's work. We will deal with bending of the Gao beam resting on the Winkler foundation. The classical Winkler model is based on the assumption of a linear force-deflection relationship and a fixed connection between the beam and the foundation. Let k_F is the foundation modulus, which will be supposed constant. Then, with respect to (15), the requested equation is

$$EI\, w^{IV} - E\alpha\, (w')^2 w'' + k_F w = f \qquad \forall x \in (0, \mathrm{L}). \tag{29}$$

Very easy is obtaining the variational formulation, because the potential energy of the Winkler foundation is

$$\Pi_F(v) = \frac{1}{2} k_F \int_0^{\mathrm{L}} v^2 \, \mathrm{d}x \tag{30}$$

and regarding (21) consequently for the total energy holds

$$\Pi_{B+F}(v) = \frac{1}{2} EI \int_0^{\mathrm{L}} (v'')^2 \, \mathrm{d}x + \frac{1}{12} E\alpha \int_0^{\mathrm{L}} (v')^4 \, \mathrm{d}x - \int_0^{\mathrm{L}} fv \, \mathrm{d}x + \frac{1}{2} k_F \int_0^{\mathrm{L}} v^2 \, \mathrm{d}x. \tag{31}$$

The variational formulation afterwards reads as follows:

Find $w \in V$ such that

$$\Pi_{B+F}(w) = \min_{v \in V} \Pi_{B+F}(v) \tag{32}$$

and, since the strict convexity still holds, this can be equivalently expressed as

$$EI \int_0^{\mathrm{L}} w''v'' \, \mathrm{d}x + \frac{1}{3} E\alpha \int_0^{\mathrm{L}} (w')^3 v' \, \mathrm{d}x + k_F \int_0^{\mathrm{L}} wv \, \mathrm{d}x = \int_0^{\mathrm{L}} fv \, \mathrm{d}x \qquad \forall v \in V. \tag{33}$$

Next our attention will be focused on the so-called *unilateral case*, when the foundation and the beam are not interconnected. This case was studied e.g. in [6] and [12] for the linear

Euler-Bernoulli beam model. We will assume that the vertical axis is turned down. Applying then the technique from the mentioned works, we can rewrite (33) as follows

$$EI \int_0^L w''v'' \, dx + \frac{1}{3}E\alpha \int_0^L (w')^3 v' \, dx + k_F \int_0^L w^+ v \, dx = \int_0^L fv \, dx \qquad \forall v \in V, \quad (34)$$

where $w^+(x) = \frac{1}{2}(w(x) + |w(x)|) = \max\{0, w(x)\}$. Of course, we are able to write the variational formulation for the unilateral problem in the form:

Find $w \in V$ such that

$$\widetilde{\Pi}_{B+F}(w) = \min_{v \in V} \widetilde{\Pi}_{B+F}(v), \qquad (35)$$

where

$$\widetilde{\Pi}_{B+F}(v) = \frac{1}{2}EI \int_0^L (v'')^2 \, dx + \frac{1}{12}E\alpha \int_0^L (v')^4 \, dx + \frac{1}{2}k_F \int_0^L (v^+)^2 \, dx - \int_0^L fv \, dx. \quad (36)$$

But from now we are going to follow a different way compared to the cited papers.

Let us define a *problem decomposition* using a linear relationship, which in general has the form

$$Bv = q \qquad v \in V, q \in Q. \qquad (37)$$

B is a linear continuous operator from V into Q. The decomposition naturally split our problem into two pieces: the beam and the foundation. For our case we choose $Q = L^2((0, L))$ and B as the identity, more precisely the canonical mapping from V into Q. Thereby we get a new variable q joined with the foundation and defined by

$$v = q \qquad v \in V, \ q \in Q, \qquad (38)$$

while the beam will be described by the old variable v. After that we have the new functional

$$\widehat{\Pi}_{B+F}(v, q) = \frac{1}{2}EI \int_0^L (v'')^2 \, dx + \frac{1}{12}E\alpha \int_0^L (v')^4 \, dx + \frac{1}{2}k_F \int_0^L (q^+)^2 \, dx - \int_0^L fv \, dx \quad (39)$$

defined on the set

$$W = \{\{v, q\} \in V \times Q \colon v = q\} \qquad (40)$$

and the variational formulation of the problem with unilateral foundation is then as follows:

Find $\{w, p\} \in W$ such that

$$\widehat{\Pi}_{B+F}(w, p) = \min_{\{v, q\} \in W} \widehat{\Pi}_{B+F}(v, q). \qquad (41)$$

It is evident that (41) is equivalent to (35). This way we follow the main idea from [7] and this is a part of a more general strategy called the *decomposition-coordination method* from [5].

But there is some inconvenience in (41). The new formulation represents a *constrained optimization* problem. To handle it right, we must define the *Lagrangian* for our problem by

$$\mathcal{L}(v, q, \mu) = \widehat{\Pi}_{B+F}(v, q) + \int_0^L \mu(v - q) \, dx \qquad v \in V, \ q \in Q, \ \mu \in \Lambda, \qquad (42)$$

where μ is the *Lagrange multiplier* associated with the constraint $v = q$ and $\Lambda = L^2((0, L))$. It can be proved (see e.g. [2]), that our problem (41) can be reformulated as the so-called *saddle-point problem* for the Lagrangian (42):

Find $\{w, p, \lambda\} \in V \times Q \times \Lambda$ such that

$$\mathcal{L}(w, p, \mu) \leq \mathcal{L}(w, p, \lambda) \leq \mathcal{L}(v, q, \lambda) \quad \forall v \in V,\ q \in Q,\ \mu \in \Lambda. \tag{43}$$

In our case (43) can be equivalently expressed as follows

$$\mathcal{L}(w, p, \lambda) = \inf_{\{v,q\} \in V \times Q} \sup_{\mu \in \Lambda} \mathcal{L}(v, q, \mu) = \sup_{\mu \in \Lambda} \inf_{\{v,q\} \in V \times Q} \mathcal{L}(v, q, \mu). \tag{44}$$

From here we can observe, that it is possible to obtain the unknowns w, p by some minimization. Therefore, by this way we transformed the constrained problem (41) into an *unconstrained* one at the cost of the additional unknown, i.e. the Lagrange multiplier λ.

By means of Gâteaux derivatives (or first variations) of the Lagrangian $\mathcal{L}$ with respect to q and μ we obtain at the point $\{w, p, \lambda\}$ the following results

$$w = p, \quad \lambda = k_F\, p^+ \quad \text{a.e. in } L^2((0, \mathrm{L})). \tag{45}$$

The first one was expected regarding (38), the second one gives us the interpretation of the Lagrange multiplier λ.

Finally, we must mention the question of the existence of a saddle point $\{w, p, \lambda\}$. In *infinite dimensions* this question coincide with the question of the existence of a Lagrange multiplier λ and it is, however, somewhat problematical. Sufficient conditions to assure the existence of the multiplier λ would be found e.g. in [2]. The problem considered in this article fulfills these conditions and (44) has therefore a solution.

5. Solution of the given problem

Now we consider some possibilities how to solve our problem (44). There are two principal ways to this objective. The first one consists in transformation our problem into the system of nonlinear equations. The second way is based on using of optimization methods to find a saddle point of (42). We can recognize the situation is in a certain manner similar to that we encountered by finding solution for bending of the Gao beam.

The first way uses a transformation to a mixed complementarity problem and will be omitted in this article as it would be rather extensive (for Euler-Bernoulli beam this approach was realized e.g. in [7]). So that we will concern our attention to the second possibility for solution of our problem (44) which consists in taking advantage of optimization methods. Despite the fact that the saddle-point problem is not a true optimization problem, we have a good opportunity in combining two methods. The first one is the *Uzawa algorithm* for finding saddle points and the second one is the so-called *augmented Lagrangian method*. Hereafter we will mainly follow [5].

The *augmented Lagrangian* $\mathcal{L}_r$ is defined in our case for any $r > 0$ by

$$\mathcal{L}_r(v, q, \mu) = \mathcal{L}(v, q, \mu) + \frac{r}{2} \int_0^{\mathrm{L}} (v - q)^2 \, \mathrm{d}x \tag{46}$$

with $\mathcal{L}$ given by (42). Next we can introduce the saddle-point problem for this augmented Lagrangian:

Find $\{w, p, \lambda\} \in V \times Q \times \Lambda$, with V from (17) and $Q, \Lambda = L^2((0, \mathrm{L}))$, such that

$$\mathcal{L}_r(w, p, \mu) \leq \mathcal{L}_r(w, p, \lambda) \leq \mathcal{L}_r(v, q, \lambda) \quad \forall v \in V,\ q \in Q,\ \mu \in \Lambda. \tag{47}$$

An advantageousness of the augmented Lagrangian method is given by the fact that we can state the following result (for the proof see [5]):

Suppose $\{w, p, \lambda\}$ is a saddle point of $\mathcal{L}$ on $V \times Q \times \Lambda$. Then $\{w, p, \lambda\}$ is a saddle point of $\mathcal{L}_r$ for every $r > 0$, and vice versa. Furthermore w is a solution of the original problem (35), (36) and we have $p = w$.

Hence we can interchange the problems (43) and (47) and from the computational point of view the second one will be much more convenient. We have to notice, that in *finite dimensions* the existence of a saddle point is assured, since we minimize under a linear equality constraint.

Considering all things, to solve the problem (35), (36) we need to determine the saddle points of the Lagrangian $\mathcal{L}$ from (42) and consequently the saddle points of the augmented Lagrangian $\mathcal{L}_r$ from (46). This can be attained with the help of a variant of the Uzawa algorithm. The rather complicated problem in the implementation of such an algorithm presents the solution of the minimization problem for $\mathcal{L}_r$ with respect to $\{v, q\}$ at each iteration. A frequently used solution procedure consists of using the block relaxation method which leads to the following algorithm

$p^0 \in Q, \lambda^1 \in \Lambda$ are given,
$\qquad$ then for $n = 1, 2, \ldots$
$\qquad\qquad$ determine w^n, p^n as follows:
$\qquad\qquad\qquad$ find $w^n \in V$ such that
$\qquad\qquad\qquad$ $\mathcal{L}_r(w^n, p^{n-1}, \lambda^n) \leq \mathcal{L}_r(v, p^{n-1}, \lambda^n) \qquad \forall v \in V$,
$\qquad\qquad\qquad$ find $p^n \in Q$ such that
$\qquad\qquad\qquad$ $\mathcal{L}_r(w^n, p^n, \lambda^n) \leq \mathcal{L}_r(w^n, q, \lambda^n) \qquad \forall q \in Q$,
$\qquad\qquad$ determine λ^{n+1} as follows:
$\qquad\qquad\qquad$ $\lambda^{n+1} = \lambda^n + \rho(w^n - p^n) \qquad \rho > 0$.

Under quite general assumptions we have convergence of this algorithm under the condition $0 < \rho < ((1 + \sqrt{5})/2)r$. The proof may be found in [5]. Let us remark that for our functional (36) aforementioned assumptions are fulfilled. The good choice for ρ seems to be in most cases $\rho = r$. Moreover, then we are able to implement some modification into our algorithm. From the equation for the minimization of $\mathcal{L}_r$ with respect to q we get

$$r(p^n - w^n) = \lambda^n - k_F(p^n)^+ \qquad \text{a.e. in} L^2((0, L)). \tag{48}$$

This result helps us to adjust the Uzawa step as follows

$$\lambda^{n+1} = \lambda^n + r(w^n - p^n) = \lambda^n + k_F(p^n)^+ - \lambda^n = k_F(p^n)^+ \tag{49}$$

and the last row of our algorithm can now be rewritten according to (49).

Finally, for computational purposes we must define suitable approximations of the infinite-dimensional spaces V, Q and Λ. Let us denote their finite-dimensional subspaces as V_h, Q_h and Λ_h. V_h will be the same as in (24), Q_h and Λ_h can be chosen as

$$Q_h = \Lambda_h = \{q_h \in L^2((0, L)) \colon q_h|_{K_i} \in P_0(K_i) \qquad \forall i = 1, \ldots, n\}, \tag{50}$$

i.e. these spaces consist of piecewise constant functions.

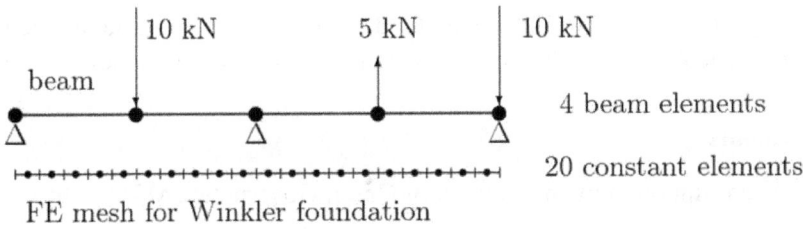

Fig. 1. Sketch for the example

6. Example

Here we want to illustrate the above explained theory and methods on a simple example. Let us consider a beam of the length L = 4 m with three supports at $x = 0$ m, $x = 2$ m and $x = 4$ m and resting on a Winkler foundation. Data for the beam and the foundation are given as follows: $EI = 2 \times 10^7$ N $\cdot$ m^2, h = 0.25 m, b = 0.4 m, $\nu = 0.3$, $k_F = 2 \times 10^7$ N $\cdot$ m^{-2}. Three isolated forces are acting at $x = 1$ m, $x = 3$ m and $x = 4$ m as it can be seen from Fig. 1, where is also an example how finite element meshes are constructed (dots denote element nodes).

Table 1. Results for the example

number of elements		EB linear beam				Gao beam			
		classical WF		unilateral WF		classical WF		unilateral WF	
beam	found.	u max	u min	u max	u min	u max	u min	u max	u min
4	40	6.237	−4.525	6.433	−5.036	5.639	−4.093	5.805	−4.544
4	100	6.236	−4.524	6.431	−5.035	5.637	−4.091	5.803	−4.542
4	400	6.236	−4.524	6.431	−5.035	5.636	−4.090	5.802	−4.541
8	40	6.238	−4.526	6.433	−5.036	5.639	−4.092	5.805	−4.544
8	104	6.237	−4.525	6.432	−5.036	5.637	−4.091	5.803	−4.542
8	400	6.237	−4.525	6.432	−5.035	5.637	−4.091	5.803	−4.542

Results for extreme displacement values are given in the Table 1, presented numbers should be multiplied by the scaling factor 10^{-5} m. The table contains results for the Euler-Bernoulli (abbr. EB) beam and for the Gao beam, both with the classical Winkler foundation (abbr. WF) and unilateral foundation. Different meshes give the quite similar numbers and we can observe something like convergence of the numerical values. The nonlinear beam proves itself as more stiff, which we could expect e.g. from (25) due to an additional stiffness matrix K_2.

7. Conclusion

We proposed here the new way how to formulate and solve the problem of bending of the non-linear Gao beam while the beam is resting on the unilateral Winkler foundation, which is not connected with the beam. The beam and the foundation have their own finite elements and element meshes which are closer to their physical fundamentals as it is in contact problems. But we are not forced to solve a contact problem. Our solution uses a saddle-point formulation and represents some compromise between a contact solution technique and a standard practice. It can be realized either through the application of methods for a system of nonlinear nondifferentiable equations, namely the nonsmooth Newton method, or by the help of the augmented Lagrangian method. The numerical example demonstrated some possibilities of the new solution method.

Acknowledgements

The work has been supported by the Council of Czech Government MSM 6198959214.

References

[1] Cea, J., Optimization: Theory and algorithms, Lectures on mathematics and physics, vol. 53, Tata Institute of Fundamental Research, Bombay, 1978.

[2] Ekeland, I., Témam, R., Convex analysis and variational problems, SIAM, Philadelphia, 1999.

[3] Gao, D. Y., Nonlinear elastic beam theory with application in contact problems and variational approaches, Mech. Research Communication, 23 (1) (1996) 11–17.

[4] Gao, D. Y., Finite deformation beam models and triality theory in dynamical post-buckling analysis, Intl. J. Non-Linear Mechanics, 35 (2000) 103–131.

[5] Glowinski, R., Numerical methods for nonlinear variational problems, Springer-Verlag, Berlin, Heidelberg, 1984.

[6] Horák, J. V., Netuka, H., Mathematical model of nonlinear foundations of Winkler's type: I. Continous problem, Proceedings of 21st conference with international participation Computational Mechanics 2005, Hrad Nečtiny, November 7–9, 2005, published by UWB in Pilsen, 2005, pp. 235–242 (in Czech).

[7] Machalová, J., Netuka, H., A new approach to the problem of an elastic beam resting on a foundation, Beams and Frames on Elastic Foundation 3, VŠB – Technical University of Ostrava, Ostrava, 2010, pp. A99–A113.

[8] Netuka, H., Horák, J. V., Mathematical model of nonlinear foundations of Winkler's type: II. Discrete problem, Proceedings of 21st conference with international participation Computational Mechanics 2005, Hrad Nečtiny, November 7–9, 2005, published by UWB in Pilsen, 2005, pp. 431–438 (in Czech).

[9] Nocedal, J., Wright, S. J., Numerical optimization, Second edition, Springer, New York, 2006.

[10] Reddy, J. N., An introduction to the finite element method, McGraw-Hill Book Co., New York, 1984.

[11] Reddy, J. N., An introduction to nonlinear finite element analysis, Oxford University Press, Oxford, 2004.

[12] Sysala, S., Unilateral elastic subsoil of Winkler's type: Semi-coercive beam problem, Applications of Mathematics 53 (4) (2008) 347–379.

[13] Washizu, K., Variational methods in elasticity and plasticity, Second edition, Pergamon Press, New York, 1975.

12

Modelling of the mechanical behaviour of porcine carotid artery undergoing inflation-deflation test

J. Vychytil[a,*], F. Moravec[a], P. Kochová[a], J. Kuncová[b], J. Švíglerová[b]

[a] *Faculty of Applied Sciences, University of West Bohemia, Univerzitní 22, 306 14 Plzeň, Czech Republic*
[b] *Faculty of Medicine in Pilsen, Charles University in Prague, Lidická 1, 301 66 Plzeň, Czech Republic*

Abstract

Samples of porcine carotid artery are examined using Tissue bath MAYFLOWER, Perfusion of tubular organs Version, Type 813/6. Pressure-diameter diagrams are obtained for fixed axial extension and volumetric flow rate. Finite element analysis of the experiment, performed using COMSOL software, indicates a negligible effect of given flow rate on the mechanical response of the tested sample. Also the effect of clamped ends is shown to be local only. Hence, static analysis in MATLAB software is performed considering the arterial segment as an incompressible hyperelastic axisymmetric tube. Residual stress at the load-free configuration is taken into account resulting in the overall stiffening of the model. Comparison of theoretical and experimental pressure-diameter curves results in the identification of material parameters using the least square method. In addition to classical hyperelastic models, such as the neo-Hookean and the Fung's exponential, two-scale model mimicking arrangement of soft tissue is considered.

Keywords: carotid artery, inflation-deflation test, modelling, hyperelasticity, parameters identification

1. Introduction

The arterial wall mechanics and its interaction with blood flow have been an object of extensive research during past decades. Many experiments have been performed to investigate the material properties of animal and human arteries, see e.g. [8, 12, 13]. Earlier works on this topic date back to the end of the 19th century, such as the work [27] proving mechanical response of arteries to be nonlinear with strain-hardening. The common assumption of incompressibility is confirmed in [5] observing volume change to be very small even for deformations greater than those *in vivo*. The anisotropy of the arterial wall is an evident fact due to the enormous complexity of its microstructure. However, experiments performed in [24] suggest that arterial segments can be considered as cylindrically orthotropic tubes for modelling purposes. According to [15, 22] material constants are found to be of the order ~ 1 MPa for elastic modulus and its fraction of approximately 1/10 for loss modulus. However, mechanical properties depend on many factors, such as age and volume fraction of constituents, see e.g. [1, 2].

Together with experiments, numerical modelling plays an important role in understanding arterial mechanics. Since the blood flow through an artery represents a complex problem of fluid-solid interaction in the framework of continuum mechanics, simplifications concerning both arterial wall and flow are applied. In [29], for instance, the bypass model is proposed considering arterial walls as rigid and the blood flow as incompressible Newtonian. More often,

---

*Corresponding author. e-mail: vychytil@kme.zcu.cz.

arterial walls are considered as hyperelastic [28] or linearly viscoelastic [4]. Such assumption is supported by a biaxial tension test performed with porcine coronary artery tissue in [21]. Although the anisotropy of the tissue is demonstrated, isotropic models are sufficient to describe *in vivo* conditions according to authors. On the other hand, isotropic hyperelastic (viscoelastic) models cannot be capable of describing all aspects of the mechanical response of the arterial wall resulting from its complex microstructure and from the fact that artery is a living tissue. According to [26], relevant features that should not be omitted in the modelling of arteries include anisotropy, residual stresses and remodeling. Therefore, the latest works deal with these effects and possible modelling approaches.

An anisotropy of the arterial wall can be described using the notion of fiber-reinforced materials [18]. This approach is applied in [14, 19] to propose a hyperelastic model of arterial wall. The fibrous nature of the arterial wall is included in the formula of strain-energy function via two unit vectors representing the orientations of collagenous fibers. Extension for the case of viscoelastic response is done in [20]. A different approach is proposed in [31] by defining the eight-chain orthotropic unit element that represents the microstructure. Employing the assumption of affine deformations at the micro-scale, an orthotropic hyperelastic material model of the arterial wall is obtained.

It is known fact that unloaded arteries are not stress-free due to the certain growth mechanisms of the different layers. Also, they are subjected to axial prestretch *in vivo*, see e.g. [7, 9, 19]. Theoretical framework for dealing with residual stress in arteries is provided in [19]. It consists in introducing three configurations: reference (stress-free), unloaded and current. The simplification of geometry of the arterial ring leads to a description of the residual stress with a single parameter. A similar approach is used in [9] to propose a finite element (FE) model of the carotid segment. Here, residual stress causes the resulting stress field to be much more uniform. In [23], the model of prestrained cytoskeleton is embedded into a macroscopic model of arterial wall using the method of homogenization. Although the model of arterial ring opens, which is in agreement with observations, author admits that it does not reflect a realistic situation. A similar approach of including prestress in the material model (using the model of prestressed cytoskeleton and the notion of representative volume element) is described in [30].

Finally, recent works on numerical modelling of arterial walls deal also with the so-called remodeling, i.e. change of either geometry or mechanical properties in order to adapt to applied load [26]. In [16], for instance, the reorientation of collagen fibers within arterial wall is modelled upon the assumption of their alignment with the directions of principal stresses.

The aim of this work is to investigate the mechanical response of the porcine carotid artery experimentally and to propose a suitable model representing the artery under experimental conditions. The paper is organized as follows. Section 2 describes an experiment which consists in loading the arterial segment with inner pressure under constant flow rate and axial prestretch. Pressure-diameter diagrams for both inflation (loading) and deflation (unloading) parts are provided. In section 3, preliminary estimates and an FE analysis of the artery under experimental conditions is performed. The simplified geometry of axisymmetric tube is considered using the hyperelastic material models of the neo-Hookean and the Fung's type, the fluid is considered to be incompressible Newtonian. Results indicate a negligible effect of flow rate substantiating the static analysis performed in MATLAB software in section 4. Here the arterial segment is represented as a hyperelastic tube loaded with inner pressure considering residual stress at the load-free configuration. Apart of the neo-Hookean and the Fung's model, the two-scale hyperelastic model introduced in [17] is employed. A comparison with experiment is provided using the least square method which leads to the identification of material parameters.

2. Experimental investigation of the artery

2.1. Experimental setting

The porcine carotid artery is obtained from domestic pig. The animal is firstly anesthetized and after dissection of the carotid artery, it is euthanized by KCl injection. The sample is prepared by cutting the segment of the artery with the length L and clamped into a measurement device (Tissue bath MAYFLOWER, Perfusion of tubular organs version, type 813/6, Hugo Sachs Electronik, Germany), see fig. 1. The sample is prestretched axially within the device to mimic *in vivo* conditions with $\lambda_z = 1.5$ so that its current length is $l = \lambda_z L$. This value is chosen according to the studies [11] and [25] where the value of 1.4 is used for common arteries in rabbits and 1.5 for porcine carotid arteries, respectively. In [19] the values between 1.1 and 1.9 are chosen depending on the kind of artery. The sample is embedded in the tissue bath and perfused intraluminarly with a constant flow rate $Q = 2$ ml/min. The Tyrode's solution with the temperature of $36\,°C$ is used. The measurement device provides for setting the outlet pressure, denoted with p_2 in fig. 1. At the same time, the outer diameter of the arterial segment is determined as follows. For each value of p_2, the middle segment of the artery of the length $l_m = 21$ mm (corresponding to $L_m = 14$ mm at the load-free state) is captured using stereomicroscope (Olympus SZ60) and camera (Olympus E440). Three values of diameter are measured from each photograph in the middle and both ends of measured segment. The resultant diameter is calculated as the arithmetic mean, i.e. $d = (d_1 + d_2 + d_3)/3$.

At first, the preconditioning of 4 cycles from 0 up to 200 mm Hg is performed to ensure unambiguous mechanical response. After that, the sample is loaded from 0 up to 200 mm Hg using the step of 10 mm Hg and then unloaded to 0 mm Hg using the same step. In fact, it is a

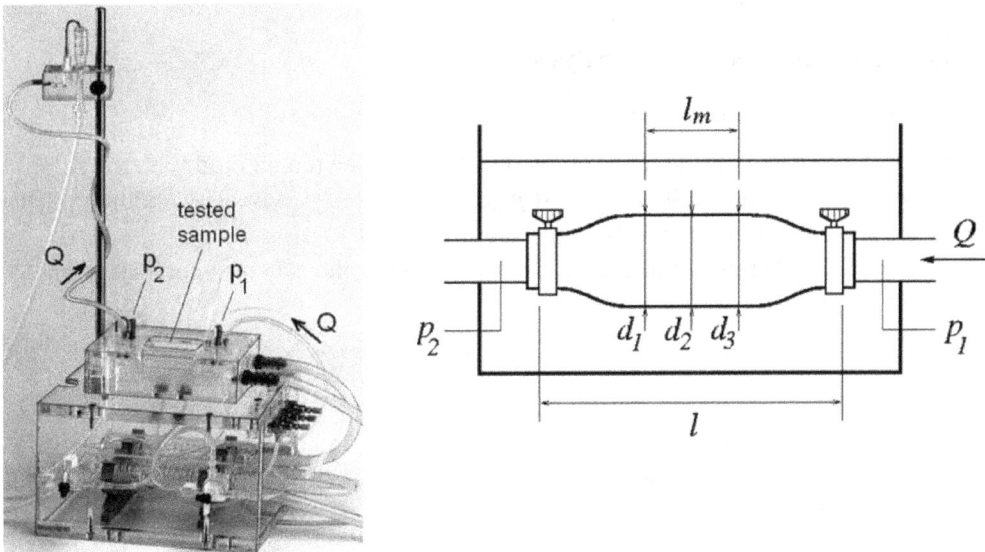

Fig. 1. The measurement device (left, taken with modifications from [32]) and the detailed scheme of clamped sample (right). The Tyrode's solution is pumped through the tested sample with constant flow rate of Q. The outlet pressure, p_2, is controlled by the device mounted on the stool. The inlet pressure is denoted with p_1. The arterial sample is clamped to the ends of two rigid pipes and prestretched axially to the current length of l. Outer diameters, d_1, d_2 and d_3 are measured in the middle segment of the length l_m

quasi-static process as setting each loading step and taking a photograph is done manually. The time of whole cycle is approximately $t \approx 6$ min providing $t_{step} \approx 9$ s for each step.

2.2. Results

The pressure-diameter diagram for whole inflation-deflation cycle is depicted in fig. 2. Both inflation and deflation parts exhibit the strain hardening as it is characteristic for soft living tissues, the hysteresis corresponds to the viscoelastic behaviour of the arterial wall. For convenience, the units of pressure are converted as 1 mm Hg $= 133.322$ Pa.

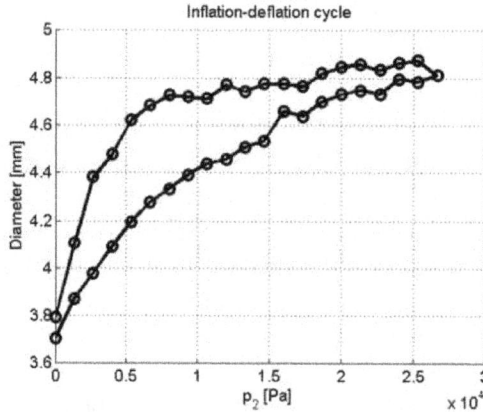

Fig. 2. The pressure-diameter diagram of the porcine carotid artery. The lower curve corresponds to inflation, the upper curve to deflation part of the cycle, respectively

3. Preliminary estimates and FE analysis

3.1. Fluid pressure, clamping effect

The Tyrode's solution is considered as an incompressible Newtonian fluid with the same characteristics as water, i.e. $\eta = 0.001$ Pa.s and $\rho = 1\,000$ kg $\cdot$ m^{-3} (viscosity and mass density). The flow obeys the incompressible Navier-Stokes equations.

Considering the arterial segment as a cylindrical tube, the pressure corresponding to the laminar flow exhibits a linear decrease,

$$p(z) = \frac{p_1 - p_2}{l} z + p_2 , \tag{1}$$

see e.g. [3]. Here, z denotes the coordinate aligned with the longitudinal axis of the tube, $z = 0$ and $z = l$ refer to the outlet and the inlet region, respectively. The pressure difference can be estimated using the Poiseuille's formula for a laminar flow,

$$Q = \frac{\pi r_{in}^4}{8} \frac{p_1 - p_2}{\mu l} , \tag{2}$$

where r_{in} denotes the inner radius of the tube. In our case, the estimate of $(p_1 - p_2) \sim 1$ Pa is obtained. Comparing to the applied load (each step $\sim 10^2$ Pa), the pressure corresponding to the flow is negligible and thus the arterial segment can be considered as statically loaded with the inner pressure $\Delta P \approx p_2$.

Although the arterial segment can be represented with a cylindrical tube at the unloaded configuration, its shape at the current configuration is in fact more complex. However, for loading consisting of the inner pressure and the longitudinal stretch, the representation with the cylindrical tube is appropriate for the middle part, far enough from clamped ends.

3.2. FE analysis

To confirm the assumptions resulting from the preliminary estimates, the FE analysis is performed. The arterial segment is represented as a cylindrical tube at the reference configuration and residual stresses are not taken into account. Considering the axial symmetry, the fluid occupies the rectangle $[0, R_{in}] \times [0, L]$ and the solid occupies the rectangle $[R_{in}, R_{out}] \times [0, L]$ at the reference configuration. Here, $R_{in} = 1.07$ mm, $R_{out} = 2.38$ mm and $L = 14$ mm.

The solid part (i.e. the arterial wall) is considered as a hyperelastic incompressible material. The neo-Hookean and the Fung's energy functions are employed, their expressions as well as particular choice of material constants are detailed in section 4. The fluid part is considered as an incompressible Newtonian fluid, as it is mentioned in previous section.

Boundary conditions of the model are defined as follows. For the fluid, the constant inlet flow rate of $Q = 2$ ml/min and the outlet pressure p_2 are prescribed. The solid part is stretched along the z-axis at first with $\lambda_z = 1.5$ and then the left ($Z = 0$) and the right ($Z = L$) faces are fixed to represent clamping of the sample.

The results are depicted in figs. 3 to 6. In fig. 3, the distribution of the Von Mises stress at the current configuration of the solid part is depicted. The profile of the tube at the current

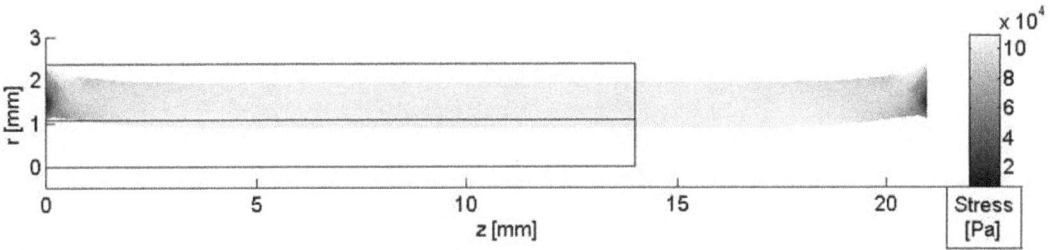

Fig. 3. Deformation and Von Mises stress distribution in the arterial wall involved by the stretch ratio of $\lambda_z = 1.5$ and the flow rate $Q = 2$ ml/min. Reference geometry is also indicated

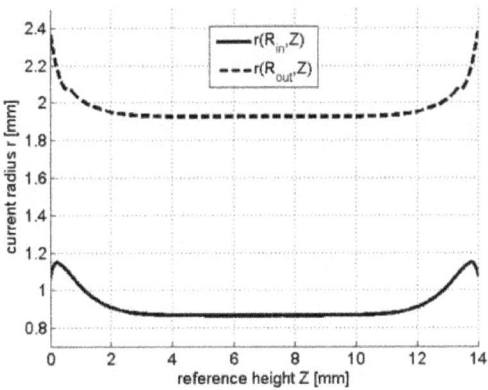

Fig. 4. Pipe profile at the current configuration for outlet pressure $p_2 = 0$ Pa

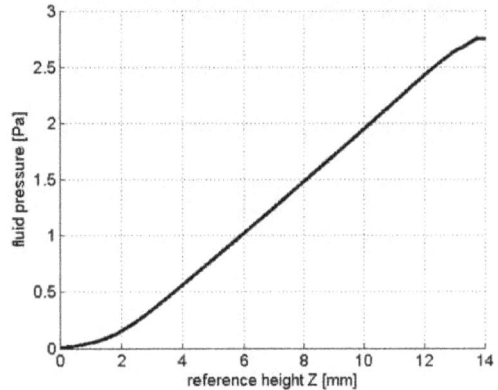

Fig. 5. Dependence of the fluid pressure on the height Z for outlet pressure $p_2 = 0$ Pa

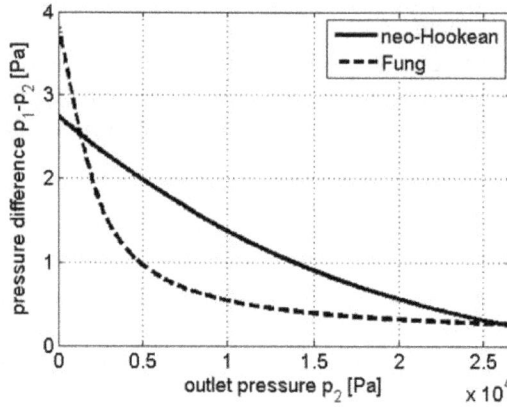

Fig. 6. Dependence of the difference between the inlet and outlet pressure on the outlet pressure with constant inner flow $Q = 2$ ml/min

configuration is also depicted in fig. 4. The distribution of the fluid pressure along the longitudinal axis is plotted in fig. 5. The dependence is nearly linear according to (1) except for the regions influenced by clamped ends. The pressure difference between the inlet and outlet region is small compared to stresses within the arterial wall and it even decreases with increasing outlet pressure, see fig. 6. Clearly, the assumptions resulting from preliminary estimations are confirmed. The clamping effect is only local and the pressure corresponding to the flow of the Tyrode's solution is negligible.

4. Static analysis

4.1. Basic relations

Description of deformations of the arterial wall follows the approach presented in [19]. To take into account the residual stresses, three configurations are introduced, see fig. 7. The reference configuration, Ω_0, corresponds to cut arterial segment that is supposed to be stress-free. The position of any material point is described using the cylindrical coordinate system $\{R, \Theta, Z\}$, where $R \in [R_{in}, R_{out}]$, $\Theta \in [0, 2\pi - \alpha]$ and $Z \in [0, L]$. Here, α denotes the opening angle. At the current configuration, the spatial coordinates are denoted with $\{r, \theta, z\}$, where $r \in [r_{in}, r_{out}]$, $\theta \in [0, 2\pi]$ and $z \in [0, l]$.

No shear occurs at the current state, hence the deformation gradient at the cylindrical coordinates takes the form

$$\mathbf{F} = \begin{pmatrix} r'(R) & 0 & 0 \\ 0 & \frac{hr}{R} & 0 \\ 0 & 0 & \lambda_z \end{pmatrix}. \tag{3}$$

Here, h is a constant parameter related to the opening angle and λ_z is a constant axial stretch,

$$h = \frac{2\pi}{2\pi - \alpha}, \quad \lambda_z = \frac{l}{L}. \tag{4}$$

The arterial wall is assumed to be incompressible providing the expression for current radius as

$$r(R) = \sqrt{\frac{R^2}{h\lambda_z} + C}. \tag{5}$$

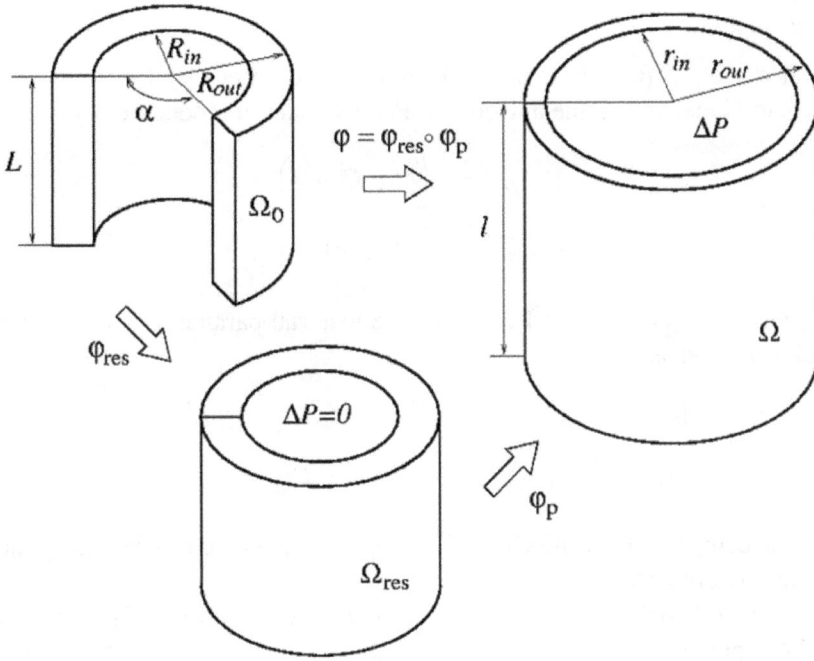

Fig. 7. An arterial segment at the stress-free reference configuration Ω_0, the load-free configuration Ω_{res} and the current configuration Ω. Redrawn with modifications from [19]

Here, C is to be determined from the boundary conditions. Applying the momentum balance,

$$\frac{\partial \sigma_{rr}}{\partial r} + \frac{1}{r}(\sigma_{rr} - \sigma_{\theta\theta}) = 0, \tag{6}$$

the constitutive equation of hyperelasticity,

$$\boldsymbol{\sigma} = \frac{\partial \hat{W}}{\partial \mathbf{F}}\mathbf{F}^T - p\mathbf{I}, \tag{7}$$

and the boundary conditions,

$$\sigma_{rr}(r_{in}) = -\Delta P, ,\quad \sigma_{rr}(r_{out}) = 0, \tag{8}$$

we obtain the relationship between ΔP and C,

$$\Delta P = \int_{R_{in}}^{R_{out}} \frac{r'}{r}\left(\frac{r}{R}\hat{W}_2 - r'\hat{W}_1\right) \mathrm{d}R. \tag{9}$$

Here, $\boldsymbol{\sigma}$ is the Cauchy stress tensor, p is the hydrostatic pressure, ΔP is the pressure applied to the inner face of the tube and $\hat{W}_i$ denotes the partial derivative of $\hat{W}$ with respect to the corresponding diagonal component of the deformation gradient,

$$\hat{W}_i = \frac{\partial \hat{W}}{\partial \lambda_i},\quad \lambda_i = F_{ii}. \tag{10}$$

4.2. Material models

Three material models defined by various formulas of strain-energy functions are employed. The isotropic neo-Hookean and the anisotropic Fung's material model are defined as

$$
\begin{aligned}
\hat{W}_{NH} &= \frac{\mu}{2}\left(\mathrm{tr}\,\hat{\mathbf{C}} - 3\right), \\
\hat{W}_{Fung} &= \frac{c}{2}\left[\exp(q) - 1\right],
\end{aligned}
\tag{11}
$$

where $\hat{\mathbf{C}} = (\det \mathbf{C})^{-1/3}\mathbf{C}$, $\mathbf{C} = \mathbf{F}^T\mathbf{F}$, μ and c are material parameters. The function q in the Fung's model is defined as

$$
\begin{aligned}
q = &\, b_1 \hat{E}_{\theta\theta}^2 + b_2 \hat{E}_{zz}^2 + b_3 \hat{E}_{rr}^2 + 2b_4 \hat{E}_{\theta\theta}\hat{E}_{zz} + 2b_5 \hat{E}_{zz}\hat{E}_{rr} + \\
&\, 2b_6 \hat{E}_{rr}\hat{E}_{\theta\theta} + b_7 \hat{E}_{\theta z}^2 + b_8 \hat{E}_{rz}^2 + b_9 \hat{E}_{r\theta}^2 .
\end{aligned}
\tag{12}
$$

Here, $\hat{E}_{ij}$ are the components of the Green-Lagrange strain tensor referred to cylindrical coordinates, b_i's are non-dimensional material parameters.

Also, the so-called "balls and springs" (BS) model introduced in [17] is employed. It is a two-scale orthotropic hyperelastic model motivated by the arrangement of the microstructure of soft tissues. Its mechanical response can be described by an approximative analytical formula of strain-energy function,

$$
W_{bs} = W_s + W_m ,
\tag{13}
$$

where

$$
W_s = \frac{1}{2}\sum_{\substack{i=1 \\ i\neq j\neq k}}^{3} K_i \frac{k_i}{1+k_i} l_{ij}l_{ik}\left[(F_{ii}-1)+\gamma_i P_i\right]^2 ,
\tag{14}
$$

and

$$
W_m = \frac{K_1 K_2 K_3 (1+k_1)(1+k_2)(1+k_3)\left[1 - F_{11}^{eff}F_{22}^{eff}F_{33}^{eff}\right]^2}{\sum_{i=1,i\neq j\neq k}^{3} 2K_j K_k (1+k_j)(1+k_k)\frac{l_{ji}l_{ki}}{\gamma_i^2}\left(F_{jj}^{eff}F_{kk}^{eff}\right)^2} .
\tag{15}
$$

Here, K_i's, k_i's, l_{ij}'s and γ_i's are material parameters related to the microstructure (stiffness of matrix and cells, geometrical anisotropy, relative sizes of cells), see [17, 30] for details. The effective stretches are defined as

$$
F_{ii}^{eff} = \frac{F_{ii}-1}{\gamma_i(1+k_i)} + 1 .
\tag{16}
$$

In general, this orthotropic model contains 11 material parameters. However, this number is reduced for the case of transverse isotropy (7 material parameters) or isotropy (3 material parameters).

4.3. Influence of residual stresses

To obtain pressure-diameter curves, equations derived in this section are implemented in MATLAB software. For a given pressure ΔP, the unknown constant C is determined from (9) using the Newton-Raphson iteration scheme. The diameter is then calculated using (5) as

$$
d = 2r(R_{out}) .
\tag{17}
$$

The influence of residual stresses on the mechanical response is studied by controlling the parameter α (the opening angle, see fig. 7). If $\alpha = 0$, no residual stress is involved, i.e. the load-free configuration is also stress-free. Increasing value of α then accounts for increasing residual stress. The pressure-diameter curves for various values of α are plotted in fig. 8 for the neo-Hookean and the Fung's model. The geometry of the segment is characterized with $R_{in} = 1.07$ mm, $R_{out} = 2.38$ mm, and the axial stretch is $\lambda_z = 1.5$. The material parameters are chosen as those resulting from identification in the following section. For both material models, the tube stiffens with increasing residual stress, i.e. certain value of inner pressure ΔP causes smaller deformation of diameter d when opening angle α increases.

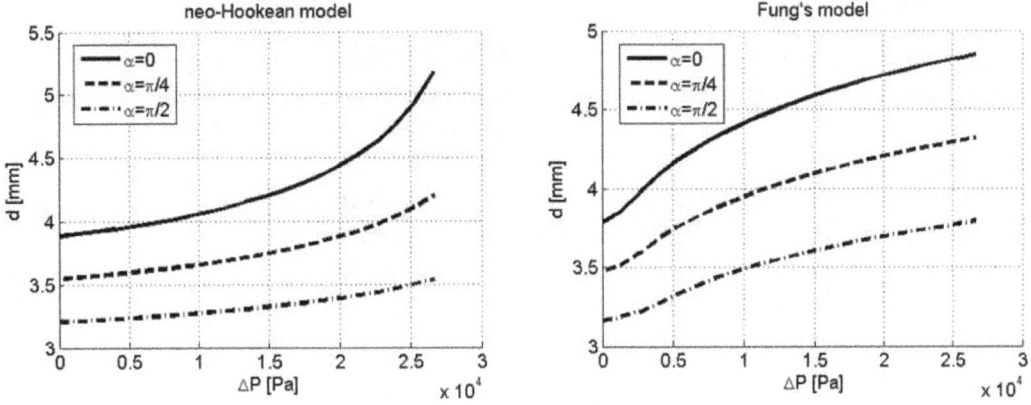

Fig. 8. Influence of the residual stress on the mechanical response of the arterial segment. The neo-Hookean (left) and the Fung's model (right) are employed with different values of opening angle α. Curves describe the dependence of outer diameter d on the applied inner pressure ΔP

4.4. Comparison with experiment

Theoretical pressure-diameter curves are compared with the inflation part of the experimental cycle depicted in fig. 2. The geometry of the model is obtained from measurements of the cut arterial segment. It is $R_{in} = 1.58$ mm, $R_{out} = 2.77$ mm, opening angle $\alpha = 0.41\,\pi$ and the axial stretch is $\lambda_z = 1.5$. A comparison with experimental data is provided using the least

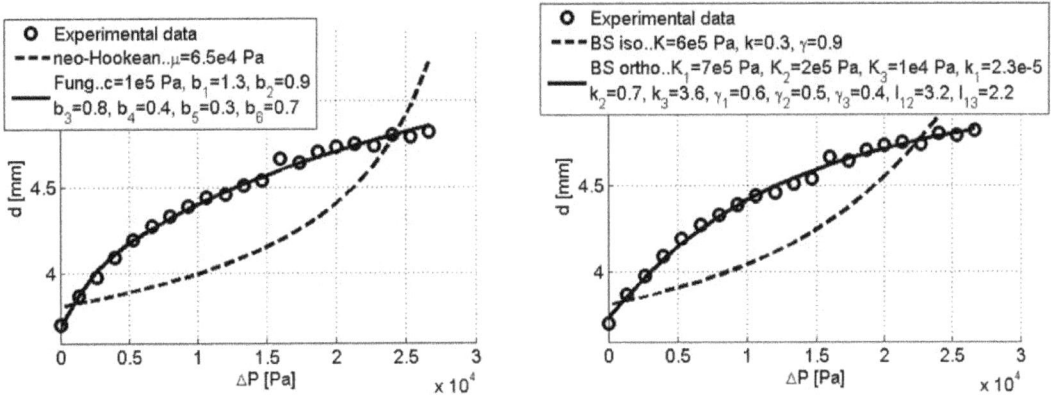

Fig. 9. Comparison of theoretical pressure-diameter curves with experimental data involving the inflation part of the cycle. The neo-Hookean and the Fung's model are depicted on the left, the BS approximative model (both orthotropic and isotropic) on the right figure

square method which leads to the identification of material parameters. Results for the neo-Hookean, the Fung's and the BS model are depicted in fig. 9. Approximate values of material parameters of each model resulting from identification are listed in boxes.

The neo-Hookean as well as the isotropic restriction of the BS model exhibit opposite pattern compared to the experiment (strain-softening instead of strain-hardening). Therefore, theoretical curves of these models do not fit experimental data. The Fung's exponential and the BS orthotropic models, on the other hand, exhibit a good agreement with experiment proving their applicability for the modelling of arterial walls.

5. Conclusion

The mechanical response of the porcine carotid artery exhibits typical patterns of soft tissues. After preconditioning, both strain-hardening and hysteresis are observed during inflation and deflation phases. In the experiment, the sample is prestretched axially to mimic *in vivo* conditions.

The preliminary estimates and the following FE analysis indicate that the arterial segment under the experimental conditions can be represented as a cylindrical tube loaded statically with the inner pressure and the axial stretch. The analysis is performed considering the load-free configuration to be stress-free for simplicity. However, the same conclusions can be expected when the residual stress is taken into account as its contribution to the mechanical response is the overall stiffening.

Adopting these assumptions, the static analysis is performed considering the arterial segment as an incompressible hyperelastic tube. Here, both axial prestretch and residual stress are taken into account. Influence of the residual stress consists in the stiffening of the model, i.e. it can be understood as a subsidiary mechanism of arteries to bear loads caused by pulsatile blood flow and to reduce inadequate deformations *in vivo*. However, residual stress is taken into account via a single parameter in this model considering the open arterial segment as stress-free. In real arteries, residual stress is a more complex problem resulting from certain growth mechanisms of the different layers.

Comparison of theoretical pressure-diameter curves with experimental data is provided for the neo-Hookean, the Fung's and the so-called "balls and springs" (BS) material model. Both the Fung's and the BS models provide a good agreement with experiment due to their anisotropy and a large number of material constants (7 and 11, respectively). The neo-Hookean and the isotropic restriction of the BS model, on the other hand, exhibit an opposite pattern (strain-softening) in the mechanical response when compared to the experiment. Therefore, these models do not seem to be suitable for the modelling of arterial walls. Values of the material parameters found for the Fung's model seem to be reasonable although in [6] slightly different values for a rabbit carotid artery were obtained. Namely, $c \sim 10^4$ Pa, b_1 and $b_2 \sim 10^{-1}$, $b_3 \sim 10^{-3}$ and b_4 to $b_6 \sim 10^{-2}$. Apart of different tissue, material parameters are obtained upon an assumption of a stress-free state at the load-free configuration which may also lead to different results. Material constant found for the neo-Hookean model corresponds to the Young's modulus of $\approx 2 \times 10^5$ Pa. It does not correspond to the Young's modulus of the examined tissue since the mechanical response of the neo-Hookean model differs significantly from the experimental observation. However, the same order of value for dynamic elastic moduli are obtained in [15, 10] for carotid human arteries ($0.49 - 6.08 \times 10^5$ Pa) and arteries of dogs ($\sim 10^5$ Pa).

Concerning the BS model, the approximative analytical formula of the strain energy function is employed in this work for simplicity. Although it exhibits a good agreement with experiment, the approximative formula is in fact not accurate for the values of material parameters obtained in the identification. In other words, it does not represent the material with its microstructure as proposed in [17]. Rather, it may be understood as a phenomenological model.

Aims for the future are to consider accurate formula of the BS model capturing the microstructure and including prestress with the possible application on the study of activation and contraction of living tissue. To consider various experimental and modelling conditions such as the variation of flow rate, dynamic and cyclic loading and to describe hysteresis in the mechanical response as well as time dependent behaviour, viscoelastic models must be employed. Possible approach for creating such models, for instance, is the notion of internal variables. Challenging topic is also the degradation of tissue at the microlevel caused by the collagenase and elastase treatment. Modelling of the mechanical response of the degraded tissue using the BS model is of further interest.

Acknowledgements

The work has been supported by the grant project GAČR 106/09/0734 and by the research project MSM 4977751303.

References

[1] Apter, J. T., Rabinowitz, M., Cummings, D. H., Correlation of visco-elastic properties of large arteries with microscopic structure, Circulation Research 19 (1966) 104–121.

[2] Bader, H., Dependence of wall stress in the human thoracic aorta on age and pressure, Circulation Research 20 (1967) 354–361.

[3] Brdička, M., Samek, L., Sopko, B., Mechanika kontinua, 2nd Edition, Academia, Praha, 2000.

[4] Čanić, S., Hartley, C. J., Rosenstrauch, D., Tambača, J., Guidoboni, G., Mikelić, A., Blood flow in compliant arteries: An effective viscoelastic reduced model, numerics, and experimental validation, Annals of Biomedical Engineering 34 (4) (2006) 575–592.

[5] Carew, T. E., Vaishnav, R. N., Patel, D. J., Compressibility of the arterial wall, Circulation Research 23 (1968) 61–68.

[6] Chuong, C. J., Fung, Y. C., Three-dimensional stress distribution in arteries, Journal of Biomechanical Engineering 105 (1983) 268–274.

[7] Chuong, C. J., Fung, Y. C., On residual stresses in arteries, Journal of Biomechanical Engineering 108 (1986) 189–192.

[8] Cox, R. H., Passive mechanics and connective tissue composition of canine arteries, American Journal of Physiology 234 (1978) 533–541.

[9] Delfino, A., Stergiopulos, N., Moore, J. E., Meister, J. J., Residual strain effects on the stress field in a thick wall finite element model of the human carotid bifurcation, Journal of Biomechanics 30 (8) (1997) 777–786.

[10] Either, C. R., Simmons, C. A., Introductory biomechanics. From cells to organisms, Cambridge university press, Cambridge, 2008.

[11] Fonck, E., Prod'hom, G., Roy, S., Augsburger, L., Rüfenacht, D. A., Stergiopulos, N., Effect of elastin degradation on carotid wall mechanics as assessed by a constituent-based biomechanical model, American Journal of Physiology – Heart and Circulatory Physiology 292 (2007) H2754–H2763.

[12] Fridez, P., Makino, A., Miyazaki, H., Meister, J. J., Hayashi, K., Stergiopulos, N., Short-term biomechanical adaptation of the rat carotid to acute hypertension: Contribution of smooth muscle, Annals of Biomedical Engineering 29 (1) (2001) 26–34.

[13] Fridez, P., Zulliger, M., Bobard, F., Montorzi, G., Miyazaki, H., Hayashi, K., Stergiopulos, N., Geometrical, functional, and histomorphometric adaptation of rat carotid artery in induced hypertension, Journal of Biomechanics 36 (2003) 671–680.

[14] Gasser, T. C., Ogden, R. W., Holzapfel, G. A., Hyperelastic modelling of arterial layers with distributed collagen fibre orientation, Journal of the Royal Society Interface 3 (2006) 15–35.

[15] Gow, B. S., Taylor, M. G., Measurement of viscoelastic properties of arteries in the living dog, Circulation Research 23 (1968) 111–122.

[16] Hariton, I., de Botton, G., Gasser, T. C., Holzapfel, G. A., Stress-driven collagen fiber remodeling in arterial walls, Biomechanics and Modeling in Mechanobiology 6 (2007) 163–175.

[17] Holeček, M., Moravec, F., Hyperelastic model of a material which microstructure is formed of "balls and springs", International Journal of Solids and Structures 43 (2006) 7 393–7 406.

[18] Holzapfel, G. A., Nonlinear solid mechanics, Wiley, Chichester, 2000.

[19] Holzapfel, G. A., Gasser, T. C., Ogden, R. W., A new constitutive framework for arterial wall mechanics and a comparative study of material models, Journal of Elasticity 61 (2000) 1–48.

[20] Holzapfel, G. A., Gasser, T. C., Stadler, M., A structural model for the viscoelastic behavior of arterial walls: Continuum formulation and finite element analysis, European Journal of Mechanics A/Solids 21 (2002) 441–463.

[21] Lally, C., Reid, A. J., Prendergast, P. J., Elastic behavior of porcine coronary artery tissue under uniaxial and equibiaxial tension, Annals of Biomedical Engineering 32 (10) (2004) 1 355–1 364.

[22] Learoyd, B. M., Taylor, M. G., Alterations with age in the viscoelastic properties of human arterial walls, Circulation Research 18 (1966) 278–292.

[23] Lukeš, V., Two-scale computational modelling of soft biological tissues, Ph.D. thesis, University of West Bohemia in Pilsen, Pilsen, 2007.

[24] Patel, D. J., Fry, D. L., The elastic symmetry of arterial segments in dogs, Circulation Research 24 (1969) 1–8.

[25] Perrée, J., van Leeuwen, T. G., Kerindongo, R., Spaan, J. A. E., VanBavel, E., Function and structure of pressurized and perfused porcine carotid arteries, Effects of in vitro balloon angioplasty, American Journal of Pathology 163 (2003) 1 743–1 750.

[26] Rodríguez, J., Goicolea, J. M., García, J. C., Gabaldón, F., Finite element models for mechanical simulation of coronary arteries, Proceedings of the 2nd International Workshop on Functional Modeling of the Heart, Lyon, Springer, Lecture Notes in Computer Science 2674, 2003, pp. 295–305.

[27] Roy, C. S., The elastic properties of the arterial wall, The Journal of Physiology 3 (1880) 125–159.

[28] Valencia, A., Solis, F., Blood flow dynamics and arterial wall interaction in a saccular aneurysm model of the basilar artery, Computers and Structures 84 (2006) 1 326–1 337.

[29] Vimmr, J., Jonášová, A., Analysis of blood flow through three-dimensional bypass model, Applied and Computational Mechanics 1 (2) (2007) 693–702.

[30] Vychytil, J., Holeček, M., Two-scale hyperelastic model of a material with prestress at cellular level, Applied and Computational Mechanics 2 (1) (2008) 167–176.

[31] Zhang, Y., Dunn, M. L., Drexler, E. S., McCowan, C. N., Slifka, A. J., Ivy, D. D., Shandas, R., A microstructural hyperelastic model of pulmonary arteries under normo- and hypertensive conditions, Annals of Biomedical Engineering 33 (8) (2005) 1 042–1 052.

[32] Instruction manual: Tissue bath MAYFLOWER. Version: Perfusion of tubular organs Type 813/6. Hugo Sachs Electronik – Harvard apparatus GmbH, Germany. http://www.hugo-sachs.de

13

Thermoelastic wave propagation in laminated composites plates

K. L. Verma[a],*

[a]*Department of Mathematics, Government Post Graduate College, Hamirpur, (H.P.) 177005, India*

Abstract

The dispersion of thermoelastic waves propagation in an arbitrary direction in laminated composites plates is studied in the framework of generalized thermoelasticity in this article. Three dimensional field equations of thermoelasticity with relaxation times are considered. Characteristic equation is obtained on employing the continuity of displacements, temperature, stresses and thermal gradient at the layers' interfaces. Some important particular cases such as of free waves on reducing plates to single layer and the surface waves when thickness tends to infinity are also discussed. Uncoupled and coupled thermoelasticity are the particular cases of the obtained results. Numerical results are also obtained and represented graphically.

Keywords: dispersion, thermoelasticity, laminated, thermal relaxation times, composites plates

1. Introduction

Increasing use of advanced composites as important structural components in modem high speed aircraft, missile, marine vehicles, and other aerospace structures, and various other applications has led to widespread research activities in the field of composite materials. Composites consist of different materials, so they are inhomogeneous and anisotropic. Different mechanical and thermal properties between constituents of such composites structures, like temperature changes, can generate residual stresses, which may lead to interface de-bonding. A possible failure of the system has intensified the need to study the thermoelastic wave propagation, especially in the form of precise numerical calculations. Consequently, it is of interest to investigate the feasibility of nondestructively, monitoring thermal, mechanical and aging in composites.

Extensive review on the dynamic behavior of anisotropic plate theories can be found in [1] and [14] and problems of wave propagation in periodically layered anisotropic media have been considered and studied in [16,28] and [3]. Dynamic behavior of the problems on the theories of laminated and composite plates have been investigated by authors [12] and [18–23]. Reasonable number of investigations of such advanced materials and their analysis also have been reported in [10,19]. In [15] a transfer matrix technique to obtain the dispersion relation curves of elastic waves propagating in multilayered anisotropic media i.e., composite laminate is developed and detailed review on the wave propagation in layered anisotropic media is studied in [11]. In [9], general problem of thermoelastic waves in anisotropic periodically laminated composites in thermoelasticity is studied.

Theory of thermoelasticity is well established, one can see the works in references [17] and [5]. Literature in this field is rather large to account for the phenomena involving the finite propagation velocity of the thermal wave, and can confer with the reference [4]. These modified

---

*Corresponding author. e-mail: klverma@aol.com.

coupled theories of thermoelasticity are based on hyperbolic-type equations for temperature and are closely connected with the theories of *second sound*, which consider heat propagation as a wavelike phenomenon. In the literature, addressing *linear* theories with relaxation time, most attention is given to the models formulated in [13] and [8]. Theory in [13] called Lord and Shulman (LS) theory is based on a modified Fourier's Law of heat conduction with one relaxation time to dictate the relaxation of thermal propagation, as well as the rate of change of strain rate and the rate of change of heat generation. Green and Lindsay (GL) theory is based on a rigorous treatment of thermodynamics, and a form of the entropy inequality. The literature dedicated to hyperbolic thermoelastic models is quite large and its detailed review can be found in [6, 7].

Theory of generalized thermoelasticty [13] is extended to anisotropic heat conducting elastic materials by [2], and hence it is valid for both isotropic and anisotropic bodies. The propagation of harmonic waves in a laminated composite plate consisting of an arbitrary number of layeres is studied in [9]. Various problems of infinite plates in the context of generalized theories thermoelasticity and the propagation of waves in layered anisotropic media in generalized thermoelasticity is investigated [24–27]. Yamada and Nasser [29] have studied harmonic wave's propagation direction in orthotropic composites.

In this article propagation of thermoelastic waves in layered laminated composites, where the direction of the corresponding harmonic waves makes an arbitrary angle with respect to the layers is examined in the context of generalized thermoelasticity with two thermal relaxation times. Three dimensional field equations of thermoelasticity are considered for this study and the corresponding characteristic equation is obtained on employing the continuity of displacements, temperature, thermal stresses and thermal gradient at the layers' interface. Some important particular cases such as of free waves on reducing plates to single layer and the surface waves when thickness tends to infinity are also discussed. Numerical results are also obtained and represented graphically.

2. Formulation

Consider a set of Cartesian coordinate system $x_i = (x_1, x_2, x_3)$ in such a manner that x_3-axis is normal to the layering. The basic field equations of generalized thermoelasticity for an infinite generally anisotropic thermoelastic medium at uniform temperature T_0 in the absence of body forces and heat sources are

$$\frac{\partial \sigma_{ij}}{\partial x_j} = \rho \frac{\partial^2 u_i}{\partial t^2}, \tag{1}$$

$$K_{ij} \frac{\partial^2 T}{\partial x_i \partial x_j} - \rho C_e \left(\frac{\partial T}{\partial t} + \tau_0 \frac{\partial^2 T}{\partial t^2} \right) = T_0 \beta_{ij} \frac{\partial}{\partial x_j} \left(\frac{\partial u_i}{\partial t} \right). \tag{2}$$

Constitutive relations for anisotropic materials in the context of generalized thermoelasticity are following:

$$\sigma_{ij} = c_{ijkl} e_{kl} - \beta_{ij}(T + \tau_1 \dot{T}), \tag{3}$$

$$\beta_{ij} = c_{ijkl} \alpha_{kl}, \quad i, j, k, l = 1, 2, 3, \tag{4}$$

where ρ is the density of the nth layer, t is time, u_i is the displacement in the x_i direction, K_{ij} are the thermal conductivities, σ_{ij} and e_{ij} are the stress and strain tensor respectively, C_e is the specific heat at constant strain, β_{ij} are thermal moduli, α_{ij} is the thermal expansion tensor, T

is temperature, and c_{ijkl} is the fourth order tensor of the elasticity. The quantities c_{ijkl}, α_{ij}, β_{ij} satisfy the symmetry conditions

$$c_{ijkl} = c_{klij} = c_{ijlk} = c_{jikl}, \qquad \alpha_{ij} = \alpha_{ji}, \qquad \beta_{ij} = \beta_{ji}. \tag{5}$$

The parameter τ_1 and τ_0 are the thermal-mechanical relaxation time and the thermal relaxation time of the GL theory and satisfy the inequality $\tau_1 \geq \tau_0 \geq 0$. Strain-displacement relation is

$$e_{ij} = \frac{1}{2}\left(\frac{\partial u_i}{\partial x_j} + \frac{\partial u_j}{\partial x_i}\right). \tag{6}$$

In addition, at the interface between two layers the tractions, temperature gradient, displacements and temperature must be continuous.

3. Analysis

For harmonic waves propagating in an arbitrary direction, the displacements components u_1, u_2, u_3 and temperature T are written as

$$(u_1, u_2, u_3, T) = \{U_1(x_3), U_2(x_3), U_3(x_3), U_4(x_3)\}e^{i\xi(l_1x_1+l_2x_2+l_3x_3-ct)}, \tag{7}$$

where ξ is the wave number, $c = \omega/\xi$ is the phase velocity, $i = \sqrt{-1}$, ω is the circular frequency, l_1, l_2 and l_3 are the direction cosine defining the propagation direction as in Fig. 1.

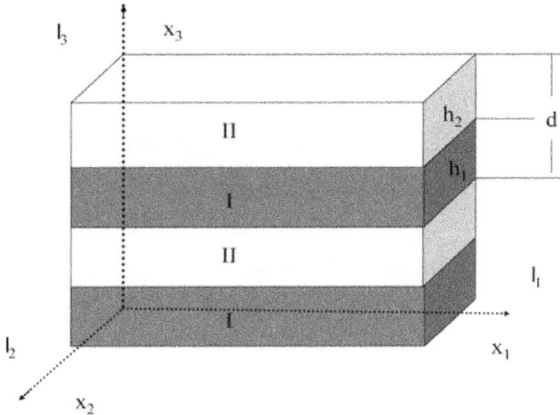

Fig. 1. Two-phase orthotropic layered thermoelastic composite plate. The direction of the propagation vector are denoted as l_1, l_2 and l_3

U_j and T are the constants related to the amplitudes of displacement and temperature, Floquet's theory requires functions U_j ($j = 1, 2, 3$ and 4) to have the same periodicity as the layering. Hence the problem is reduced to that of one pair of layers, where

$$U_j = \bar{U}_j e^{-i\xi(l_3+\alpha)x_3}, \quad j = 1, 2, 3, 4, \tag{8}$$

where $\bar{U}_j$ are constants. On substitution of Eq. (8) into Eqs. (1)–(2), via (3)–(6) and specializing the equations for orthotropic media, it follows that

$$M_{mn}(\alpha)\bar{U}_n = 0, \quad m, n = 1, 2, 3, 4, \tag{9}$$

where

$$M_{11} = (l_1^2 + l_2^2 \bar{c}_{66} + \alpha^2 \bar{c}_{55} - \zeta^2), \quad M_{12} = (\bar{c}_{12} + \bar{c}_{66})l_1 l_2, \tag{10}$$

$$M_{13} = -(\bar{c}_{13} + \bar{c}_{55})l_1 \alpha, \quad M_{14} = l_1,$$

$$M_{22} = (l_1^2 \bar{c}_{66} + l_2^2 \bar{c}_{22} + \alpha^2 \bar{c}_{44} - \zeta^2), \quad M_{23} = -(\bar{c}_{23} + \bar{c}_{44})l_2 \alpha, \quad M_{24} = \bar{\beta}_2 l_2,$$

$$M_{33} = (l_1^2 \bar{c}_{55} + l_2^2 \bar{c}_{44} + \alpha^2 \bar{c}_{33} - \zeta^2), \quad M_{34} = -\bar{\beta}_3 \alpha, \quad M_{41} = \varepsilon \omega^* \zeta^2 l_1 \tau_g,$$

$$M_{42} = \varepsilon \omega^* \zeta^2 l_2 \bar{\beta}_2 \tau_g, \quad M_{43} = -\varepsilon \omega^* \zeta^2 \alpha \bar{\beta}_3 \tau_g, \quad M_{44} = l_1^2 + \bar{K}_2 l_2^2 + \bar{K}_3 \alpha^2 - \omega^* \zeta^2 \tau, \tag{11}$$

where $\zeta^2 = \frac{\rho c^2}{c_{11}}$, $\omega^* = \frac{c_{11} C_e}{K_1}$, $\varepsilon = \frac{\beta_1^2 T_0}{\rho C_e c_{11}}$, and $\tau_g = \tau_1 + i/\omega$, $\tau = \tau_0 + i\omega$. The existence of nontrivial solutions for $\bar{U}_j$ ($j = 1, 2, 3, 4$) demands the vanishing of the determinant in Eqs. (9), and yields the eighth degree polynomial equation

$$\alpha^8 + A_1 \alpha^6 + A_2 \alpha^4 + A_3 \alpha^2 + A_4 = 0, \tag{12}$$

where the coefficients A_1, A_2, A_3 and A_4 are

$$A_1 = [Q_1 \omega^* \varepsilon \tau_g \zeta^2 + P_1 \bar{K}_3 + c_{33} c_{44} c_{55}(l_1^2 + l_2^2 \bar{K}_2 - \omega^* \tau \zeta^2)]/\Delta,$$

$$A_2 = [Q_2 \omega^* \varepsilon \tau_g \zeta^2 + P_2 \bar{K}_3 + P_1(l_1^2 + \bar{K}_2 l_2^2 - \omega^* \tau \zeta^2)]/\Delta,$$

$$A_3 = [Q_3 \omega^* \varepsilon \tau_g \zeta^2 + P_3 \bar{K}_3 + P_2(l_1^2 + l_2^2 \bar{K}_2 - \omega^* \tau \zeta^2)]/\Delta,$$

$$A_4 = [Q_4 \omega^* \varepsilon \tau_g \zeta^2 + P_3(l_1^2 + l_2^2 \bar{K}_2 - \omega^* \tau \zeta^2)]/\Delta,$$

$$P_1 = [(c_{22} c_{33} - 2c_{23} c_{44} - c_{23}^2)c_{55} + c_{33} c_{44} c_{66}]l_2^2 + [(c_{33} - 2c_{13} c_{55} - c_{13}^2)c_{44} + c_{33} c_{55} c_{66}]l_1^2 - (c_{33} c_{44} + c_{33} c_{55} + c_{44} c_{55})\zeta^2,$$

$$P_2 = [(c_{33} - 2c_{13} c_{55} - c_{13}^2)c_{66} + c_{44} c_{55}]l_1^4 + [(c_{22} c_{33} - 2c_{23} c_{44} - c_{23}^2)c_{66} + c_{22} c_{55} c_{44}]l_2^4 +$$
$$[-c_{12}^2 c_{33} - 2(c_{33} c_{44} - c_{66} c_{23} c_{55} - c_{12} c_{44} c_{55} + c_{13} c_{22} c_{55} - 2c_{44} c_{55} c_{66} - c_{13} c_{44} c_{66} +$$
$$c_{12} c_{33} c_{66} - c_{12} c_{13} c_{44} - c_{13} c_{23} c_{66} - c_{12} c_{23} c_{55} - c_{12} c_{13} c_{23}) - c_{13}^2 c_{22} + c_{22} c_{33} - c_{23}^2]l_1^2 l_2^2 +$$
$$[(2c_{13} c_{55} - c_{66} c_{33} - c_{55} c_{44} - c_{44} - c_{33} - c_{66} c_{55} + c_{13}^2)l_1^2 +$$
$$(2c_{23} c_{44} + c_{23}^2 - c_{22} c_{33} - c_{22} c_{55} - c_{66} c_{44} - c_{55} c_{44} - c_{33} c_{66})l_2^2 + (c_{33} + c_{44} + c_{55})\zeta^4]\zeta^2,$$

$$P_3 = (c_{55} l_1^2 + c_{44} l_2^2 - \zeta^2)\{[(1 + c_{66})l_1^2 + (c_{22} + c_{66})l_2^2]\zeta^2 - \zeta^4 +$$
$$[(2c_{22} c_{66} + c_{12}^2 - c_{22})c_{55}]l_1^2 l_2^2 - c_{22} c_{66} l_2^4 - c_{66} l_1^4\},$$

$$\Delta = c_{33} c_{44} c_{55} \bar{K}_3, \quad Q_1 = -c_{44} c_{55} \bar{\beta}_3^2,$$

$$Q_2 = (c_{55} + c_{44})\bar{\beta}_3^2 \zeta^2 + [2(c_{13} c_{44} + c_{44} c_{55})\bar{\beta}_3 - c_{33} c_{44} - (c_{44} + c_{66} c_{55})\bar{\beta}_3^2]l_1^2 +$$
$$[2(c_{23} c_{55} + c_{44} c_{55})\bar{\beta}_2 \bar{\beta}_3 - c_{33} c_{55} \bar{\beta}_2^2 - (c_{22} c_{55} + c_{44} c_{66})\bar{\beta}_3^2]l_2^2,$$

$$Q_3 = ((1 + c_{66})\bar{\beta}_3^2 + (c_{44} + c_{33}) - 2(c_{55} + c_{13})\bar{\beta}_3)l_1^2 \zeta^2 +$$
$$((c_{22} + c_{66})\bar{\beta}_3^2 + 2(c_{33} + c_{55})\bar{\beta}_2^2 - 2(c_{23} + c_{44})\bar{\beta}_2 \bar{\beta}_3)l_2^2 \zeta^2 +$$
$$[(-\bar{\beta}_3^2 + 2(c_{55} + c_{13})\bar{\beta}_3 - c_{33})c_{66} - c_{44} c_{55}]l_1^4 +$$
$$[-c_{22} c_{66} \bar{\beta}_3^2 - (c_{33} c_{66} + c_{44} c_{55})\bar{\beta}_2^2 + 2(c_{44} c_{66} + c_{66} c_{23})\bar{\beta}_2 \bar{\beta}_3]l_2^4 +$$
$$[(c_{23}^2 + 2c_{23} c_{44} - c_{22} c_{33}) + 2(c_{22} c_{55} + c_{13} c_{22} - c_{12} c_{44} - c_{44} c_{66} - c_{12} c_{23} - c_{23} c_{66})\bar{\beta}_3 +$$
$$2(c_{12} c_{13} + c_{33} c_{66} - c_{44} c_{55} - c_{13} c_{23} - c_{23} c_{55} - c_{13} c_{44})\bar{\beta}_2 +$$
$$(c_{13}^2 - c_{33} + 2c_{13} c_{55})\bar{\beta}_2^2 + (c_{12}^2 - c_{22} + 2c_{12} c_{66})\bar{\beta}_3^2 +$$
$$2(c_{23} - c_{13} c_{66} + c_{44} - c_{12} c_{55} - c_{66} c_{55} - c_{12} c_{13})\bar{\beta}_3 \bar{\beta}_2]l_1^2 l_2^2 - \bar{\beta}_3 \zeta^4,$$

$$Q_4 = (c_{55} l_1^2 + c_{44} l_2^2 - \zeta^2)[(l_1^2 + \bar{\beta}_2^2 l_2^2)\zeta^2 + (-\bar{\beta}_2^2 + 2\bar{\beta}_2 c_{66} - c_{22} + 2\bar{\beta}_2 c_{12})l_1^2 l_2^2 -$$
$$c_{66} \bar{\beta}_2^2 l_2^4 - c_{66} l_1^4].$$

Eqs. (8) using Eq. (7) are rewritten as

$$(U_1, U_2, U_3, U_4) = \sum_{q=1}^{8} (\bar{U}_{1q}, \bar{U}_{2q}, \bar{U}_{3q}, \bar{U}_{4q}) e^{-i\xi(l_3 + \alpha_q)x_3}. \tag{13}$$

For each α_q, $q = 1, 2, \ldots, 8$, using the Eqs. (9) and express the displacements ratios as

$$\frac{D_1(\alpha_q)}{D(\alpha_q)} = \frac{\bar{U}_{2q}}{\bar{U}_{1q}} = \gamma_q, \qquad \frac{D_2(\alpha_q)}{D(\alpha_q)} = \frac{\bar{U}_{3q}}{\bar{U}_{1q}} = \delta_q, \tag{14}$$

$$\frac{D_3(\alpha_q)}{D(\alpha_q)} = \frac{\bar{U}_{4q}}{\bar{U}_{1q}} = \Theta_q \quad \text{for} \quad q = 1, 2, \ldots, 8,$$

where

$$\begin{aligned}
D_1(\alpha_q) &= M_{23}(\alpha_q)M_{34}(\alpha_q)M_{41}(\alpha_q) + M_{24}(\alpha_q)M_{33}(\alpha_q)M_{41}(\alpha_q) - \\
& \quad M_{13}(\alpha_q)M_{24}(\alpha_q)M_{43}(\alpha_q) + M_{12}(\alpha_q)M_{34}(\alpha_q)M_{43}(\alpha_q) + \\
& \quad M_{13}(\alpha_q)M_{23}(\alpha_q)M_{44}(\alpha_q) - M_{12}(\alpha_q)M_{33}(\alpha_q)M_{44}(\alpha_q), \\
D_2(\alpha_q) &= M_{23}(\alpha_q q)M_{24}(\alpha_q)M_{41}(\alpha_q) + M_{12}(\alpha_q)M_{23}(\alpha_q)M_{44}(\alpha_q) + \\
& \quad M_{13}(\alpha_q)M_{24}(\alpha_q)M_{42}(\alpha_q) + M_{22}(\alpha_q)M_{34}(\alpha_q)M_{41}(\alpha_q) - \\
& \quad M_{13}(\alpha_q)M_{22}(\alpha_q)M_{44}(\alpha_q) - M_{12}(\alpha_q)M_{34}(\alpha_q)M_{42}(\alpha_q), \\
D_3(\alpha_q) &= M_{23}^2(\alpha_q)M_{41}(\alpha_q) - M_{22}(\alpha_q)M_{33}(\alpha_q)M_{41}(\alpha_q) - \\
& \quad M_{12}(\alpha_q)M_{23}(\alpha_q)M_{43}(\alpha_q) + M_{13}(\alpha_q)M_{22}(\alpha_q)M_{43}(\alpha_q) + \\
& \quad M_{12}(\alpha_q)M_{33}(\alpha_q)M_{42}(\alpha_q) - M_{13}(\alpha_q)M_{23}(\alpha_q)M_{42}(\alpha_q), \\
D(\alpha_q) &= M_{23}(\alpha_q)M_{34}(\alpha_q)M_{42}(\alpha_q) - M_{24}(\alpha_q)M_{33}(\alpha_q)M_{42}(\alpha_q) - \\
& \quad M_{22}(\alpha_q)M_{34}(\alpha_q q)M_{43}(\alpha_q) + M_{22}(\alpha_q)M_{33}(\alpha_q)M_{44}(\alpha_q) - \\
& \quad M_{23}^2(\alpha_q)M_{44}(\alpha_q) + M_{23}(\alpha_q)M_{24}(\alpha_q)M_{43}(\alpha_q).
\end{aligned} \tag{15}$$

Then the solution given by Eq. (13) may be rewritten as

$$(U_1, U_2, U_3, U_4) = \sum_{q=1}^{8} (1, \gamma_q, \delta_q, \Theta_q) \bar{U}_{1q} e^{-i\xi(l_3 + \alpha_q)x_3}. \tag{16}$$

In view of the continuity of the displacement components, temperature, tractions and temperature gradient across the interface of the two layers, the following conditions must be satisfied:

$$u_j^I \big|_{x_3=0^-} = u_j^{II} \big|_{x_3=0^+}, \qquad T^I \big|_{x_3=0^-} = T^{II} \big|_{x_3=0^+}, \tag{17}$$

$$\sigma_{3j}^I \big|_{x_3=0^-} = \sigma_{3j}^{II} \big|_{x_3=0^+}, \qquad T'^I \big|_{x_3=0^-} = T'^{II} \big|_{x_3=0^+}, \tag{18}$$

where $T' = \frac{\partial T}{\partial x_3}$ superscripts I and II refer to layers one and two respectively, 0^+ and 0^- are values of x_3 near zero. Because of periodicity of the deformation and thermoelastic stress fields, additional conditions obtained are

$$u_j^I \big|_{x_3=h_1^-} = u_j^{II} \big|_{x_3=-h_2^+}, \qquad T^I \big|_{x_3=h_1^-} = T^{II} \big|_{x_3=-h_2^+}, \tag{19}$$

$$\sigma_{3j}^I \big|_{x_3=h_1^-} = \sigma_{3j}^{II} \big|_{x_3=-h_2^+}, \qquad T'^I \big|_{x_3=h_1^-} = T'^{II} \big|_{x_3=-h_2^+}, \quad j = 1, 2, 3. \tag{20}$$

On substituting the displacements, temperature, stresses and temperature gradient components into Eqs. (17)–(18), sixteen linear homogeneous equations for sixteen constants $U_{11}^I, U_{12}^I, \ldots,$ U_{17}^{II} and U_{18}^{II} are obtained. For nontrivial solutions, the determinant of coefficient matrix must vanish. This yields the following characteristic equation:

$$\det \begin{pmatrix} P_{jk} & -\bar{P}_{jk} \\ Q_{jk} & -\bar{Q}_{jk} \end{pmatrix} = 0, \quad j, k = 1, 2, \ldots, 8. \tag{21}$$

The entries of 8×8 matrices $P_{jk}, \bar{P}_{jk}, Q_{jk}$ and $\bar{Q}_{jk}$ are

$$\begin{aligned}
P_{1j} &= 1, \quad P_{2j} = \gamma_j^I, \quad P_{3j} = \delta_j^I, \quad P_{4j} = \Theta_j^I, \\
P_{5j} &= b_{1j}^I c_{55}^I, \quad P_{6j} = b_{2j}^I c_{44}^I, \quad P_{7j} = b_{3j}^I, \quad P_{8j} = -b_{4j}^I, \\
\bar{P}_{1j} &= 1, \quad \bar{P}_{2j} = \gamma_j^{II}, \quad \bar{P}_{3j} = \delta_j^{II}, \quad \bar{P}_{4j} = \Theta_j^{II}, \\
\bar{P}_{5j} &= \eta b_{1j}^{II} c_{55}^{II}, \quad \bar{P}_{6j} = \eta b_{2j}^{II} c_{44}^{II}, \quad \bar{P}_{7j} = \eta b_{3j}^{II}, \quad \bar{P}_{8j} = \eta b_{4j}^{II}, \\
Q_{jk} &= P_{jk} E_k^-, \quad \bar{Q}_{jk} = \bar{P}_{jk} E_k^+,
\end{aligned} \tag{22}$$

where $E_j^- = \mathrm{e}^{-iQ(l_3 + \alpha_j^{(1)})h_1/h}$, $Q = \xi(h_1 + h_2)$, $E_j^+ = \mathrm{e}^{-iQ(l_3 + \alpha_j^{II})h_2/h}$, $\eta = c_{11}^{II}/c_{11}^I$,

$$\begin{aligned}
b_{1j}^{(m)} &= l_1 \delta_j^{(m)} - \alpha_j^{(m)}, \quad b_{2j}^{(m)} = l_2 \delta_j^{(m)} - \alpha_j^{(m)} \gamma_j^{(m)}, \\
b_{3j}^{(m)} &= \bar{c}_{13}^{(m)} l_1 + \bar{c}_{23}^{(m)} l_2 \gamma_j^{(m)} - \bar{c}_{33}^{(m)} \alpha_j^{(m)} \delta_j^{(m)} - \beta_3 \Theta_j^{(m)}, \\
b_{4j}^I &= (l_3 + \alpha_j^{(m)}) \Theta_j^{(m)} = i\xi \alpha_j^{(m)} \Theta_j^{(m)}, \quad \bar{c}_{jk}^{(m)} = c_{jk}^{(m)}/c_{11}^{(m)}, \quad m = \mathrm{I, II}. \tag{23a}
\end{aligned}$$

From Eq. (21), we have $\det[P_{jk}] \det([-\bar{Q}_{jk}] - [Q_{jk}][P_{jk}]^{-1}[-\bar{P}_{jk}]) = 0$ which implies that

$$\text{either } \det[P_{jk}] = 0, \tag{23b}$$
$$\text{or } \det([-\bar{Q}_{jk}] - [Q_{jk}][P_{jk}]^{-1}[-\bar{P}_{jk}]) = 0. \tag{23c}$$

If Eq. (23b) holds true, then the problem reduces to a free wave propagation in a single thermoelastic plate of thickness h_1, and in this case $([-\bar{Q}_{jk}] - [Q_{jk}][P_{jk}]^{-1}[-\bar{P}_{jk}])$ will not exist as P_{jk}singular. On the hand P_{jk} is nonsingular $[P_{jk}]^{-1}$ exists and accordingly

$$\det([-\bar{Q}_{jk}] - [Q_{jk}][P_{jk}]^{-1}[-\bar{P}_{jk}]) = 0. \tag{24a}$$

Similarly Eq. (21) can also be written as

$$\det[-\bar{Q}_{jk}] \det([P_{jk}] - [-\bar{P}_{jk}][-\bar{Q}_{jk}]^{-1}[Q_{jk}]) = 0, \tag{24b}$$

which implies that either

$$\det[-\bar{Q}_{jk}] = 0, \tag{24c}$$

or

$$\det([P_{jk}] - [-\bar{P}_{jk}][-\bar{Q}_{jk}]^{-1}[Q_{jk}]) = 0. \tag{24d}$$

If Eq. (24b) holds true, then again the problem reduces to a single thermoelastic plate of thickness h_2, and $([-\bar{Q}_{jk}] - [Q_{jk}][P_{jk}]^{-1}[-\bar{P}_{jk}])$ will not exists as $\bar{Q}_{jk}$ is singular.
On the hand, if $\bar{Q}_{jk}$ is non-singular, therefore

$$\det([-\bar{Q}_{jk}] - [Q_{jk}][P_{jk}]^{-1}[-\bar{P}_{jk}]) = 0. \tag{25}$$

In order to solve the problem numerically it is sufficient to consider either Eq. (24a) or Eq. (25) for composite plates and to solve for free thermoelastic plate Eq. (23b) or Eq. (24b) can be considered.

4. Particular cases

4.1. Classical case

If the coupling constant $\varepsilon = 0$, then thermal and elastic fields decoupled from each other and from Eq. (11) we have $M_{41} = M_{42} = M_{43} = 0$. In this case Eq. (12) factorised into

$$(l_1^2 + \bar{K}_2 l_2^2 + \bar{K}_3 \alpha^2 - \omega^* \zeta^2 \tau)(\Delta \alpha^6 + F_1 \alpha^4 + F_2 \alpha^2 + F_3) = 0. \tag{26}$$

One of the factor of the above equation

$$\Delta \alpha^6 + F_1 \alpha^4 + F_2 \alpha^2 + F_3 = 0 \tag{27}$$

corresponds to the characteristic equation in the uncoupled thermoelasticity, where

$$\Delta = c_{33} c_{44} c_{55},$$
$$F_1 = [(c_{22} c_{33} - 2 c_{23} c_{44} - c_{23}^2) c_{55} + c_{33} c_{44} c_{66}] l_2^2 +$$
$$[(c_{33} - 2 c_{13} c_{55} - c_{13}^2) c_{44} + c_{33} c_{55} c_{66}] l_1^2 -$$
$$(c_{33} c_{44} + c_{33} c_{55} + c_{44} c_{55}) \zeta^2,$$
$$F_2 = [(c_{33} - 2 c_{13} c_{55} - c_{13}^2) c_{66} + c_{44} c_{55}] l_1^4 + [(c_{22} c_{33} - 2 c_{23} c_{44} - c_{23}^2) c_{66} + c_{22} c_{55} c_{44}] l_2^4 +$$
$$[-c_{12}^2 c_{33} - 2(c_{33} c_{44} - c_{66} c_{23} c_{55} - c_{12} c_{44} c_{55} + c_{13} c_{22} c_{55} - 2 c_{44} c_{55} c_{66} - c_{13} c_{44} c_{66} +$$
$$c_{12} c_{33} c_{66} - c_{12} c_{13} c_{44} - c_{13} c_{23} c_{66} - c_{12} c_{23} c_{55} - c_{12} c_{13} c_{23}) - c_{13}^2 c_{22} + c_{22} c_{33} - c_{23}^2] l_1^2 l_2^2 +$$
$$[(2 c_{13} c_{55} - c_{66} c_{33} - c_{55} c_{44} - c_{44} - c_{33} - c_{66} c_{55} + c_{13}^2) l_1^2 +$$
$$(2 c_{23} c_{44} + c_{23}^2 - c_{22} c_{33} - c_{22} c_{55} - c_{66} c_{44} - c_{55} c_{44} - c_{33} c_{66}) l_2^2 + (c_{33} + c_{44} + c_{55}) \zeta^4] \zeta^2,$$
$$F_3 = (c_{55} l_1^2 + c_{44} l_2^2 - \zeta^2)\{[(1 + c_{66}) l_1^2 + (c_{22} + c_{66}) l_2^2] \zeta^2 - \zeta^4 +$$
$$[(2 c_{22} c_{66} + c_{12}^2 - c_{22}) c_{55}] l_1^2 l_2^2 - c_{22} c_{66} l_2^4 - c_{66} l_1^4\}.$$

In this case, Eqs. (14) simplify to

$$D_1(\alpha_q) = M_{13}(\alpha_q) M_{23}(\alpha_q) - M_{12}(\alpha_q) M_{33}(\alpha_q),$$
$$D_2(\alpha_q) = M_{12}(\alpha_q) M_{23}(\alpha_q) - M_{13}(\alpha_q) M_{22}(\alpha_q),$$
$$D_3(\alpha_q) = 0, \qquad D(\alpha_q) = M_{22}(\alpha_q) M_{33}(\alpha_q) - M_{23}^2(\alpha_q) \tag{28}$$

and the reduced result corresponds to the purely elastic orthotropic materials, which is obtained and studied by Yamada and Nasser [29]. On the other hand, the second factor of the Eq. (26) is

$$l_1^2 + \bar{K}_2 l_2^2 + \bar{K}_3 \alpha^2 - \omega^* \zeta^2 \tau = 0, \tag{29}$$

which corresponds to the purely thermal wave. Hence thermal wave in the generalized theory of thermoelasticity is influenced by the thermal relaxation time τ.

4.2. Thermoelastic free waves

When layer I = II and $h_1 = h_2$(say h) then the thickness of the layer is $2h$, on considering origin at mid of the plate, then the above analysis reduces to a single plate. In this case, the eight roots of Eq. (12) can be arranged in four pairs as $\alpha_{j+1} = -\alpha_j, j = 1, 3, 5, 7$.

It is observed from Eq. (11) that M_{13}, M_{23}, M_{34} and M_{43} are odd functions of α, and the other M_{ij}'s are even functions of α. On employing the thermal stresses and thermal gradient free surfaces conditions

$$\sigma_{3j} = T' = 0, \quad x_3 = \pm h, \quad j = 1, 2, 3, \tag{30}$$

and employing the relations (14), we have

$$\gamma_{q+1} = \gamma_q, \quad \delta_{q+1} = -\delta_q \text{ and } \Theta_{q+1} = \Theta_q. \tag{31}$$

Hence from (23a)

$$b_{1q+1} = -b_{1q}, \quad b_{2q+1} = -b_{2q}, \quad b_{3q+1} = b_{3q} \text{ and } b_{4q+1} = -b_{4q}, \tag{32}$$

$$b_{1j} = l_1 \delta_j - \alpha_j, \quad b_{2j} = l_2 \delta_j - \alpha_j \gamma_j,$$

$$b_{3j} = \bar{c}_{13} l_1 + \bar{c}_{23} l_2 \gamma_j - \bar{c}_{33} \alpha_j \delta_j - \beta_3 \Theta_j, \quad b_{4j} = -\mathrm{i}\xi \alpha_j \Theta_j \bar{c}_{jk} = c_{jk}/c_{11}, \tag{33}$$

$$\det [P_{jk}] = 0. \tag{34}$$

Eq. (34) is the corresponding characteristic equation for free waves in generalized thermoelasticity. Further, if thickness $d = (h_1 + h_2) \to \infty$, in Eq. (34) then the problem reduces to thermoelastic surface waves.

4.3. Coupled thermoelasticity

This is the case, when thermal relaxation times $\tau_0 = \tau_1 = 0$ and hence, $\tau = \tau_g = \mathrm{i}/\omega$. Following above, we arrived at frequency equation of the coupled thermoelasticity. When $\tau_1 = \tau_0 \neq 0$, characteristic Eq. (21) becomes the frequency equation in the LS theory of generalized thermoelasticity.

5. Numerical results and discussion

Using Eq. (24a) numerical results are presented to exhibit the dependence of dispersion on the angle of propagation and thermal relaxation time. The materials chosen for this purpose are aluminum epoxy composite as layer I ($h_1 = 0.6$) and carbon steel as layer II ($h_2 = 0.4$).

Since the distinction among the wave mode types of thermoelastic waves in anisotropic plates is somewhat artificial, as the thermal and elastic wave modes are generally coupled, they are referred to as quasilongitudinal and quasitransverse, quasishear horizontal modes and quasithermal. For wave propagation in the direction of higher symmetry (see Section 4), some wave types revert to pure modes and lead to a simple characteristic equation of lower order, and consequently to the loss of pure wave modes in the direction of general propagation. Here Fig. 2 depicts the dispersion curves for the direction cosines of propagation $l_1 = 0.259$, $l_2 = 0.542$, and $l_3 = 0.799$, whereas Fig. 3 demonstrate the dispersion behavior when the direction cosines of propagation are same but the coupling constant $\varepsilon = 0$, i.e., thermal and elastic fields are not coupled.

Similarly, dispersion curves with the direction cosines of propagation $l_1 = 0.195$, $l_2 = 0.515$, and $l_3 = 0.834$ are shown in Fig. 4, whereas when the direction cosines of propagation are same but the coupling constant $\varepsilon = 0$.

Similarly, on considering the direction cosines of propagation $l_1 = 0.125$, $l_2 = 0.707$, and $l_3 = 0.696$ dispersion curves are shown in Fig. 6, whereas when the coupling constant $\varepsilon = 0$, keeping the same direction cosines dispersion curves are shown in Fig. 7.

It is observed that in generalized thermoelasticity, at zero wave number limits, each figure (Figs. 2, 4 and 6) displays four thermoelastic wave speeds corresponding to one quasilongitudinal, two quasitransverse and one quasithermal. It is apparent that the largest value corresponds to the quasi-longitudinal and the additional mode appears is a quasi-thermal mode. At low wave number limits, modes are found to highly influenced and also vary with the direction. A small

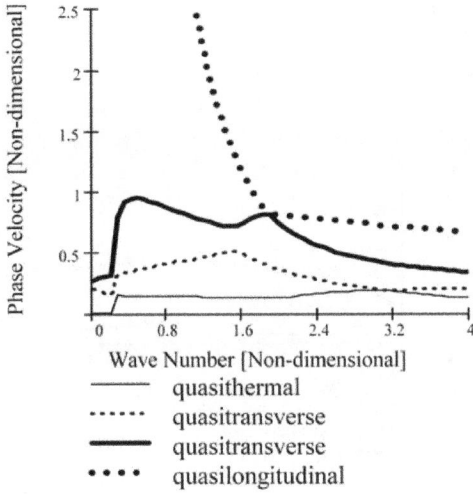

Fig. 2. Phase velocity versus wave number for the direction cosine $l_1 = 0.259$, $l_2 = 0.542$ and $l_3 = 0.799$ in generalized thermoelasticity

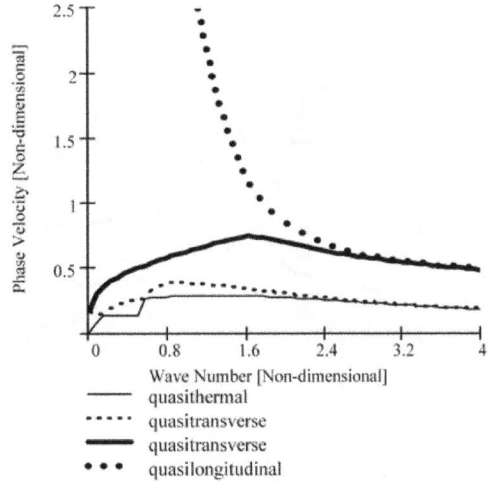

Fig. 3. Phase velocity versus wave number for the direction cosine $l_1 = 0.259$, $l_2 = 0.542$ and $l_3 = 0.799$ when the coupling parameter is zero

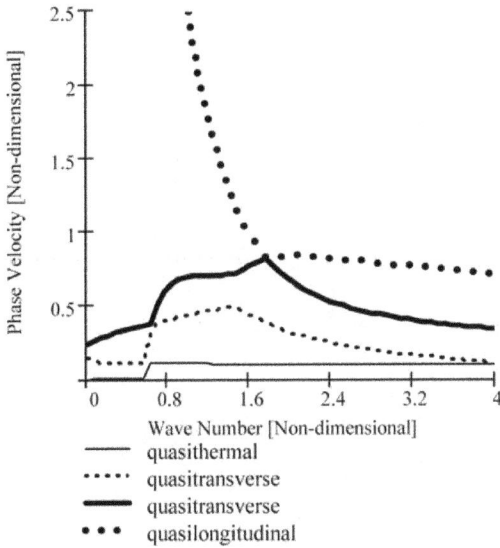

Fig. 4. Phase velocity versus wave number for the direction cosine $l_1 = 0.195$, $l_2 = 0.515$ and $l_3 = 0.834$ in generalized thermoelasticity

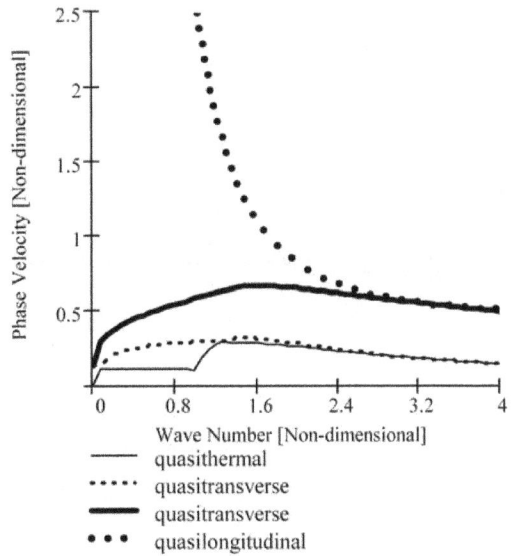

Fig. 5. Phase velocity versus wave number for the direction cosine $l_1 = 0.195$, $l_2 = 0.515$ and $l_3 = 0.834$ when the coupling parameter is zero

change is observed in these modes values as ξ increases and others higher modes appear, one of the modes seems to be associated with quick change in the slope. It is also observed that with change in direction, lower modes appear to have large influence than the higher modes where a small variation is noticed. When the when the coupling constant $\varepsilon = 0$, i.e., thermal and elastic fields are not coupled, Figs. 3, 5 and 7 demonstrate the dispersion behavior of wave speed modes with different angles of propagation. From these figures, it is observed that at low wave number limits, although wave speed modes are dispersive, but are different from the coupled case. Thus in generalized thermoelasticity, at low values of the wave number, only

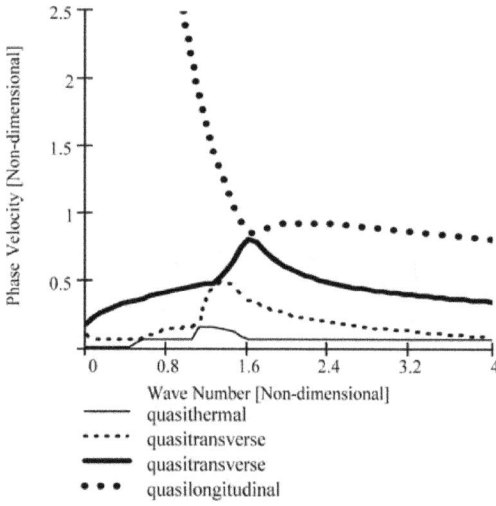

Fig. 6. Phase velocity versus wave number with direction cosine $l_1 = 0.125$, $l_2 = 0.707$ and $l_3 = 0.696$ in generalized thermoelasticity

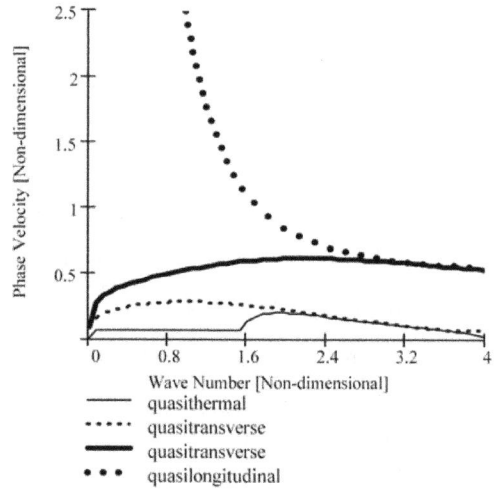

Fig. 7. Phase velocity versus wave number for the direction cosine $l_1 = 0.125$, $l_2 = 0.707$ and $l_3 = 0.696$ when the coupling parameter is zero

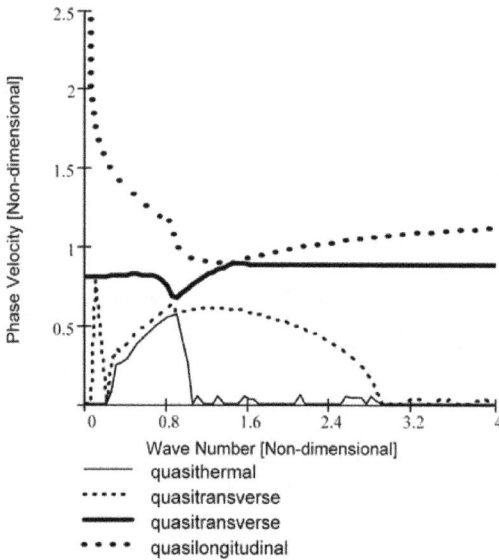

Fig. 8. Phase velocity versus wave number in GL theory of thermoelasticity with thermal relaxation times $\tau_0 = 2 \cdot 10^{-7}$, $\tau_1 = 2 \cdot 10^{-6}$

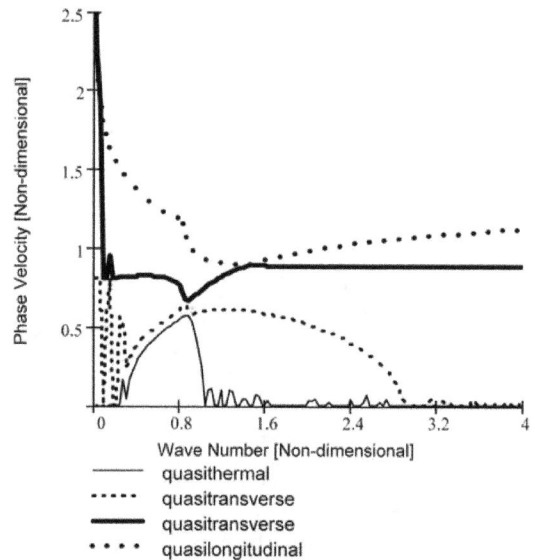

Fig. 9. Phase velocity versus wave number in GL theory of thermoelasticity with thermal relaxation times $\tau_0 = 2 \cdot 10^{-7}$, $\tau_1 = 10 \cdot 10^{-7}$

the lower modes hihgly affected and the little change is observed at the relatively high values of wave number. The low value region of the wave number is found to be of more physical interest in generalized thermoelasticity. As high wave number limits exhibit no effect on wave speeds, therefore the second sound effects are short lived in the laminated composites plates in generalized thermoelasticity.

To observe the influence of the thermal relaxations, selected values of thermal relaxation times τ_1 and τ_0 are cosidered, Figs. 8–10 demonstarte the variations of phase velocity with wave number and the dispersive character of quasilongitudinal, quasitransverse and quasither-

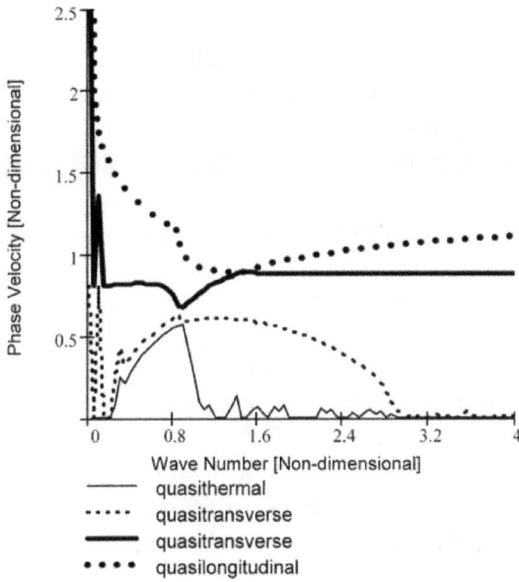

Fig. 10. Phase velocity versus wave number in GL theory of thermoelasticity with thermal relaxation times $\tau_0 = 2 \cdot 10^{-7}, \tau_1 = 4 \cdot 10^{-7}$

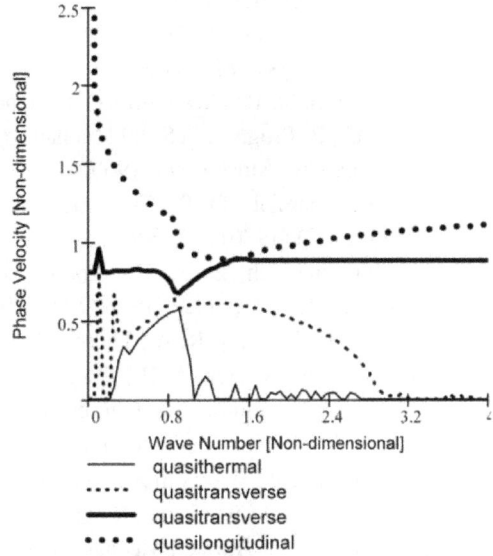

Fig. 11. Phase velocity versus wave number in LS theory of thermoelasticity with thermal relaxation times $\tau_0 = 2 \cdot 10^{-7}$

mal modes are represented. Quasilongitudinal, quasitransverse (two) and quasi-thermal waves are found coupled with each other due to the thermal and anisotropic effects, also wave-like behavior of the quasi-thermal modes is characterized in Green and Lindsay (GL) thermoelasticity theory. Also Fig. 11 is drawn by considering τ_0 only, a single time constant which represents the dispersion curve in Lord and Shulman (LS) theory.

Although the thermal relaxation times τ_1 and τ_0 are derived from distinctively different physical assumptions and physical laws, the dispersion behavior described by LS and GL theory for thermoelastic waves are remarkably similar even in laminated composites plates. It is probably due to the fact that even though the theories are entirely different in their approach to form a coupled thermoelasticity theory, they are remarkably similar in their formulation.

6. Conclusion

Dispersion of a 3D layered heat conducting composite plate in an arbitrary direction in the theory of generalized thermoelasticity is studied. Equations of motion for 3D continuum formulated for an infinite layered plate of an anisotropic thermoelastic medium with uniformly distributed temperature. The Floquet method is used for the derivation of general solution of displacements and temperature distributions. Special cases such as classical, free waves and coupled thermoelasticity are also presented and discussed. Influence of wave propagation direction on plate dispersion is analysed numerically and analytically.

References

[1] Auld, B. A., Acoustic Fields and Waves in Solids, Volume 1. John Wiley and Sons, New York, 1973.

[2] Banerjee, D. K., Pao, Y. K., Thermoelastic waves in anisotropy solids, Journal of the Acoustical Society of America 56 (1974) 1 444–1 454.

[3] Braga, A. M. B., Hermann, G., Plane waves in anisotropic layered composite, In Wave Propagation in Structural Composites, edited by A. K. Mal and T. C. Ting, ASME New York, 1988.

[4] Cattaneo, C., Form of heat equation which eliminates the paradox of instantaneous propagation, Comptes Rendus de lAcadé-mie des Sciences 247 (1958) 431–433.

[5] Chadwick, P., Progress in Solid Mechanics, (Edited by R. Hill and I. N. Sneddon) 1 North Holland Publishing Co., Amsterdam. 1960.

[6] Chandrasekharaiah, D. S., Thermoelasticity with second sound. A review, Applied Mechanics Review 39 (3) (1986) 355–376.

[7] Chandrasekharaiah, D. S., Hyperbolic thermoelasticity. A review of recent literature, Applied Mechanics Review 51(12) (1998) 705–729.

[8] Green, A. E., Lindsay, K. A., Thermoelasticity, Journal of Elasticity 2 (1972) 1–7.

[9] Hawwa, M. A., Nayfeh, A. H., The general problem of thermoelastic waves in anisotropic periodically laminated composites, Composite Engineering 5(1995) 1 499–1 517.

[10] Jones, R. M., Mechanics of composite materials. Scripta Book Co. Washington, 1975.

[11] Liu, G. R., Elastic Waves in Anisotropic Laminates Xi, Z. C. CRC Press, ACES 2002.

[12] Liu, G. R., Tani, J., Watanabe, K., Ohyoshi, T., Lamb Wave Propagation in Anisotropic Laminates, ASME Journal of Applied Mechanics, 57 (1990) 923–929.

[13] Lord, H. W., Shulman, Y. A., Generalized dynamical theory of thermoelasticity, Journal of the Mechanics and Physics of Solids 15 (1967) 299–309.

[14] Nayfeh, A. H., Wave propagation in layered anisotropic media. North-Holland, Amsterdam, 1995.

[15] Nayfeh, A. H., The general problem of elastic wave propagation in multilayered anisotropic media, Journal of Acoustical Society of America 89 (1991) 1 521–1 531.

[16] Norris, A. N., Waves in Periodically Layered Media, A comparison of two theories, SIAM Journal of Applied Mathematics 53 (5) (1993) 1 195–1 209.

[17] Nowacki, W., Dynamic problems of thermoelasticity-Noordhoff. International Publishing, Leyden, the Netherlands, 1975.

[18] Postma, G. W., Wave propagation in a stratified medium, Geophysics 20 (1955) 780–806.

[19] Reddy, J. N., Mechanics of laminated composite plates: theory and analysis. CRC Press, 1997.

[20] Reddy, J. N., On refined theories of composite laminate, Meccanica 25 (1990) 230–238.

[21] Rytov, S. M., Acoustical propagation of a thinly laminated medium, Soviet Physics Acoustics 2 (1956) 68–80.

[22] Sun, C. T., Achenbach, J. D., Herrmann, G., Continuum theory for a laminated medium, Journal of Applied Mechanics 35 (1968) 467–475.

[23] Sve, C., Time harmonic waves traveling obliquely in a periodically laminated medium, Journal of Applied Mechanics 38 (1971) 447–482.

[24] Verma, K. L., Hasebe, N., Wave propagation in plates of general anisotropic media in generalized thermoelasticity, International Journal of Engineering Science 39 (15) (2001) 1 739–1 763.

[25] Verma, K. L., Thermoelastic vibrations of transversely isotropic plate with thermal relaxations, International Journal of Solids and Structures 38 (2001) 8 529–8 546.

[26] Verma, K. L., Hasebe, N., Wave propagation in transversely isotropic plates in generalized thermoelasticity, Archives of Applied Mechanics 72 (6–7) (2002) 470–482.

[27] Verma, K. L., On the propagation of waves in layered anisotropic media in generalized thermoelasticity, International Journal of Engineering Science 40 (18) (2002) 2 077–2 096.

[28] Wang, L., Rokhlin, S. I., Floquet wave homogenization of periodic anisotropic media, Journal of the Acoustical Society of America 112(1) (2002) 38–45.

[29] Yamada, M., Nasser, S. N., Propagation direction in layered transversely isotropic elastic composite, Journal of Composite Materials 15 (1981) 531–542.

14

Using a tensor model for analyzing some aspects of mode-II loading

S. Seitl[a,*], D. Fernández-Zúñiga[b], A. Fernández-Canteli[b]

[a]*Institute of Physics of Materials, Academy of Sciences of the Czech Republic, v. v. i. Žižkova 22, 616 62 Brno, Czech Republic*

[b]*Dept. of Construction and Manufacturing Engineering, E.P.S. de Ingeniería de Gijón, University of Oviedo, Campus de Viesques, 33203 Gijón, Spain*

Abstract

When analyzing the scatter and discrepancies arising among the fracture toughness resulting for different materials and given mixity ratio K_{IIC}/K_{IC} three factors seems to be influential in contributing to the still unsatisfactory state of affairs in this field: a) the lack of established requirements as regards geometry and minimal in- and out-of-plane dimensions of specimens regulating the test for determining mode-II fracture toughness K_{IIC} or, in the more general case, its equivalent in mixed mode cases, b) the role played by the micro-cracking present in the process zone, acknowledged as a microstructural phenomenon already pointed out by Kalthoff and co-workers, needs to be experimentally investigated, and is not considered in the mainly analytical and numerical focussing pursued here, and c) the insufficient attention paid to the particularity of the stress fields around the crack front before and after the daughter crack is formed. In this work, the last question is addressed with the intention of contributing to the clarification of some points with regard to crack instability under mode-II and mixed-mode loading, in particular, why it is difficult to formulate a sufficiently simple failure model for mechanical components or real structures for which the type of load or the geometry results in stress states from which the potential of mixed mode failure arises.

Keywords: Arcan-Richard specimen, shear mode, crack tip stress field, two-parameter fracture mechanics, fracture parameters, numerical simulation

1. Introduction

The transition of a mode-I crack to a mixed-mode one with the formation of a daughter crack kinked with respect to the original mother crack implies a modification of the stress intensity tensor encompassing different constraint conditions. This emphasizes the significance of analyzing the influence of the constraint conditions before and after the formation of the daughter crack in order to interpret the instability criterion as initiation (before) and the crack growth rate (after) as propagation.

In this work a tensor approach is proposed to consider the real situation of the stress intensity field at the crack. In a certain respect it can be considered an extension of the model handled in [11], in which the existence of different stress intensity factors before and after the crack kinking is underlined indicates that the consideration of the potential mode-I situation in the prospective propagation direction before kinking does not correspond to the regular mode-I state as would be present in the mother crack subjected to mode-I loading. Although the constraint

---

*Corresponding author. e-mail: seitl@ipm.cz.

state for mode II is not thickness dependent, at least in the "before" state, it denotes different constraint conditions than in the case of regular mode-I in the prospective direction.

When analyzing the scatter and discrepancies arising among the fracture toughness resulting for different materials and given mixity ratio K_{IIC}/K_{IC} three factors seems to be influential and to contribute to the continuing, unsatisfactory state of the situation:

a) The lack of established requirements as regards geometry and minimal in- and out-of-plane dimensions of specimens regulating the test for determining mode-II fracture toughness K_{IIC} or, in the more general case, its equivalent in mixed mode cases. This fact contrasts with the mode-I case, for which linear elastic or small scale yielding fracture mechanics and a state of plane strain dominating at the crack tip are ensured by the requirements made explicit by the ASTM and ESIS standards.

b) The role played by the micro-cracking present in the process zone, being acknowledged as a microstructural phenomenon already pointed out by Kalthoff and co-workers [7], needs to be experimentally investigated, and is not considered in the mainly analytical and numerical focussing pursued here. This is the reason why the fracture toughness under mode-II K_{IIC} cannot be directly related to that under mode-I K_{IC}.

c) The insufficient attention paid to the particularity of the stress fields around the crack front before and after the daughter crack is formed. According to the tensor approach proposed in [10] the orientation of the prospective crack necessarily follows the direction predicted using the maximal tangential stress model if initiation is assumed always to succeed under model-I loading. Further, the transition between the stress fields before and after the kinked (daughter) crack is formed must be taken into consideration [11]. In fact, the stress field, represented by the stress intensity tensor, around the mother crack front before the crack kinks to the prospective direction differentiates from the regular mode-I stress intensity tensor present at the crack front of the daughter crack. The specific features of crack kinking under mode-II and mixed-mode fatigue loading are also recognized and discussed.

In this work, some aspects of the mode-II and mixed-mode problem are handled, in particular, why it is difficult to formulate a sufficiently simple failure model for mechanical components or real structures for which the type of load or the geometry result in stress states from which the potential of mixed mode failure arises.

2. Stress and strain tensors near the crack front under a general load

In the following, the general expression of the stress field in the proximity of the crack front is derived using a tensor approach and then particularized for the cases of pure mode-I and pure mode-II.

2.1. General definitions

The following tensor magnitudes are defined in the stress field [13] $\sigma_{ij}(z, r, \theta; B)$:

a) Stress intensity field tensor:

$$\phi_{ij}(r, \theta, z; B) = \sqrt{2\pi r}\sigma_{ij}(r, \theta, z; B). \qquad (1)$$

b) Spatial stress intensity tensor:

$$k_{ij}^*(\theta, z; B) = \lim_{r \to 0} \phi_{ij}(r, \theta, z; B) = \lim_{r \to 0} \sqrt{2\pi r} \sigma_{ij}(r, \theta, z; B). \tag{2}$$

c) Stress intensity tensor:

$$k_{ij}(z; B) = k_{ij}^*(\theta, z; B)|_{\theta=\theta_{cr}} = \lim_{r \to 0} \sqrt{2\pi r} \sigma_{ij}(r, \theta, z; B)|_{\theta=\theta_{cr}}. \tag{3}$$

d) Spatial constraint tensor corresponding to the second term of Williams' expansion:

$$t_{ij}^*(\theta, z; B) = \lim_{r \to 0} \left[\sigma_{ij}(r, \theta, z; B) - \frac{k_{ij}^*}{\sqrt{2\pi r}} \right]. \tag{4}$$

e) Constraint tensor:

$$t_{ij}(z; B) = t_{ij}^*(\theta, z; B)|_{\theta=\theta_{cr}} = \lim_{r \to 0} \left[\sigma_{ij}(r, \theta, z; B) - \frac{k_{ij}^*}{\sqrt{2\pi r}} \right] |_{\theta=\theta_{cr}}. \tag{5}$$

f) Constraint function:

$$\psi_{ij}(r, z; B) = \phi(r, \theta, z; B)|_{\theta=\theta_{cr}} = \sqrt{2\pi r} \sigma_{ij}(r, \theta, z; B)|_{\theta=\theta_{cr}}. \tag{6}$$

Using expressions (2) and (4), the stress tensor σ_{ij} in the proximity of a straight crack tip in a plane normal to the crack front at the point (r, θ, z) for a given specimen thickness B (see Fig. 1) can be expressed in polar coordinates as a Williams' expansion [13]:

$$\sigma_{ij}(r, \theta, z; B) = \frac{k_{ij}(\theta, z; B)}{\sqrt{2\pi}} r^{-1/2} + t_{ij}^*(\theta, z; B) r^0 + O_{ij}(r^{1/2}, \theta, z; B), \tag{7}$$

where O_{ij} represents the remaining higher terms. A justification for the extension of this formula to the component $\sigma_{zz}(r, \theta, z; B)$ is provided in the next subsection.

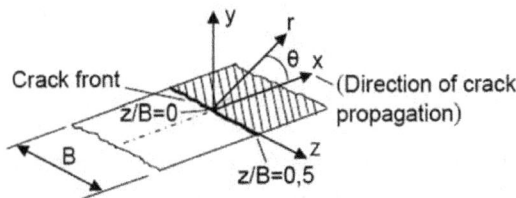

Fig. 1. Crack front and associated coordinate systems (Cartesian and polar)

By identifying (7) with the conventional formulation of Williams' expansion for the general case of mixed-mode I–II, k_{ij}^* can be expressed, indistinctly, in terms of K_I or K_{II}, but considering the preponderance of the mode-II component in the case to be handled, the reference to K_{II} is preferred.

$$\begin{aligned} k_{ij}^*(\theta, z; B) &= K_{II}(z; B) f_{ij}^{(K)}(\theta), \\ t_{ij}^*(\theta, z; B) &= T_{stress}(z; B) f_{ij}^{(T)}(\theta), \end{aligned} \tag{8}$$

where K_{II} is the stress intensity factor for mode-II, defined as $K_{II}(z; B) = \lim_{r \to 0} \sqrt{2\pi r} \tau_{xy}(z; B)$, T_{stress} is the classical T-stress (see [10, 11]) and $f_{ij}(\theta)$ are the geometric functions with super index (T) or (K) referred, respectively, to the tensors k_{ij}^* and t_{ij}^*, which can be expressed in terms of the inverse of the mixity ratio $1/\alpha = K_I/K_{II}$ as:

$$
\begin{aligned}
f_{rr}^{(K)} &= \frac{1}{4}\left[\frac{1}{\alpha}\left(5\cos\frac{\theta}{2} - \cos\frac{3\theta}{2}\right) + \left(-5\sin\frac{\theta}{2} + 3\sin\frac{3\theta}{2}\right)\right], \\
f_{r\theta}^{(K)} &= f_{\theta r}^{(K)} = \frac{1}{4}\left[\frac{1}{\alpha}\left(\sin\frac{\theta}{2} + \sin\frac{3\theta}{2}\right) + \left(\cos\frac{\theta}{2} + 3\cos\frac{3\theta}{2}\right)\right], \\
f_{\theta\theta}^{(K)} &= \frac{1}{4}\left[\frac{1}{\alpha}\left(3\cos\frac{\theta}{2} + \cos\frac{3\theta}{2}\right) + \left(-3\sin\frac{\theta}{2} - 3\sin\frac{3\theta}{2}\right)\right], \\
f_{zz}^{(K)} &= \nu\left(2\frac{1}{\alpha}\cos\frac{\theta}{2} - \sin\frac{\theta}{2}\right), \\
f_{rz}^{(K)} &= f_{zr}^{(K)} = f_{\theta z}^{(K)} = f_{z\theta}^{(K)} = 0
\end{aligned}
\tag{9}
$$

and

$$
\begin{aligned}
f_{rr}^{(T)} &= \cos^2\theta, \\
f_{\theta\theta}^{(T)} &= \sin^2\theta, \\
f_{zz}^{(T)} &= \frac{E\varepsilon_{zz}}{T_{stress}} - \nu, \\
f_{r\theta}^{(T)} &= f_{\theta r}^{(T)} = f_{rz}^{(T)} = f_{zr}^{(T)} = f_{\theta z}^{(T)} = f_{z\theta}^{(T)} = 0,
\end{aligned}
\tag{10}
$$

as can be verified from the literature [1, 5, 11, 13]. The extension of these formulae to the component $\sigma_{zz}(r, \theta, z; B)$ is justified in the next subsection.

The critical orientation can be ascertained from the assumption that failure succeeds under mode-I conditions. This condition requires that the stress intensity tensor k_{ij} becomes a diagonal one, what implies that $f_{r\theta} = 0$. The following equation in θ_{cr} is then obtained:

$$
\sin\frac{\theta_{cr}}{2} + \sin\frac{3\theta_{cr}}{2} + \alpha\left(\cos\frac{\theta_{cr}}{2} + 3\cos\frac{3\theta_{cr}}{2}\right) = 0,
\tag{11}
$$

or

$$
\alpha = \frac{\sin\theta_{cr}}{1 + \sin\theta_{cr} - 3\cos\theta_{cr}}
\tag{12}
$$

from which after some algebra the value of θ_{cr} can be found in terms of the mixity ratio α resulting from the particular loading case considered. Note that this condition is equivalent to the failure criterion controlled by the maximal tangential stress as proposed by Erdogan and Sih [2].

Accordingly, under pure mode-I loading, i.e., $\alpha = 0$, it results in $\theta_{cr} = 0$ for which

$$
k_{ij}(0; B) = K_I(0; B)\begin{vmatrix} 1 & 0 & 0 \\ 0 & 1 & 0 \\ 0 & 0 & 2\nu \end{vmatrix}
$$

and

$$
t_{ij}(0; B) = T_{stress}(0; B)\begin{vmatrix} 1 & 0 & 0 \\ 0 & 0 & 0 \\ 0 & 0 & \frac{E\varepsilon_{zz}(0;B)}{T_{stress}(0;B)} - \nu \end{vmatrix},
\tag{13}
$$

in this case K_I has been referred to by obvious reasons. For pure mode-II, the k_{ij} and t_{ij} tensors can be derived for the mother crack orientation, i.e., $\theta = 0$. In this case, $T_{stress}^{(II)} = 0$ due to the anti-symmetric load and boundary conditions at the crack so that:

$$k_{ij}(0; B) = K_{II}(0; B) \begin{vmatrix} 0 & 0 & 0 \\ 0 & 1 & 0 \\ 0 & 0 & \nu \end{vmatrix} \quad \text{and} \quad t_{ij}(0; B) = \begin{vmatrix} 0 & 0 & 0 \\ 0 & 0 & 0 \\ 0 & 0 & 0 \end{vmatrix}, \tag{14}$$

where K_{II} is now taken as the reference magnitude.

Nevertheless, according to [7] the instability conditions resulting during the failure process, i.e. for $\theta = \theta_{cr} = 70.5$, implies necessarily investigating the stress fields *before* and *after* the secondary or daughter kinked crack is formed. In fact, these two states reveal significantly different stress states.

a) *Before the daughter crack is formed (primarily for crack initiation)*

$$k_{ij}(0; B) = K_{II}(0; B) \begin{vmatrix} 0 & 0 & 0 \\ 0 & 1 & 0 \\ 0 & 0 & \nu \end{vmatrix} \quad \text{and} \quad t_{ij}(0; B) = \begin{vmatrix} 0 & 0 & 0 \\ 0 & 0 & 0 \\ 0 & 0 & 0 \end{vmatrix}, \tag{15}$$

b) *After the daughter crack is formed (primarily for crack propagation)*

$$k_{ij}(0; B) = K_I(0; B) \begin{vmatrix} 1 & 0 & 0 \\ 0 & 1 & 0 \\ 0 & 0 & 2\nu \end{vmatrix}$$

and

$$t_{ij}(0; \theta_{cr}) = T_{stress}(0; B) \begin{vmatrix} \cos^2 \theta_{cr} & 0 & 0 \\ 0 & \sin^2 \theta_{cr} & 0 \\ 0 & 0 & \frac{E\varepsilon_{zz}}{T_{stress}} - \nu \end{vmatrix}. \tag{16}$$

Thus, the tensor approach demonstrates that for pure mode-II loading and consequently for mixed-mode, the stress intensity tensor k_{ij} at the state *before* differs substantially from that corresponding to the state *after*, irrespective of the crack length a, provided the latter is small. This applies not only to the in-plane singularity controlled by k_{rr} and $k_{\theta\theta}$ but also the out-of-plane singularity controlled by k_{zz}. Accordingly, the stress state *before*, i.e., the one supposedly determining the crack instability condition or crack initiation, can be labelled as a *spurious mode-I state*, and cannot be identified with that arising from a regular mode-I failure, as is generally accepted [2]. Further, because the constant stress tensor t_{ij} for pure-mode-II is null, none or negligible in-plane and out-of-plane constraint effects due to specimen thickness or crack ratio are expected, since such an influence could be assigned only to the higher terms of the tensor expansion. This has been confirmed by earlier research performed by Kalthoff and co-workers at the University of Bochum [6, 9] performed on steel and aluminium alloys, see Fig. 2.

In the case of mixed-mode, increasing mode-I participation, i.e., diminishing mixity ratio, promotes the potential influence of the constraint effect, in particular of the specimen thickness on the crack instability.

Once the daughter crack is formed, i.e. for the state *after*, the stress intensity tensor k_{ij} as well as the constant stress tensor recover the structure of the regular mode-I implying the

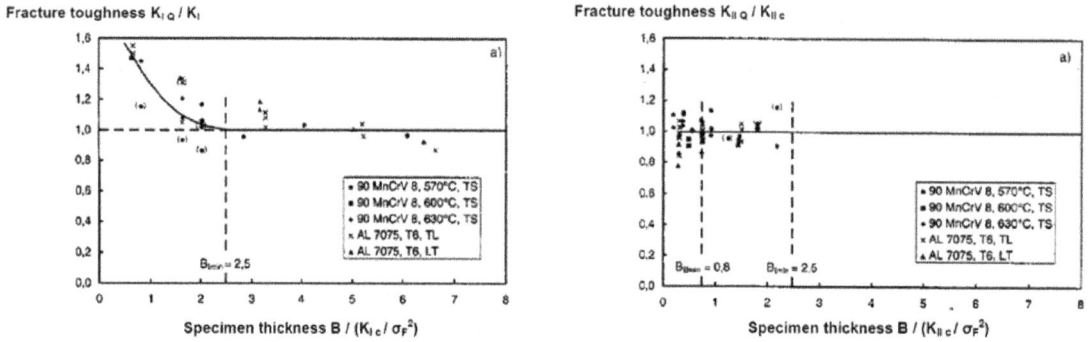

Fig. 2. Dependency and non-dependency of fracture toughness with respect to specimen thickness, respectively, for mode-I and mode-II (from [6])

presence of constraint effects during the crack propagation. In any case, in the first stages of crack propagation the relative size of crack and process zones should have an influence on the crack propagation conditions, an aspect that points out the possible influence of the material fracture properties during the failure sequence. This would explain the different ratios K_{IIC}/K_{IC} observed for different materials, see [12] and [9], and therefore the impossibility of deducing directly K_{IIC} from K_{IC} (see [7]). Since this corresponds to the propagation phase, different crack velocities should be observed for specimens with different thicknesses

2.2. Strain relations at the crack front

So far, the expression of the out-of-plane stress $\sigma_{zz}(r, \theta, z; B)$ has not been justified yet. Applying the generalized Hooke's law $\varepsilon_{ij} = \frac{1+\nu}{E}\sigma_{ij} - \frac{\nu}{E}\sigma_{kk}\delta_{ij}$, allows deriving the expression

$$\varepsilon_{zz}(z, r, \theta; B) = \frac{\sigma_{zz}(r, \theta, z; B) - \nu\left[\sigma_{rr}(r, \theta, z; B) + \sigma_{\theta\theta}(r, \theta, z; B)\right]}{E}. \tag{17}$$

According to [4] $\varepsilon_{zz}(z, r, \theta; B)$ cannot be singular in r what implies from (17)

$$\sigma_{zz}(r, \theta, z; B) - \nu\left[\sigma_{rr}(r, \theta, z; B) + \sigma_{\theta\theta}(r, \theta, z; B)\right] \neq \infty \quad \text{for} \quad r \to 0. \tag{18}$$

Accordingly, for any position at the crack front, perhaps with the exception of locations close to $z = \pm B/2$ not considered here, $\sigma_{zz}(r, \theta, z; B)$ must necessarily be singular with the same order of singularity as $\sigma_{rr}(r, \theta, z; B)$ and $\sigma_{\theta\theta}(r, \theta, z; B)$, so that the Williams' expansion is also extensible to

$$\sigma_{zz}(r, \theta, z; B) = \frac{k_{zz}^*(\theta, z; B)}{\sqrt{2\pi r}} + t_{zz}^*(\theta, z; B) + \dots, \tag{19}$$

thus

$$k_{zz}^*(\theta, z; B) = k_{zz}(z; B)f_{zz}^{(K)}(\theta) = K_{II}(z; B)f_{zz}^{(K)}(\theta), \tag{20}$$

validating (7) and (8). From above it follows

$$\varepsilon_{zz}(r, \theta, z; B) = \frac{1}{E}\left[\frac{k_{zz}^*(\theta, z; B) - \nu\left[k_{rr}^*(\theta, z; B) + k_{\theta\theta}^*(\theta, z; B)\right]}{\sqrt{2\pi r}} + t_{zz}^*(\theta, z; B) - \nu[t_{rr}^*(\theta, z; B) + t_{\theta\theta}^*(\theta, z; B)] + \dots\right], \tag{21}$$

so that the condition $\varepsilon_{zz}(r, \theta, z; B) \neq \infty$ at the crack front implies

$$k_{zz}^*(\theta, z; B) - \nu\left[k_{rr}^*(\theta, z; B) + k_{\theta\theta}^*(\theta, z; B)\right] = 0 \tag{22}$$

and by considering (8) and (9), it follows

$$K_I(z; B) f_{zz}^{(K)}(\theta) - \nu \left[K_I(z; B) f_{zz}^{(K)}(\theta) + K_I(z; B) f_{zz}^{(K)}(\theta) \right] = 0 \qquad (23)$$

from which finally results (see [10])

$$f_{zz}^{(K)}(\theta) = \nu(f_{zz}^{(K)}(\theta) + f_{zz}^{(K)}(\theta)) = 2\nu \left(\cos \frac{\theta}{2} - \alpha \sin \frac{\theta}{2} \right) \qquad (24)$$

and

$$k_{zz}(z; B) = k_{zz}^*(\theta_{cr}, z; B) = K_{II}(z; B) f_{zz}^{(K)}(\theta_{cr}) = 2\nu K_{II}(z; B) \left(\frac{1}{\alpha} \cos \frac{\theta_{cr}}{2} - \sin \frac{\theta_{cr}}{2} \right), \qquad (25)$$

as it would be expected from (20).

Since the numerator of the first term of (21) does not depend on r, the following condition must be accomplished:

$$\lim_{r \to 0} \frac{k_{zz}^*(\theta, z; B) - \nu \left[k_{rr}^*(\theta, z; B) + k_{\theta\theta}^*(\theta, z; B) \right]}{\sqrt{2\pi r}} = 0 \qquad (26)$$

so disregarding the higher terms in the Williams' expansion that results from (21)

$$
\begin{aligned}
\varepsilon_{zz}(r, \theta, z; B)|_{r \to 0} &= \frac{1}{E} \left[t_{zz}^*(\theta, z; B) - \nu(t_{rr}^*(\theta, z; B) + t_{\theta\theta}^*(\theta, z; B)) \right] = \\
&\frac{1}{E} \left[t_{zz}(z; B) f_{zz}^{(T)}(\theta) - \nu(T(z; B) f_{rr}^{(T)}(\theta) + T(z; B) f_{\theta\theta}^{(T)}(\theta)) \right] = \\
&\frac{1}{E} \left[t_{zz}(z; B) f_{zz}^{(T)}(\theta) - \nu T(z; B)(f_{rr}^{(T)}(\theta) + f_{\theta\theta}^{(T)}(\theta)) \right] = \qquad (27) \\
&\frac{1}{E} \left[t_{zz}(z; B) f_{zz}^{(T)}(\theta) - \nu T(z; B)) \right],
\end{aligned}
$$

but since $f_{rr}^{(T)}(\theta) + f_{\theta\theta}^{(T)}(\theta) = \cos^2 \theta + \sin^2 \theta = 1$ (see [10]), it results in

$$\varepsilon_{zz}(r, \theta, z; B)|_{r \to 0} = \varepsilon_{zz}(r, z; B)|_{r \to 0} = \frac{t_{zz}(z; B) - \nu T(z; B)}{E}, \qquad (28)$$

confirming that $\varepsilon_{zz}(\theta)$ is not dependent on θ for $r = 0$.

2.3. Results expected from the analytical

According to the analytical expressions derived for the different orientations of a crack subjected to pure mode-II loading conditions the following results are predicted:

For $\theta = 0$

$$
\begin{aligned}
k_{xx} &= k_{yy} = 0, \\
k_{xy} &= K_{II}(z; B) = \lim_{r \to 0} \sqrt{2\pi r} \tau_{xy}(z; B), \\
k_{zz} &= \nu(k_{xx} + k_{yy}) = 0, \\
\varepsilon_{zz}|_{r=0} &= 0, \qquad (29) \\
t_{xx} &= t_{yy} = 0, \text{ according to the load and boundary conditions,} \\
t_{zz} &= 0, \text{ according to } \varepsilon_{zz}|_{r=0} = 0 = \frac{t_{zz} - \nu t_{xx}}{E}.
\end{aligned}
$$

For $\theta = \theta_{cr} = 70.5$ (prospective crack propagation direction)

    a) *before the daughter crack is formed*

$$
\begin{aligned}
k_{rr} &= 0, \\
k_{zz} &= \nu(k_{rr} + k_{\theta\theta}) = \nu k_{\theta\theta}, \\
\varepsilon_{zz}|_{r=0} &= 0, \\
t_{rr} &= t_{\theta\theta} = 0, \text{ according to the load and boundary conditions}, \\
t_{zz} &= 0, \text{ according to } \varepsilon_{zz}|_{r=0} = 0 = \frac{t_{zz} - \nu t_{rr}}{E}.
\end{aligned}
\tag{30}
$$

    b) *after the daughter crack is formed*

$k_{rr} = k_\theta = 1.155 K_{II}$, irrespective of the specimen thickness
$k_{zz} = \nu(k_{rr} + k_{\theta\theta}) = 2 \times 1.155\, \nu K_{II}$

In close proximity to the crack front, varying as a function of the specimen thickness B, this results in
$t_{\theta\theta} = 0$, according to the load and boundary conditions

$$
\varepsilon_{zz}|_{r=0} \neq 0
\tag{31}
$$

$t_{rr} \neq 0$, according to the load and boundary conditions
$t_{zz} \neq 0$, according to $\varepsilon_{zz}|_{r=0} = \frac{t_{zz} - \nu t_{rr}}{E}$.

3. Numerical calculations

With the aim of checking the analytical expressions found above, finite element calculations were performed using the ANSYS code version for an Arcan-Richard specimen of different specimen thicknesses and crack ratios. The Arcan-Richard specimen and corresponding experimental setup are shown in Fig. 3.

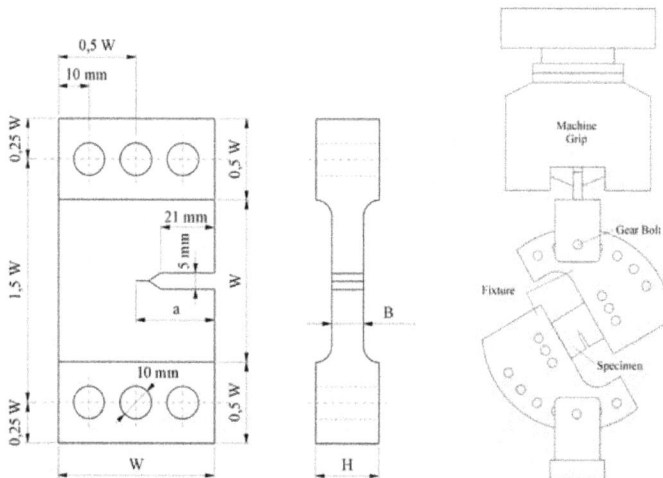

Fig. 3. Arcan-Richard – specimen and Arcan-Richard fixture system, taken from [9]

The specimen dimensions were (see Fig. 3): $W = 50$ mm, $a/W = 0.3$, 0.5 and 0.7 and specimen thickness $B = 5$, 10 and 50 mm. A remote load $P = 100$ N for 2D-plane strain and $P = 100 \times B$ N for 3D model was applied, Young's modulus is $E = 2 \times 10^5$ MPa and Poisson's ration is $\nu = 0.34$.

4. Results and discussion

The stress intensity factor K and the T-stress values were computed by means of the finite element method and using the stress difference method [14]. As a first step, the 2D finite element method solution was employed on the Arcan-Richard specimens to verify the accuracy of the numerical model used. A typical finite element mesh and the boundary conditions used in the computations are shown in Fig. 4 together with a detailed view of the small region near the crack tip. The size of the smallest element in the crack tip is 5×10^{-5} mm.

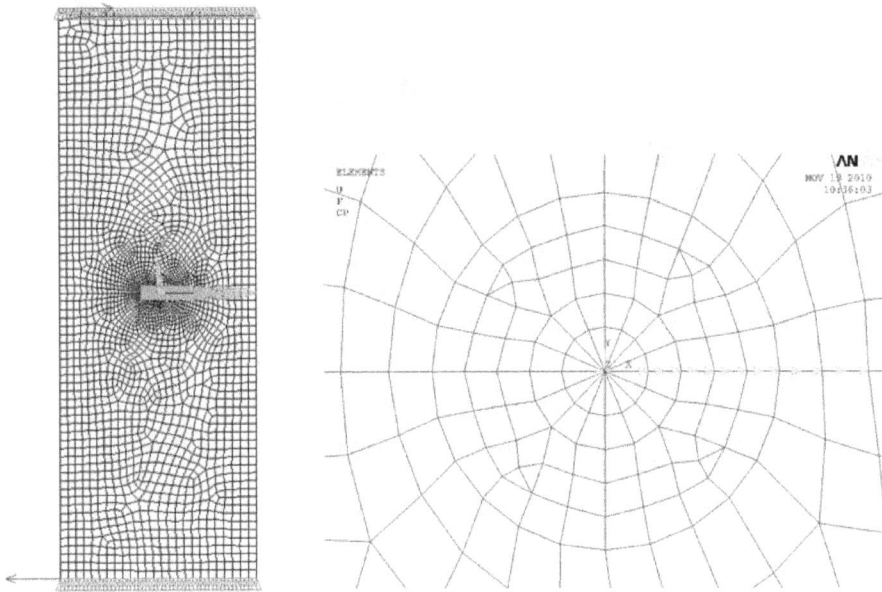

Fig. 4. Load application and finite element mesh used in the finite element calculations: detailed view of the small region near the crack tip

The analytical expressions deduced for the components of the stress intensity tensor k_{ij} and those for the constant tensor t_{ij} should be validated by the numerical calculation, first for the initial crack direction $\theta = 0$ then for the prospective crack propagation direction $\theta = \theta_{cr}$ for both of the states: *before* and *after* the daughter crack is formed. In this work, only a selected number of components have been considered, see section 4. The comparison of data from literature (e.g. [8]) and from our numerical 2D-model (plane strain) is shown in Fig. 5 the data are in good relation; the differences are smaller than 2 %. The influence of specimen thickness on the fracture toughness is shown in the Fig. 6.

The real test conditions applied the Arcan-Richard specimen do not correspond an ideal simulation of the mode-II test presented here. The influence of grips will be studied later, see Fig. 3. As a result, the out-of-plane stress intensity component for $\theta = 0$, $k^*_{zz}|_{\theta=0}$, is zero as predicted, $k^*_{zz}|_{\theta=0} = 2\nu K_I = 0.68 K_I$, i.e., $K_I = 0$ MPa m$^{1/2}$. The same conclusions are for t_{xx} and t_{zz}, as well, see Fig. 7 for t_{xx} stress component. Finally, note that critical crack orientation

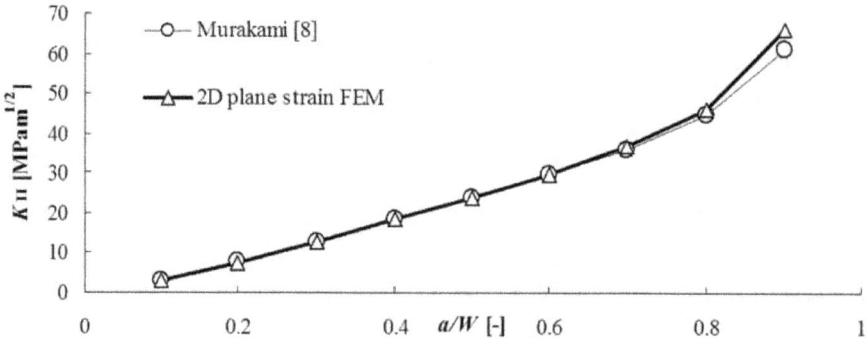

Fig. 5. Comparison of results from the used numerical model with the literature data [8]. The loading force $P = 100$ N

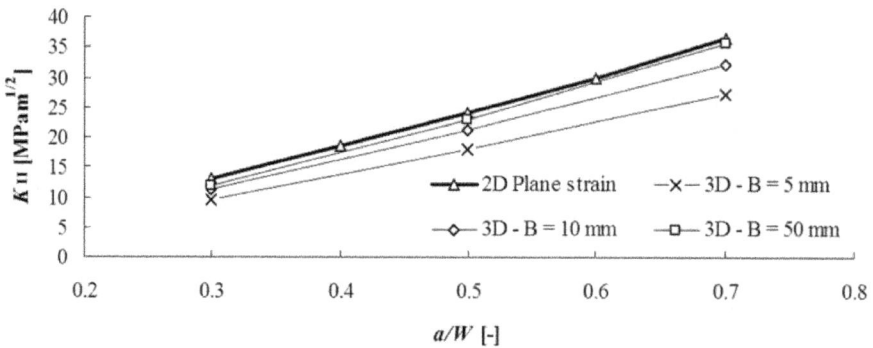

Fig. 6. Results of the stress intensity factors K_{II} for the A-R specimen under expectedly mode-II conditions for different specimen thickness and crack ratios. The loading force $P = 100 \times B$ N

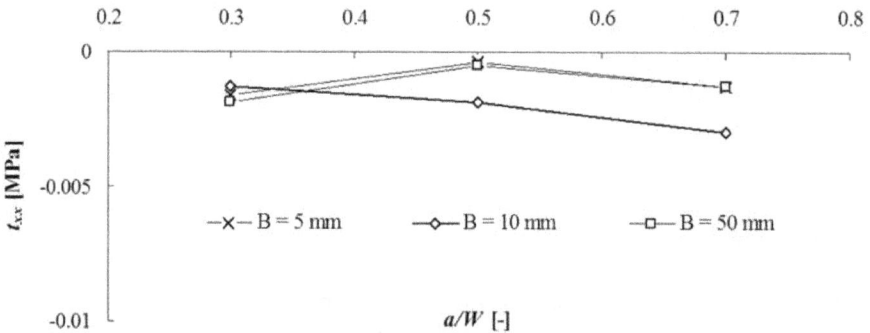

Fig. 7. Example of results of the t_{xx} stress components for the A-R specimen under mode-II conditions for different specimen thickness and crack ratios

from MTS — criteria [4] is $\theta_{cr} = 70.5$ but in reality this can be influenced by an existence of non zero mixity ratio $\alpha = K_{II}/K_I$, equation (12). Consequently, $k^*_{\theta\theta}|_{\theta=70.5}$ will be zero as predicted. A more extensive numerical calculation must be performed if a detailed checking of the analytical results expected according to the sets (29), (30) and (31) are pursued.

The role played by the micro-cracking present in the process zone, being acknowledged as a microstructural phenomenon and already pointed out by Kalthoff and co-workers, needs to be experimentally investigated, and is not considered in the mainly analytical and numerical focusing pursued here. This might be a reason why the fracture toughness under mode-II K_{IIC} cannot be directly related to that under mode-I K_{IC}. All this evidences the necessity of standardizing the specification of minimum specimen sizes for determining the true valid values of fracture toughness under mode-II and mixed-mode loading as suggested in [3,4].

5. Conclusion

The main conclusions of this work are the following:

- A tensor approach is applied to derive the general analytical expressions of the stress and strain state for mixed-mode conditions I–II at the crack front of Arcan-Richard specimens encompassing as particular cases pure mode-I, pure mode-II conditions.

- For pure mode-II, the approach confirms two different stress and strain fields at the crack front before and after the daughter crack forms implying also different in-plane and out-of-plane constraint conditions in the crack surrounding. This also applied to mixed-mode conditions.

- For the earlier state, a spurious mode-I state prevails in the prospective crack propagation direction θ_{cr} characterized by a zero stress intensity tensor component k_{rr} that one presumably governing the crack initiation conditions. Near the crack front, no influence of the specimen thickness on the constraint conditions is observed, and therefore no influence of specimen thickness on the fracture toughness is expected as long as a pure mode-II stress state prevails along the initial crack direction $\theta = 0$. This is confirmed by earlier external research performed on steel and aluminium alloys.

- As soon as the daughter crack forms, i.e., in the so-called state after, a regular mode-I stress state arises at the crack front. Constraint effects are observed as a result of the specimen thickness and the influence of specimen thickness on the fracture toughness is to be expected.

- As a result of the presence of a mode-I component, the influence of the specimen thickness B and crack ratio a/W, though small, is noticeable both in the results of the component k_{zz} and of the t_{ij} components, t_{rr} and t_{zz} that are close to zero.

- Further calculations are envisaged to analyze the stress relations in the state after, particularly in matters concerning constraint evolution during the crack growth process.

- The analytical derivations and the numerical calculations prove the utility of the tensor approach proposed in this work.

Acknowledgements

The authors acknowledge partial economical support from FYCIT, Project IB08-171 and from the Spanish Ministry of Science and Innovation, Project DPI 2007-66903, Academy of Sciences of the Czech Republic, project M10041090 and Czech Science Foundation, projects 101/09/0867 and P108/10/2049.

References

[1] Anderson, W. L., Fracture mechanics, fundamentals and applications, CRC Press, 3rd. Edition, 2004.

[2] Erdogan, F., Sih, G. C., On the crack extension in plates under plane loading and transverse shear, Journal of Basic Engineering, Transactions of ASME, 1963, pp. 519–527.

[3] Fernández-Zúñiga, D., Fernández-Canteli, A., Doblaré, M., Kalthoff, J. F., Bergmannshoff, D., Novel test criteria for determining fracture toughness under pure mode-II and mixed-mode loading, XIV European Conference of Fracture, Krakow (Polonia), Vol. I, 2002, pp. 521–530.

[4] Fernández-Canteli, A., Castillo, E., Fernández-Zúñiga, D., Linear elastic fracture mechanics based criteria for fracture including out-of-plane constraint effect, Submitted to Theoretical and Applied Fracture Mechanics.

[5] Giner, E., Fernández-Zúñiga, D., Fernández-Sáez, J., Fernández-Canteli, A., On the Jx1-integral and the out-of-plane constraint in a 3D elastic cracked plate loaded in tension. Int. J. of Solids and Structures 47, 2010, pp. 934–946.

[6] Hiese, W., Gültigkeitskriterien Zur Bestimmung Von Scherbruchzähigkeiten, Doctoral Thesis, Ruhr-Universität Bochum, 2000.

[7] Kalthoff, J. F., Fernández-Canteli, A., Blázquez, A., Fernández-Zúñiga, D., Singular stress fields and instability Conditions for mode II and mixed-mode loaded cracks, Strength, Fracture and Complexity 4, 2006, pp. 141–160.

[8] Murakami, Y. et al. Stress Intensity factors, Handbook, Pergamon Press, 1998.

[9] Podleschny, R., Untersuchungen zum Instabilitätsverhalten scherbeanspruchten Risse, Doctoral Thesis, Institut für Mechanik, Ruhr-Universität Bochum, 1995.

[10] Seitl, S., Knésl, Z., Two parameter fracture mechanics: fatigue crack behaviour under mixed mode conditions. Engng. Fract. Mech. 72, 2008, pp. 857–865.

[11] Smith, D. J., Ayatollahi, M. R., Pavier, M. J., On the consequences of T-stress in elastic brittle fracture, Proc. R. Soc. 462, 2006, pp. 2 415–2 437.

[12] Tenhaeff, D., Untersuchungen zum Ausbreitungsverhalten von Rissen bei überlagerter Normal- and Schubbeanspruchung, Doctoral Thesis, Universität Kaiserslautern, 1987.

[13] Williams, M. L., On the stress distribution at the base of a stationary crack, J. Appl. Mech. 24, 1957, pp. 109–114.

[14] Yang, B., Ravi-Chandar, K., Evaluation of elastic T-stress by the stress difference method, Engineering Fracture Mechanics 64, 1999, pp. 589–605.

15

Experimental and numerical analysis of in- and out- of plane constraint effects on fracture parameters: Aluminium alloy 2024

S. Seitl[a,*], P. Hutař[a], T. E. García[b], A. Fernández-Canteli[b]

[a] *Institute of Physics of Materials, Academy of Sciences of the Czech Republic, v. v. i. Žižkova 22, 616 62 Brno, Czech Republic*

[b] *Oviedo University, Dept. of Construction and Manufacturing Engineering; E.P.S. de Ingeniería de Gijón, University of Oviedo, Campus de Viesques, 33203 Gijón, Spain*

Abstract

The influence of in- and out- of plane constraints on the behaviour of a crack under mode I loading conditions is studied. The independence of the stress intensity tensor, with respect to the specimen thickness B shows that under loss of constraint conditions higher order members of the Williams' tensor expansion must be considered if the experimental results for increasing apparent fracture toughness resulting from decreasing specimen thickness are to be explained. This is achieved using the constraint curves that define the intensity field tensor along the crack propagation direction and can be alternative to the T-stress approach. This approach is then applied to crack instability assessment for program compact tension (CT — positive values of T-stress) and three point bending (3PB — from negative to positive values of T-stress) specimens with different thicknesses. The theoretical results are compared with experimental ones obtained from the research program on aluminium alloy 2024.

Keywords: LEFM, stress intensity tensor, constraint, aluminium alloy, plane strain, plane stress

1. Introduction

The character of the stress fields near the crack front has been extensively studied for years [17]. The classical linear elastic fracture mechanics approach stems from the influence of the singular term of the asymptotic expression, determined by its amplitude, or stress intensity factor (SIF, [8]). More accurate two-parameter approaches such as SIF — T-stress [1] or SIF — T_{zz} in thin elastic plates [7, 14] have been developed to describe the stress field in the vicinity of the crack tip. Such approaches have been applied successfully in engineering fracture design though they are unable to describe the effect of the out-of-plane constraint on the crack-tip field and the fracture toughness. In fact, it is well known that the fracture toughness highly depends on the thickness of the test specimen, so that considering variable fracture toughness, according to the 3D out-of-plane stress level, is inconceivable in practical engineering applications.

The 3D crack-front fields have been studied by different authors [9, 10, 13], as have the effect of T_{zz} on the 3D crack front fields and fracture toughness [7], proving their application to fracture and fatigue problems.

In particular, the importance of the stress intensity tensor is acknowledged. The effect of in- and out-of-plane constraint is investigated on the basis of previous analytical and numerical results [3–6, 11, 12] supported by a tensor approach based on the Williams' tensor expansion [17]. Apparently, increasing fracture toughness resulting from decreasing specimen thickness can be explained. The main aim of this work is to investigate the influence of specimen thicknesses on the values of the fracture mechanical parameters using constraint local fracture mechanics. This approach is then applied to crack instability assessment for program compact tension (CT — positive values of T-stress) and three point bending (3PB — from negative to positive values of T-stress) specimens made from aluminium alloy 2024 with different thicknesses to explain the change of Al 2024 fracture toughness values obtained from the experiment.

2. Tensor description at the crack front

In the following the general expression of the stress field in the proximity of the crack front is derived using a tensor description particularized for the case of pure mode I (the crack propagates in x direction, $\theta_{\mathrm{cr}} = 0$). It comprises the definition of the following magnitudes related to the stress fields $\sigma_{ij}(r, \theta, z; B)$ in front of the crack front, see Fig. 1.

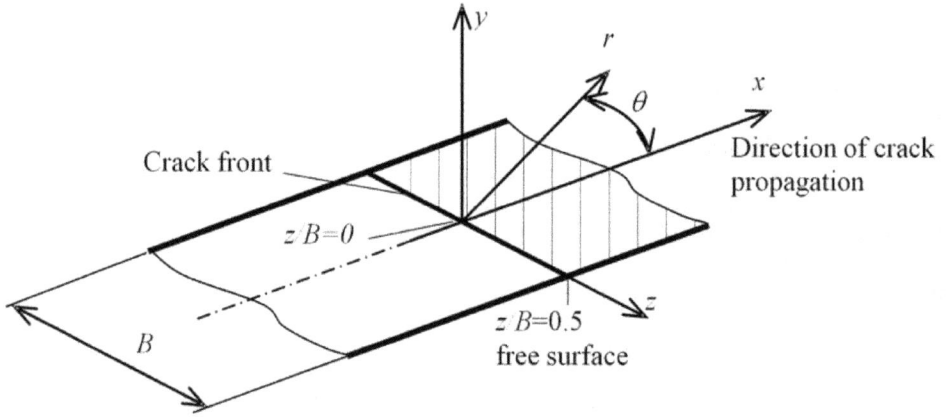

Fig. 1. Crack front and associated coordinate systems (Cartesian and polar)

1. Stress intensity field tensor:

$$\phi_{ij}(r, \theta, z; B) = \sqrt{2\pi r}\, \sigma_{ij}(r, \theta, z; B). \tag{1}$$

2. Spatial stress intensity tensor:

$$k_{ij}^*(\theta, z; B) = \lim_{r \to 0} \phi_{ij}(r, \theta, z; B) = \lim_{r \to 0} \sqrt{2\pi r}\, \sigma_{ij}(r, \theta, z; B). \tag{2}$$

3. Stress intensity tensor:

$$k_{ij}(z; B) = k_{ij}^*(\theta, z; B)|_{\theta=0} = \lim_{r \to 0} \sqrt{2\pi r}\, \sigma_{ij}(r, \theta, z; B)|_{\theta=0}. \tag{3}$$

4. Spatial constraint tensor, defined as that corresponding to the second term of the Williams' expansion:

$$t_{ij}^*(\theta, z; B) = \lim_{r \to 0} \sqrt{2\pi r}\, \sigma_{ij}(r, \theta, z; B) - \frac{k_{ij}^*}{\sqrt{2\pi r}}. \tag{4}$$

5. Constraint tensor:

$$t_{ij}(z; B) = t_{ij}^*(\theta, z; B)|_{\theta=0} = \lim_{r \to 0}\left[\sigma_{ij}(r, \theta, z; B) - \frac{k_{ij}^*}{\sqrt{2\pi r}}\right]|_{\theta=0}. \tag{5}$$

6. Constraint function:

$$\psi_{ij}(r, \theta, z; B) = \frac{k_{ij}^*(\theta, z; B)}{\sqrt{2\pi}}\, r^{-1/2} + t_{ij}^*(\theta, z; B)\, r^0 + O_{ij}(r, \theta, z; B). \tag{6}$$

Using the equations (2) and (4), the stress tensor σ_{ij} in the proximity of a straight crack tip in a plane normal to the crack front at the point $(r, \theta, z; B)$ for a given specimen thickness B, see Fig. 1, can be expressed in polar coordinates as the Williams' expansion:

$$\sigma_{ij}(r, \theta, z; B) = \frac{k_{ij}^*(\theta, z; B)}{\sqrt{2\pi}}\, r^{-1/2} + t_{ij}^*(\theta, z; B)\, r^0 + O_{ij}(r^{1/2}, \theta, z; B), \tag{7}$$

where O_{ij} represents the remaining higher order terms as a function of the polar coordinates r, θ, z and the thickness B.

For the Cartesian system in Fig. 1, the stress intensity tensor k_{ij} for a specimen of the thickness B at the mid-plane under mode I loading ($\theta_{cr} = 0$) is invariably given by:

$$k_{ij}(z; B) = \begin{pmatrix} K_I(z; B) & 0 & 0 \\ 0 & K_I(z; B) & 0 \\ 0 & 0 & 2\nu K_I(z; B) \end{pmatrix} = K_I(z; B)\begin{pmatrix} 1 & 0 & 0 \\ 0 & 1 & 0 \\ 0 & 0 & 2\nu \end{pmatrix}. \tag{8}$$

The tensor t_{ij} for a specimen of the thickness B in the proximity of the crack front, not close to the crack edge, under mode I loading ($\theta_{cr} = 0$) is given by

$$t_{ij}(z; B) = \begin{pmatrix} t_{xx}(z; B) & 0 & 0 \\ 0 & 0 & 0 \\ 0 & 0 & t_{zz}(z; B) \end{pmatrix} = \begin{pmatrix} T(z; B) & 0 & 0 \\ 0 & 0 & 0 \\ 0 & 0 & E\varepsilon_{zz}(z; B) + \nu T(z; B) \end{pmatrix}. \tag{9}$$

By identifying (8)–(9) with the conventional formulation of the Williams' expansion for the case of mode I, k_{ij}^* can be expressed in terms of K_I and t_{ij} can be expressed in term of T-stress:

$$\begin{aligned} k_{ij}(\theta, z; B) &= K_I(z; B)f_{ij}^{(K)}(\theta), \\ t_{ij}(\theta, z; B) &= T(z; B)f_{ij}^{(T)}(\theta), \end{aligned} \tag{10}$$

where K_I is the stress intensity factor for mode I and T is the classical T-stress and $f_{ij}(\theta)$ are geometric functions with super index (T) or (K) referred, respectively. See more in [3–6].

3. Experimental program and its results

In order to verify the suitability of the tensor description of the stress field, an experimental program was launched.

3.1. *Specimens and methods*

For this purpose, three point bending (3PB, see schematic diagram in Fig. 5) and compact tension (CT, see schematic diagram in Fig. 6) specimens made of aluminium alloy 2024 with different thicknesses ($B = 5$ mm and 20 mm) and different crack length ratios $a/W \in (0.2; 0.7)$ were tested (more than 30 specimens). The specimen width was $W = 50$ mm in all cases. The material parameter K_{IC} (fracture toughness) was determined according to the British standard method [2] and ASTM E1820 [15].

First, the specimens were notched in different notch lengths and thereafter they were cyclically loaded until approximately 5 mm pre-crack was achieved. After the test the total crack length was measured and the corresponding ratio a/W was calculated. The crack surface of the fractured specimens is shown in Fig. 2. There are three parts of the surface: i) notch surface, ii) fatigue pre-crack surface and iii) brittle surface. The left and right parts of Fig. 2 show the typical specimen surfaces after the measurement for $B = 20$ mm and 5 mm, respectively.

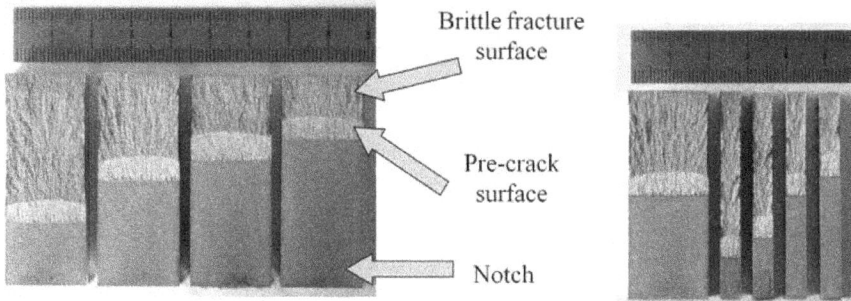

Fig. 2. Crack surfaces of specimens for various ratios a/W after failure for thicknesses $B = 20$ mm (left) and 5 mm (right)

3.2. *Experimental results*

For the used material Al 2024, the experimental results are indicated below. The obtained results are presented by values of the fracture toughness K_{IC} (for mode I) that is divided by K_{ICnorm} (the value for fracture toughness for plane strain condition $B \to \infty$).

The ratios of fracture toughness for the 3PB and CT specimens (see Figs. 5 and 6, respectively) with various thicknesses B (5 mm and 20 mm) K_{IC}, is determined for the various total crack length ratios $a/W \in (0.2; 0.7)$, see Figs. 3 and 4.

The experimentally obtained results for 3PB specimen are shown in Fig. 3. The thin specimens ($B = 5$ mm) provide fracture toughness K_{IC} values higher in the first part of interval $a/W \in (0.2; 0.4)$ than those provided by wide specimens ($B = 20$ mm). The remaining part of both curves $a/W \in (0.5; 0.8)$ are similar.

The results for CT specimens are shown in Fig. 4. For thin specimens (thickness $B = 5$ mm), the values of fracture toughness K_{IC} are higher in the whole interval $a/W \in (0.2; 0.8)$ than those provided by wide specimens (thickness $B = 20$ mm).

In the following the experimental results are correlated with constraint characteristics.

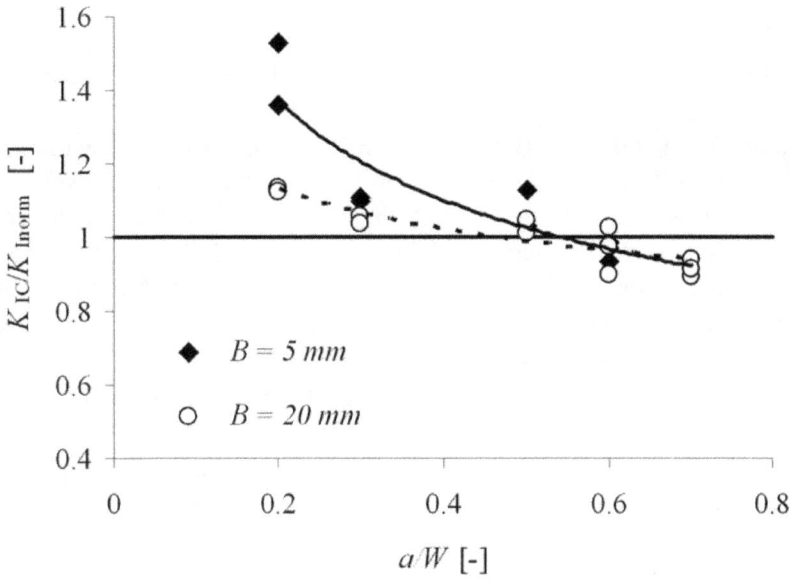

Fig. 3. Dependence of K_{IC}/K_{ICnorm} with respect to a/W determined for three point bending specimen

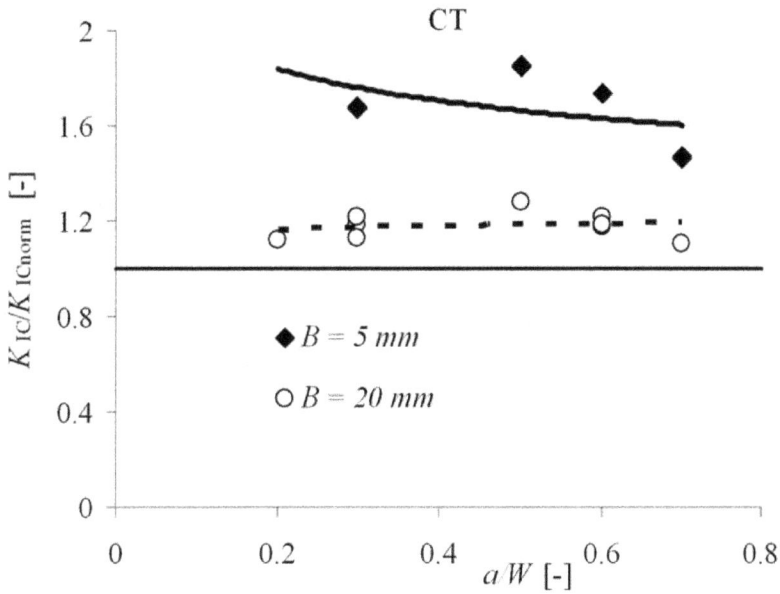

Fig. 4. Dependence of K_{IC}/K_{ICnorm} with respect to a/W determined for compact tension specimen

4. Numerical calculation and its results

With the aim of checking the validity of the preceding theoretical derivations and experimental results, linear elastic numerical calculations were performed for material parameters Al 2024 using the finite element code ANSYS.

4.1. Numerical models

The numerical models were in line with the experimental program prepared for two basic geometries: three points bend (3PB) (see Fig. 4) and compact tension (CT) specimens (see Fig. 5) with different crack ratios $a/W = 0.2$; 0.41 and 0.7. The ratio $a/W = 0.41$ has been selected because there is for 3PB specimen the value of T-stress equals 0 for 2D solution. The rank of specimen thicknesses were extended about the thickness $B = 2$ mm to cover theoretically possible slim specimens, so the values of thicknesses $B = \{2, 5, 20\}$ mm were used for numerical study. The crack ratios a/W were selected to cover all cases of constraint level for 3PB specimen that is valid for 2D solution ($T < 0, T = 0, T > 0$).

A homogenous linear elastic material model with the Young's modulus $E = 0.72$ GPa and the Poisson's ratio $\nu = 0.34$ was taken into account (the material parameters corresponding to Al 2024).

The Fig. 5 shows a schematic diagram of the 3PB model, the used dimensions and its boundary conditions. The Fig. 6 shows schematic diagram of the CT model, the used dimensions and its load.

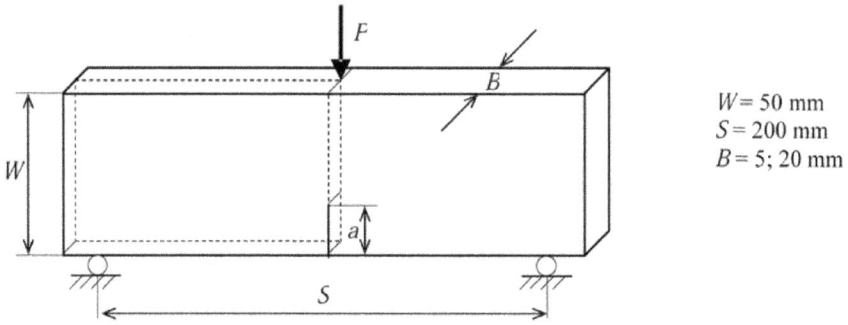

$W = 50$ mm
$S = 200$ mm
$B = 5; 20$ mm

Fig. 5. Schematic diagram of three points bend specimen (3PB)

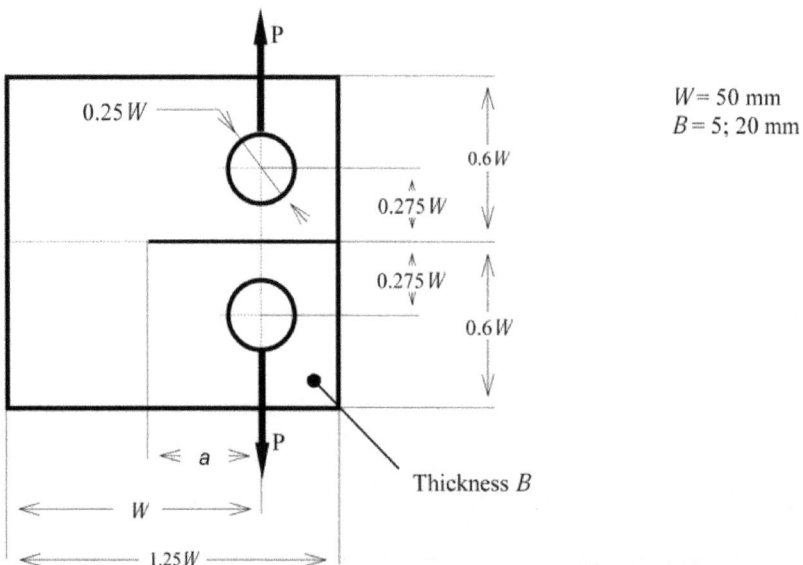

$W = 50$ mm
$B = 5; 20$ mm

Fig. 6. Schematic diagram of compact tension specimen (CT)

4.2. Numerical results

a) Validation of presented model

As the first step, the 2D numerical models presented for the assessment of the fracture mechanics parameters were compared with the data published in [16] for the 2D solution. As the second step, the numerical models were extended to 3D and the Poisson's ratio $\nu = 0$ was used for the basic validation of this 3D model.

b) Numerical results for 3D models

The aim of this work is to find a correlation between the experiment results (see section 3) and the parameters describing the stress field in front of the crack. To that end, in this section 3PB and CT numerical models with various thicknesses are studied. The values of loading P corresponding to K_{ICnorm} (the fracture toughness of Al 2024 for plane strain condition $B \to \infty$) were estimated as functions of the ratio a/W and were used for K_I estimation from 3D calibration curves. Theoretically, the same stress fields around the crack tip should be obtained.

As representatives' parameters from the Williams' tensor description curves of K_I, T_{xx}, ε_{zz} and T_{zz}, were selected; see section 2.

As the first parameters, the stress intensity factors estimated for various specimen thicknesses are shown in Fig. 7. The dependence of K_I with respect to the thickness B is dominant

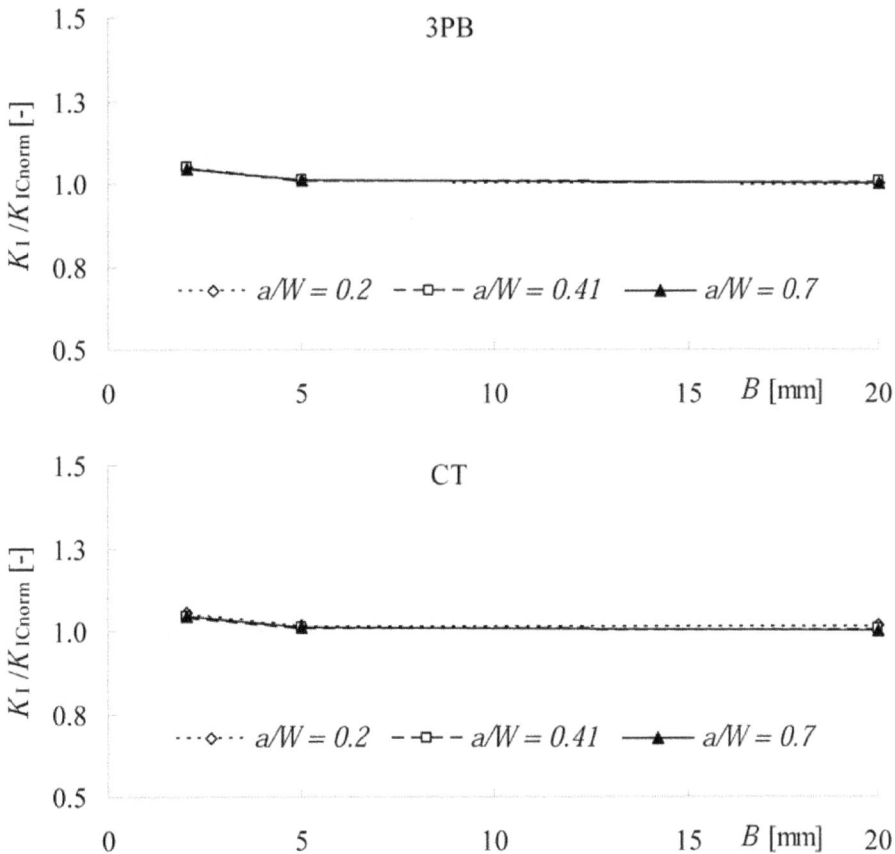

Fig. 7. Change of the values of the stress intensity factors (K_I/K_{ICnorm}) along the 3PB and CT specimen's thickness

for the change of values for both cases 3PB and CT specimens. For the infinity thickness, the value will be equal to 1. The influence of the relative crack length (a/W) on the values of K_I is not apparent.

The shapes of the curves of the second parameter, the T_{xx}-stress, significantly depend on the specimen thickness, as shown in Fig. 8. The value of T_{xx} grows as the specimen thickness decreases. The difference between the two dimensional (plane strain, $B \to \infty$) and three dimensional solutions, especially for the thin structures, is essential. According to the results obtained, we can say that the parameter characterizing the constraint (T_{xx}) is much more sensitive to the effect of the free surface than the stress intensity factor values and the two dimensional solution of the T_{xx} can lead to significant discrepancies in comparison with the 3D solution that is more close to reality. Especially for the relative crack length $a/W = 0.2$ in Fig. 8. For 3PB specimen, according to the 2D solution, there is a negative value of the T_{xx}, but for the 3D solution the value of T_{xx} is positive.

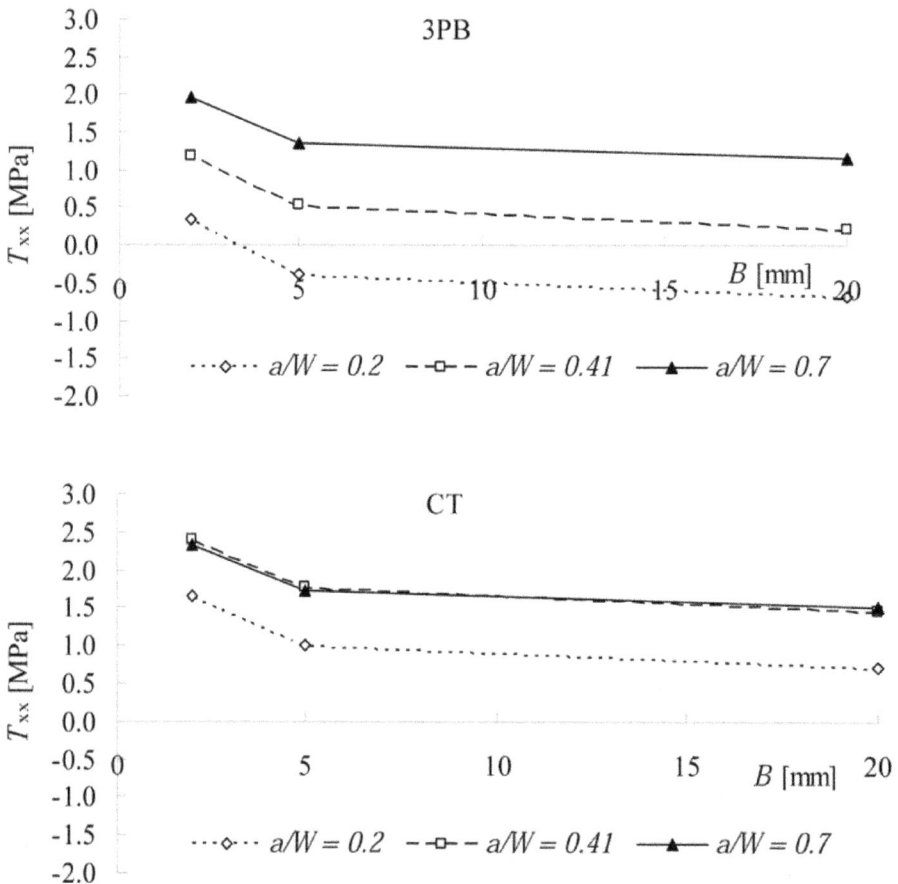

Fig. 8. Change of the values of T-stress (xx) along the 3PB and CT specimen's thickness

For determining the second parameter T_{zz} of the Williams' tensor description, the knowledge of the ε_{zz} values is required, see eq. (9). The values of out of plane strain ε_{zz} are shown in Fig. 9. The curves of ε_{zz} are plotted as functions of the specimen thickness B. The results for the different specimen thicknesses corroborate that the out of plane strain ε_{zz} is not zero as

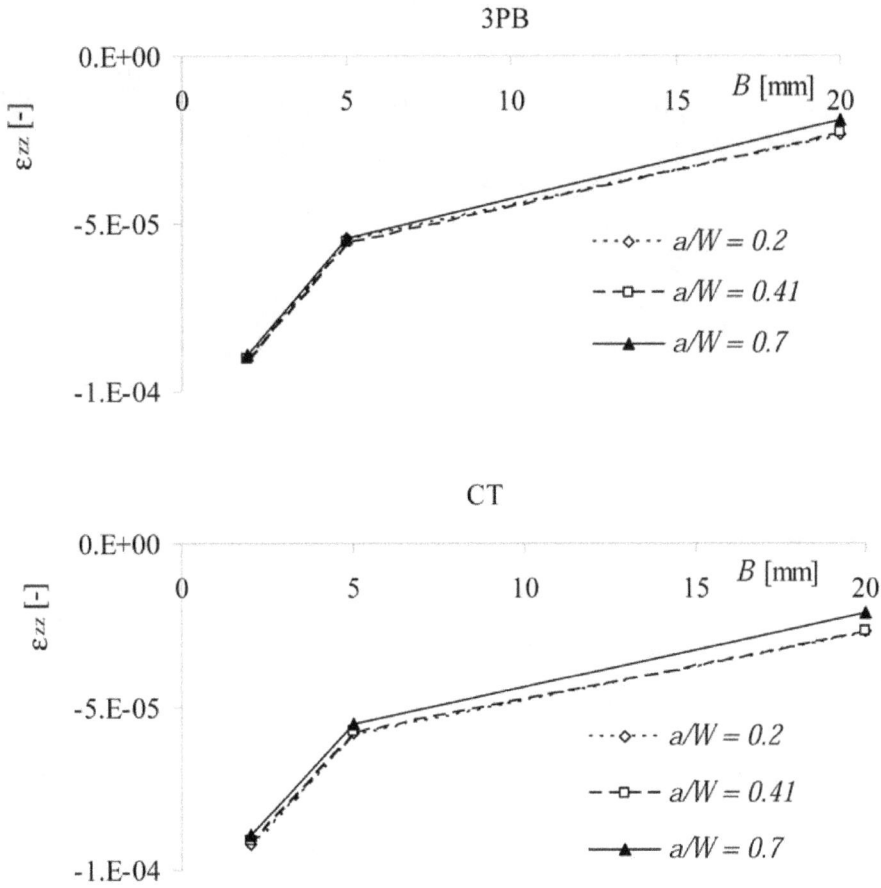

Fig. 9. Change of the values of ε_{zz} along the 3PB and CT specimen's thickness

supposed in the 2D solution. The ε_{zz} curves tend to zero when the specimen thickness is going to infinity ($B \rightarrow \infty$). The trend of both ε_{zz} curves for 3PB and CT specimens is the same due mainly to the influence of the specimen thickness.

As the second parameter of the Williams' tensor description, the change of T_{zz} is shown in Fig. 9. According to theory the values of T_{zz} for the thickness $B \rightarrow \infty$ tends towards zero. For decreasing values of thickness B the effect of the out of plane constraint increases.

It follows from the performed calculations that the in-plane and the out-of-plane constraint effects mix in 3D cases and that the elastic crack front fields become much more complicated than the 2D one. Two common types of fracture test specimens were examined: 3PB and CT specimens. These specimens represent two typical loading configurations in fracture analyses. To find out the characteristic effects of the specimen thickness, we experimentally tested different thicknesses of both specimen types in the present paper. The trend of the experimental results can be attributed to different level of constraint in both cases.

3PB

CT

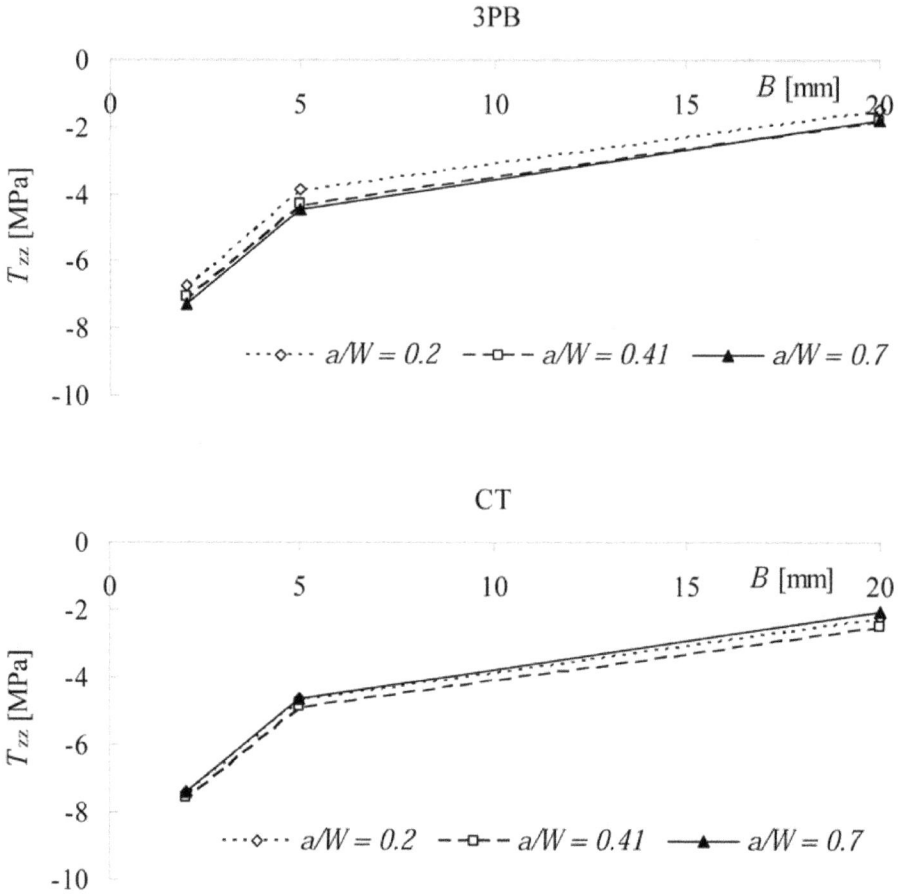

Fig. 10. Change of the values of T_{zz} along the 3PB and CT specimen's thickness

5. Conclusion

The results from the numerical solution and the experimental program are presented. These results were obtained for Al 2024 specimens. Two different geometries (3PB and CT) and two different thicknesses ($B = 20$ mm and 5 mm) were used. The following conclusions can be made:

- To explain the dependence of K_{IC} values on specimen thickness, the constraint characteristics (in- and out-of-plane) have to be used.

- Moreover, especially for thin specimens with $B < 5$ mm, the numerical support for experiment has to be done on the base of a 3D model.

- The experimental results support the analytical derivations found by using the tensor approach.

- The trend of the resulting K_{IC} curves for different a/W values can be attributed to simultaneous influence of both T_{xx} and T_{zz} stresses.

- Further calculations are envisaged to analyse the stress relations in the state after, particularly in what concerns the constraint evolution during the crack growth process.

- The analytical derivations and the numerical calculations prove the utility of the tensor approach proposed in this work.

Acknowledgements

This work was supported by the Academy of Sciences of the Czech Republic, projects M10041090 and M100411204, and by the Spanish Ministry of Science and Innovation, Project BIA2010-19920.

References

[1] Betegon, C., Hancock, J., Two parameter characterization of elastic-plastic crack-tip fields, Journal of Applied Mechanics, 113 (1991) 104–110.

[2] British standard method for determination of the rate of fatigue crack growth in metallic materials 1998.

[3] Fernández-Zúñiga, D., Giner, E., Fernández-Sáez, J., Fernández-Canteli, A., Modelo tensorial para el análisis tridimensional de placas fisuradas solicitadas en modo II y modo mixto, Anales de Mecánica de la Fractura 28 (2011) 631–636. (in Spanish)

[4] Fernández-Canteli, A., Castillo, E., Fernández-Zúñiga, D., Lateral constraint index to assess the influence of specimen thickness, Proceedings of the 16th European Conference of Fracture, Alexandropoulis, Greece, 2006, pp. 1–8.

[5] Fernández-Canteli, A., Giner, E., Fernández-Zúñiga, D., Fernández-Sáez, J., Considerations concerning plane stress and plane strain at the crack front, Proceedings of the XIII Portuguese Conference on Fracture, Porto, Portugal, 2012, pp. 1–6.

[6] Fernández-Canteli, A., Giner, E., Fernández-Zúñiga, D., Fernández-Sáez, J., A unified analysis of the in-plane and out-of-plane constraints in 3-D linear elastic fracture mechanics, Proceedings of the 19th European Conference on Fracture, Kazan, Russia, 2012, pp. 1–8.

[7] Guo, W., Recent advances in three-dimensional fracture mechanics, Key Engineering Materials 183 (2000) 193–198.

[8] Irwin, G. R., Fracture, Handbuch der Physik 6. Springer-Verlag, Heidelberg (1958) 551–590.

[9] Kwon, S. W., Sun, C. T., Characteristics of three-dimensional stress fields in plates with a through the thickness crack, International Journal of Fracture 104 (2000) 291–315.

[10] Nakamura, T., Parks, D. M., Three-dimensional stress field near the crack front of a thin elastic plate, Journal of Applied Mechanics 55 (1988) 805–813.

[11] Seitl, S., Fernández-Zúñiga, D., Fernández-Canteli, A., Using a tensor model for analyzing some aspects of mode-II loading, Applied and Computational Mechanics 5 (1) (2011) 55–66.

[12] Seitl, S., García, T. E., Fernández-Canteli, A., In- and out-of plane constraint effects on the fracture parameters of Aluminum alloy 2024: Finite element modeling and experimental results, Proceedings of the 27th Conference on Computational Mechanics, Plzen, Czech Republic, 2011.

[13] Ševčík, M., Hutař, P., Zouhar, M., Náhlík, L., Numerical estimation of the fatigue crack front shape for a specimen with finite thickness, International Journal of Fatigue 39 (2012) 75–80.

[14] She, Ch., Guo, W., The out-of-plane constraint of mixed-mode cracks in thin elastic plates, International Journal of Solids and Structure 44 (2007) 3 021–3 034.

[15] Standard test method for measurement of fracture toughness, ASTM E1820-05, 2011.

[16] Tada, H., Paris, P. C., Irwin, G. R., The stress analysis of cracks handbook (3rd ed.). American Society of Mechanical Engineers 2000.

[17] Williams, M. L., On the stress distribution at the base of a stationary crack, Journal of Applied Mechanics 24 (1957) 109–114.

16

New implicit method for analysis of problems in nonlinear structural dynamics

A. A. Gholampour[a,*], M. Ghassemieh[a]

[a]School of Civil Engineering, University of Tehran, Tehran, Iran

Abstract

In this paper a new method is proposed for direct time integration of nonlinear structural dynamics problems. In the proposed method the order of time integration scheme is higher than the conventional Newmark's family of methods. This method assumes second order variation of the acceleration at each time step. Two variable parameters are used to increase the stability and accuracy of the method. The result obtained from this new higher order method is compared with two implicit methods; namely the Wilson-θ and the Newmark's average acceleration methods.

Keywords: direct time integration, nonlinear structural dynamics, second order acceleration, implicit method

1. Introduction

Problems in the theory of vibration are divided in two categories; wave propagation problems and inertia problems, which the latter is called structural dynamics. In these problems, governing field equation is a second order differential equation [1, 2]. For nonlinear systems, it is usually expected to solve equations of motion numerically [2]. In the time integration methods, time is divided to several time steps and an algorithm is used to predict the values of displacement, velocity or acceleration at each time based on previous value. The algorithm is based on an assumption for variation of displacement in each time step and satisfying the equation of motion in selected discrete times. In fact it is a form of finite difference solution for differential equations [2–11].

In nonlinear analysis, stiffness is calculated at the beginning of each time step and then response is calculated at the end of this time step with assuming that stiffness is constant throughout the step. Therefore nonlinearity is considered by continuously updating the stiffness. Calculated responses will be considered at the end of each time step as the initial conditions for next time step. Therefore system nonlinearity behavior is replaced with a series of consecutive approximate linear characteristics [1, 2, 5, 8].

In some of algorithms, in each time step, equation of motion is written at the beginning of the time step and the unknown values at the end of time step is explicitly calculated, these methods are called explicit methods. In some other methods, to calculate the unknown values at the end of time step it is required to write the equation of motion at this point, these methods are called implicit methods [2–9]. A method is called convergent if its error for a specific time is decreased, by decreasing time step length. Also, a method is consistent if the upper bound of

---

*Corresponding author. e-mail: aagholampour@ut.ac.ir.

its residue (error in satisfying the equation of motion), is a constant power of time step length. In accuracy evaluation of the time integration methods, usually two quantities are determined, numerical damping (dissipation) and periodic error (dispersion).

In unconditionally stable methods, instability never happens, no matter how the long time step is [1–5]. Newmark, [12], presented a one-step algorithm with two parameters and he noted that γ should be taken as 0.5, because for values more than 0.5 positive numerical damping will exist and for values less than that, it will have negative numerical damping (numerical instability). The average acceleration form appears to be the most popular one. After him, lots of researches have worked on his idea. Wilson presented a modified form of linear acceleration method, called Wilson-θ method [13], and improved it to an unconditionally stable method. He also proposed the concept of collocation to develop dissipative algorithms, which were further generalized in [13]. The Wilson-θ method is unconditionally stable for $\theta = 1.37$. This method is subject to both phase and amplitude errors depending on the time step used.

Classical methods such as the Newmark's method [12] or the Wilson-θ method [13] assume a constant or linear expression for the variation of acceleration at each time step. In conditionally stable methods, the time step must be smaller than a critical time step as a constant times the smallest period of the system, consequently often entails using time steps that are much smaller than those needed for accuracy [7]. In this paper, we illustrate how to derive equations of proposed method from the Taylor series expansion in which algorithmic parameters are inserted. In this new implicit method, it is assumed that the acceleration varies quadratically within each time step. Considering this assumption and employing the two parameters δ and α, the proposed method is derived.

2. Proposed Method

The governing nonlinear equation of motion is expressed as:

$$M\ddot{x} + C\dot{x} + K(x)x = P, \tag{1}$$

where M is a constant mass matrix, C is a constant damping matrix, and $K(x)$ is a nonlinear stiffness matrix; P is vector of applied forces; x, $\dot{x}$ and $\ddot{x}$ are the displacement, velocity and acceleration vectors respectively.

By Applying the Taylor series expansions of $x_{t+\Delta t}$ and $\dot{x}_{t+\Delta t}$ about time t and truncating the equations, the following forms of equations are obtained:

$$x_{t+\Delta t} = x_t + \Delta t\dot{x}_t + \frac{\Delta t^2}{2}\ddot{x}_t + \frac{\Delta t^3}{6}\dddot{x}_t + \alpha\Delta t^4\,\ddddot{x}_t, \tag{2}$$

$$\dot{x}_{t+\Delta t} = \dot{x}_t + \Delta t\ddot{x}_t + \frac{\Delta t^2}{2}\dddot{x}_t + \delta\Delta t^3\,\ddddot{x}_t. \tag{3}$$

If the acceleration variation is assumed to be second order within time $t - \Delta t$ to $t + \Delta t$, the Eqs. (2) and (3) can be written as:

$$x_{t+\Delta t} = x_t + \Delta t\dot{x}_t + \left[\left(\alpha - \frac{1}{12}\right)\ddot{x}_{t-\Delta t} + \left(\frac{1}{2} - 2\alpha\right)\ddot{x}_t + \left(\alpha + \frac{1}{12}\right)\ddot{x}_{t+\Delta t}\right]\Delta t^2, \tag{4}$$

$$\dot{x}_{t+\Delta t} = \dot{x}_t + \left[\left(\delta - \frac{1}{4}\right)\ddot{x}_{t-\Delta t} + (1 - 2\delta)\ddot{x}_t + \left(\delta + \frac{1}{4}\right)\ddot{x}_{t+\Delta t}\right]\Delta t. \tag{5}$$

Eqs. (4) and (5) can be used to approximate the displacement and velocity at time $t + \Delta t$ respectively. It can be proven that this strategy guarantees the second-order accuracy for any choice of δ and α. The parameters δ and α are introduced in order to improve accuracy and stability. Special case $\delta = 1/4$, $\alpha = 1/12$ leads to the linear acceleration method.

Consider equation of motion in time $t + \Delta t$ as following:

$$M\ddot{x}_{t+\Delta t} + C\dot{x}_{t+\Delta t} + K_t x_{t+\Delta t} = P_{t+\Delta t}. \tag{6}$$

By substituting Eqs. (4) and (5) into the equation of motion Eq. (6), $\ddot{x}_{t+\Delta t}$ is calculated. Note that x_0 and $\dot{x}_0$ are known and $\ddot{x}_0$ can be calculated using Eq. (1) at time $t = 0$. We need the solution at time Δt before we can begin to apply Eqs. (4) and (5). It can be computed by using any one step methods such as the linear acceleration or the average acceleration methods. Now, we can obtain $x_{2\Delta t}$ from Eqs. (4) and (5), then $x_{3\Delta t}$, and so on.

3. Examples

In order to see the result of the proposed method and to see its advantages over the other implicit existing methods, let's consider two examples which the results obtained from the proposed method are compared with the Wilson-θ and average acceleration (Newmark's) methods.

Example 1 [2]: Consider a single degree of freedom system in Fig. 1. Fig. 2 shows the equivalent spring force (f_s) versus displacement diagram with elastoplastic behavior. k_{el} from Fig. 1 is the slope of the linear part of the Fig. 2. This structure is under acting force (p) as Fig. 3. The initial conditions are $x(0) = \dot{x}(0) = 0$ that $0 \leq t \leq 9$ sec and $\Delta t = 0.1$.

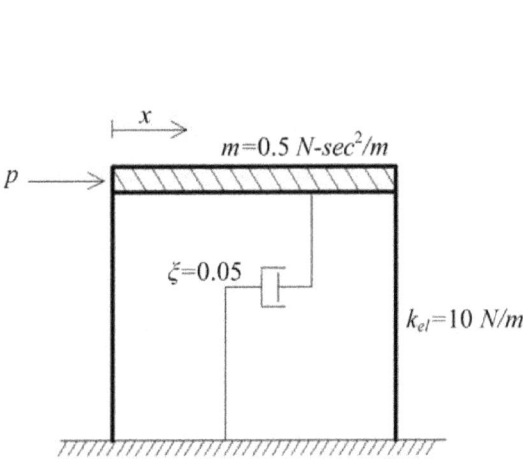

Fig. 1. Frame of structure [2]

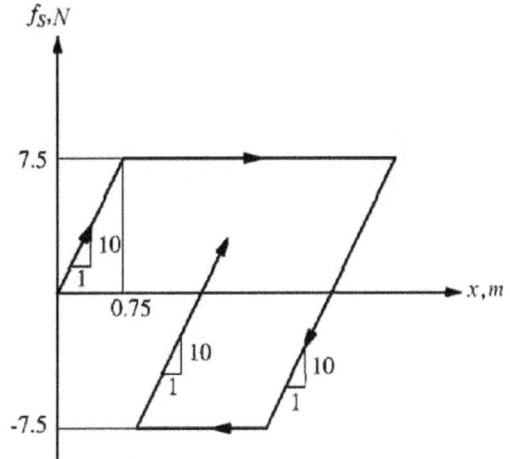

Fig. 2. Force-displacement relationship [2]

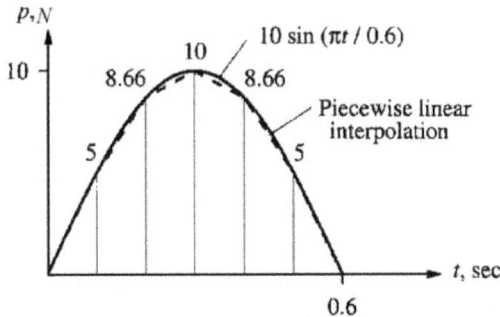

Fig. 3. Exciting force [2]

Table 1. Numerical responses using the Wilson-θ, average acceleration, and proposed methods

Time [sec]	Displacement Responses [m]		
	Wilson-θ ($\theta = 1.4$)	Average acceleration method	Proposed method ($\delta = 1/3, \alpha = 1/6$)
0 0	0 000 0	0 000 0	0 000 0
0.1	0.036 8	0.043 7	0.043 7
0.2	0.183 3	0.232 6	0.219 5
0.3	0.483 0	0.612 1	0.590 9
0.4	0.900 7	1.082 5	1.061 6
0.5	1.322 6	1.527 9	1.482 2
0.6	1.682 8	1.837 7	1.739 4
0.7	1.978 3	1.889 3	1.742 2
0.8	1.862 3	1.671 6	1.482 6
0.9	1.301 1	1.280 1	1.065 6

In Table 1 displacement results of this system due to the applied loading $P(t)$ (see Fig. 3) are given. The results obtained using of Wilson-θ, average acceleration, and proposed methods are compared.

Also the displacement responses versus time are shown in Fig. 4.

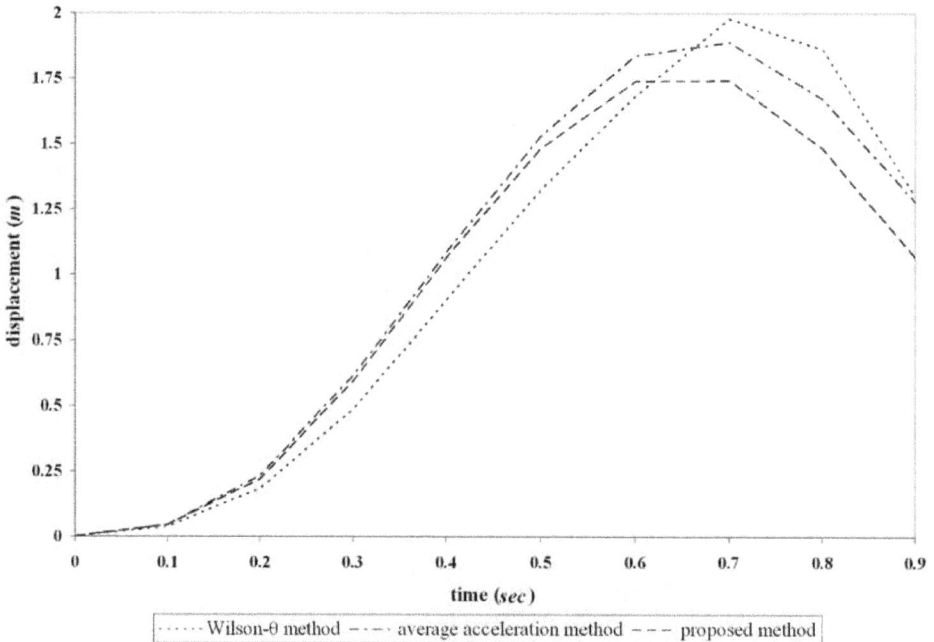

Fig. 4. Displacement responses versus time diagram for example 1

Example 2 [14]: Consider the second order nonlinear differential equation as following:

$$\ddot{x} + \sin x = 0, \tag{7}$$

with initial conditions $x(0) = \pi/2$ and $\dot{x}(0) = 0$ that $0 \le t \le 20$. Let's select $\Delta t = 0.1$ and define the error at time t as following:

$$e_t = |x_t - x_{t(exact)}|, \tag{8}$$

in which $x_{t(exact)}$ is the exact solution and x_t is the numerical solution (angle (degree)) at time t. The values obtained by the Wilson-θ, average acceleration, and proposed methods can be compared to each other in Fig. 5.

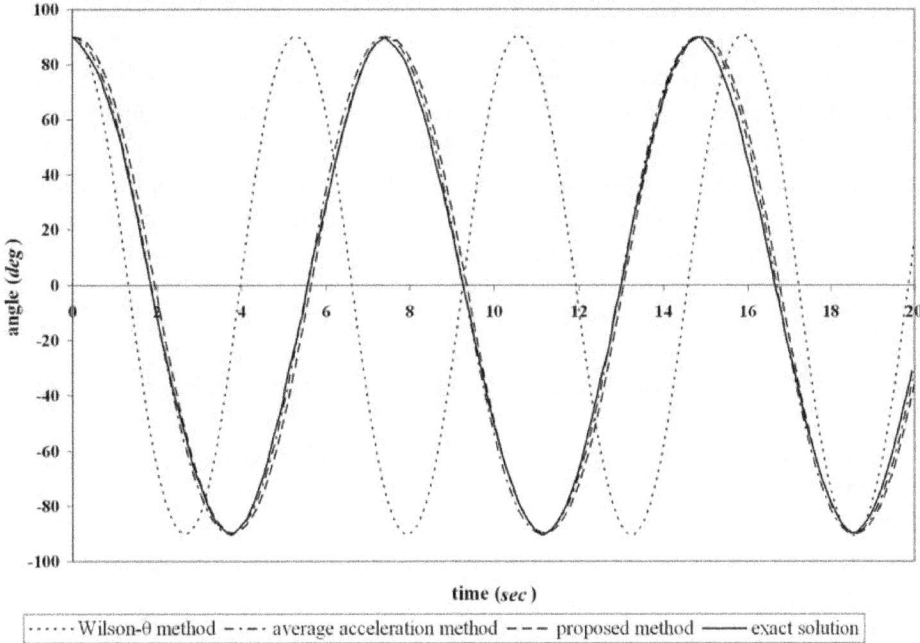

Fig. 5. Angle responses versus time diagram from example 2

The numerical solution calculated from mentioned methods and their error respect to the exact solution of Eq. (7) have been shown in Table 2 for $t = 6$ sec to $t = 7$ sec.

Table 2. Angle responses using the Wilson-θ, average acceleration, and proposed methods and their error respect to the exact solution

Time [sec]	Wilson-θ ($\theta = 1.4$)		Average acceleration method		Proposed method ($\delta = 1/3, \alpha = 1/6$)	
	x_t	e_t	x_t	e_t	x_t	e_t
6	62.1602	33.3314	34.3660	5.5372	28.1380	0.6908
6.1	53.8810	18.2972	41.5566	5.9728	35.5922	0.0084
6.2	44.6907	1.9390	48.3691	5.6174	42.7140	0.0376
6.3	34.7041	14.5329	54.7518	5.5149	49.4463	0.2093
6.4	24.0700	31.2963	60.6705	5.3042	55.7431	0.3768
6.5	12.9718	47.4454	66.0907	5.6736	61.5644	1.1472
6.6	1.6215	64.6319	70.9838	4.7304	66.8814	0.6280
6.7	−9.7689	81.3105	75.3383	3.7967	71.6713	0.1298
6.8	−20.9645	96.2283	79.1427	3.8789	75.9170	0.6531
6.9	−31.7591	111.718	82.3799	2.4214	79.6068	0.3517
7	−41.9635	125.158	85.0499	1.8557	82.7294	0.4647

Table 2 shows that the numerical values of x_t calculated using the proposed method are more accurate than those for the Wilson-θ and average acceleration methods. In this example, we presented only angle responses, whereas the angular velocity and angular acceleration responses calculated using the proposed method are also more accurate than the other methods.

4. Conclusion

A new implicit step by step integration technique for problems in structural dynamics was illustrated. A second order polynomial as a function of time was used in order to approximate the variation of acceleration during the time steps. Therefore the proposed method was shown more accurate values than the Wilson-θ and average acceleration methods. This method was a two parameter method (δ and α). Proposed method allows numerical damping while retaining second order accuracy. The new method can be used for either linear or nonlinear problems, although in this paper, we have discussed only nonlinear problems.

References

[1] Paz, M., Structural Dynamics: Theory and Computation, 4th ed., Chapman & Hall, New York, 1997.

[2] Chopra, A., Dynamics of Structures: Theory and Applications to Earthquake Engineering, 3rd ed., Prentice-Hall, Upper Saddle River, New Jersey, 2007.

[3] Hughes, T. J. R., Belytschko, T., A precis of developments in computational methods for transient analysis, Journal of Applied Mechanics, vol. 50, Dec. 1983, pp. 1 033–1 041, doi: 10.1115/1.3167186.

[4] Subbaraj, K., Dokainish, M. A., A survey of direct time integration methods in computational structural dynamics. II. Implicit methods, Computers & Structures, vol. 32, 1989, no. 6, pp. 1 387–1 401, doi: 10.1016/0045-7949(89)90315-5.

[5] Humar, J. L., Dynamics of Structures, Prentice-Hall, Englewood Cliffs, New Jersey, 1990.

[6] Bathe, K. J., Finite Element Procedures, Prentice-Hall, Englewood Cliffs, New Jersey, 1996.

[7] Crisfield, M. A., Non-Linear Finite Element Analysis of Solids and Structures, John Wiley & Sons, Vol. 2, 1997.

[8] Clough, R. W., Penzien, J., Dynamics of Structures, McGraw Hill, 1983.

[9] Park, K. C., Practical aspects of numerical time integration, Computers & Structures, vol. 7, jun. 1977, pp. 343–353, doi: 10.1016/0045-7949(77)90072-4.

[10] Belytschko, T., Liu, W. K., Moran, B., Nonlinear Finite Elements for Continua and Structures, 3rd ed., John Wiley & Sons, Chichester, UK, 2000.

[11] Chen, S., Hansen, J., Tortorelli, D., Unconditionally energy stable implicit time integration: application to multibody system analysis and design, International Journal for Numerical Methods in Engineering, 48, pp. 791–822, 2000.

[12] Newmark, N. M., A method of computation for structural dynamics, Journal of the Engineering Mechanics Division, 85(3), pp. 67–94, 1959.

[13] Wilson, E. L., Farhoomand, I., Bathe, K. J., Nonlinear Dynamic Analysis of Complex Structures, International Journal of Earthquake Engineering and Structural Dynamics, 1(3), pp. 241–252, 1973.

[14] Bert, C. W., Striklin, J. D., Comparative evaluation of six different numerical integration methods for non-linear dynamic systems, Journal of Sound and Vibration, vol. 127, 1988, no. 2, pp. 221–229, doi: 10.1016/0022-460X(88)90298-2.

17

Application of patch test in meshless analysis of continuously non-homogeneous piezoelectric circular plate

P. Staňák[a,*], V. Sládek[a], J. Sládek[a], S. Krahulec[a], L. Sátor[a]

[a]*Institute of Construction and Architecture, Slovak Academy of Sciences, Dúbravská cesta 9, 845 03 Bratislava, Slovakia*

Abstract

Proposed paper presents application of the patch test for meshless analysis of piezoelectric circular plate with functionally graded material properties. Functionally graded materials (FGM) are the special class of composite materials with continuous variation of volume fraction of constituents in predominant direction. Patch test analysis is an important tool in numerical methods for addressing the convergence. Meshless local Petrov-Galerkin (MLPG) method together with moving least-squares (MLS) approximation scheme is applied in the analysis. No finite elements are required for approximation or integration of unknown quantities. Circular plate is considered as a 3-D axisymmetric piezoelectric solid. Considering the axial symmetry, the problem is reduced to a 2-dimensinal one. Displacement and electric potential fields are prescribed on the outer boundaries in order to reach the state of constant stress field inside the considered plate as required by the patch test and the governing equations. Values of prescribed mechanical and electrical fields must be determined in order to comply with applied FGM gradation rule. Convergence study is performed to assess the considered meshless approach and several conclusions are finally presented.

Keywords: patch test, meshless method, functionally graded materials, MLS approximation

1. Introduction

Piezoelectric materials have found wide range of applications as sensors and actuators in variety of advanced engineering systems and structures. Structures which incorporate piezoelectric or another smart material are usually called adaptive structures. In this way various possibilities of vibration suppression, structural health monitoring and shape control are available [1, 8]. Smart materials are characterized by ability of converting energy form one form into another by response to an external impulse, for piezoelectric materials it is conversion of mechanical energy to electrical one and opposite. Many piezoelectric structures have plate-like shapes. Such devices can be found in acoustic ultrasound resonators or certain types of accelerometers and sensors. Active piezoelectric elements laminated into multilayer composite plates can significantly improve its structural properties [7]. Important investigations in the field of piezoelectric plates were first introduced by Tiersten [19].

Recently also the functionally graded materials (FGMs) [18] have been widely applied in engineering applications because of their excellent properties. FGMs are multi-component composite materials in which the volume fraction of the material constituents is continuously varying in a predominant direction. This feature can be used to tune the selected properties into desired value. Originally these materials have been introduced to benefit from the ideal

---

*Corresponding author. e-mail: peter.stanak@savba.sk.

performance of its constituents, e.g. for thermal shielding applications where high thermal resistance of the ceramics on the one side and mechanical strength of metals on the other side
is utilized. Similarly also piezoelectric parameters can be enhanced if various piezoelectric
ceramic materials are combined.

Design of structures incorporating piezoelectric and functionally graded materials requires
advanced computational techniques for analysis of these materials under various operational
conditions. Development of new computational methods for solving such complex problems
requires tools for checking the quality of computed results. The patch test can be considered as
one of such numerical procedures to address the convergence of the applied method. Convergence, in sense of numerical computing, describes how the results, under specific conditions,
approach the exact solution. The patch test was developed primary for the finite element method
(FEM) however can be used also to other computational methods.

The test, in its original form, is based on selecting a patch (group) of finite elements and
imposing upon it nodal displacements corresponding to any state of constant strain. Rigid body
condition is also satisfied by the patch test if displacement field corresponding to zero strain is
imposed. If the FE model passes the patch test, one can expect that the solution will converge to
exact values as the mesh is refined. The patch test was first introduced by Irons and Razzaque [8]
to examine the soundness of a nonconforming plate element. In this original form, the patch
test was primarily a test for polynomial completeness, i.e. the ability to reproduce exactly a
polynomial of order k [4].

Even though the finite element method has encountered wide acceptance and success on
commercial market, it possess some drawbacks such as locking of elements, stress disconti-
nuity across elements or costly remeshing in large problems with moving boundaries. The
meshless methods, an attractive option to solve these drawbacks were developed in the last
decade. Among many meshless or meshfree methods available the meshless local Petrov-
Galerkin (MLPG) method [2, 3] has received considerable scientific attention. Focusing on
nodes instead of finite elements have certain advantages. High computational demands of
remeshing in large scale problems or possible shear locking of elements appearing in the FEM
can be efficiently eliminated. MLPG was recently applied to broad field of engineering problems including laminated plates [12], rectangular piezoelectric plates [13] or 3-D axisymmetric
piezoelectric solids [10]. Thermal bending of plates with functionally graded material properties analyzed by MLPG was presented in [11]. Recently, FGM circular plates were also
analyzed [14] and solutions for composite circular plates with piezoelectric layers were presented [16, 17]. The patch test of the MLPG solution was performed for the Laplace equation
on the square domain in [3] with successful results. Sladek et al. [15] used the patch test to
examine the accuracy of various meshless interpolations in non-homogeneous media. If the
MLPG method recovers the prescribed exact solution at all interior nodes, then the method
passes the patch test [2].

Presented paper is adressing the modified patch test for the functionally graded piezoelectric
circular plate analyzed by the MLPG method. The circular plate can be considered as a 3-D
axisymmetric body with axis of symmetry passing through the center of the plate coinciding
with vertical z-axis. With use of cylindrical coordinates the original 3-D axisymmetric problem
can be reduced to 2-D problem located on the cross-section of the plate as shown in Fig. 1.

The coupled electro-mechanical fields are described by constitutive relations and governing
partial differential equations (PDEs). Nodal points are spread on the analyzed domain without
any restrictions. Small local circular subdomain is introduced around each nodal point. Local
integral equations (LIEs) constructed from governing PDEs are defined over these circular sub-

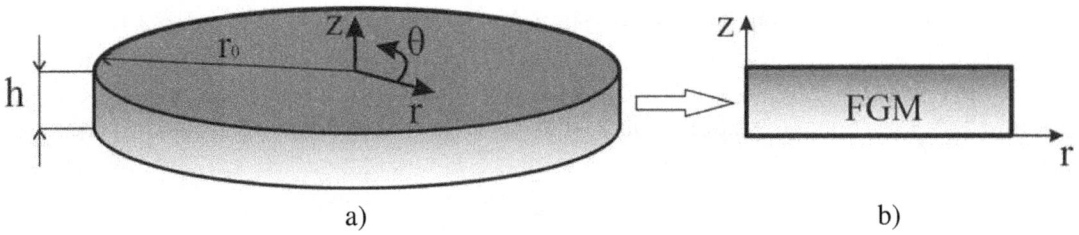

Fig. 1. Geometry of the circular plate: a) original 3-D problem, b) assumed 2-D geometry

domains. For a simple shape of subdomains e. g. circles, numerical integration of LIEs can be easily carried out. Moving Least-Squares (MLS) approximation scheme [2,6] is used to approximate the spatial variations of electric and mechanical fields. Since position of nodal points is not restricted by any finite elements, random distribution of nodes is possible, in general. There is also no restriction on the shape of analysed domain, however one have to keep in mind that sufficient ammount of nodes must be applied to adequately capture the geometry and mantain an accurate approximation of nodal quantities.

An exponential variation of material properties is assumed for the FGM plate. In order to obtain the state of constant stress satisfying the governing equations the axial displacement and electric potential must be specified on the outer boundaries in a form that comply well with prescribed FGM gradation. Compared to homogeneous materials the non-constant strain field is present, thus the original conditions of the patch test must be modified. The essential boundary conditions and prescribed mechanical and electrical fields are satisfied by the collocation of MLS approximation expressions on boundary nodes.Convergence study for increasing number of nodal points is finally conducted and results are discussed.

2. Governing equations

Piezoelectric materials are characterized by mutual coupling of elastic field and electric field. Quasi-static approximation of electric field is assumed, thus mechanical and electrical forces are balanced at any time instant. Governing equations for general piezoelectric body under static loading are given by the elastostatic equations and the first Maxwell's equation for the vector of electric displacements as

$$\sigma_{ij,j}(\mathbf{x}) + X_i(\mathbf{x}) = 0, \tag{1}$$

$$D_{i,i}(\mathbf{x}) - R(\mathbf{x}) = 0, \tag{2}$$

where σ_{ij}, D_i, X_i, R are stresses, electric displacements, vector of body forces and volume density of free charges, respectively. The linear constitutive equations for functionally graded piezoelectric material can be expressed as

$$\sigma_{ij}(\mathbf{x}) = C_{ijkl}(\mathbf{x})\varepsilon_{kl}(\mathbf{x}) - e_{kij}(\mathbf{x})E_k(\mathbf{x}), \tag{3}$$

$$D_i(\mathbf{x}) = e_{ikl}(\mathbf{x})\varepsilon_{kl}(\mathbf{x}) + h_{ik}(\mathbf{x})E_k(\mathbf{x}), \tag{4}$$

where $C_{ijkl}(\mathbf{x})$, $e_{kij}(\mathbf{x})$, $h_{ik}(\mathbf{x})$ represent spatially dependent elastic, piezoelectric and dielectric material coefficients, respectively. The strain tensor ε_{ij} and electric field vector E_k are related to elastic displacements u_i and electric potential ψ by

$$\varepsilon_{ij} = \frac{1}{2}(u_{i,j} + u_{j,i}), \tag{5}$$

$$E_k = -\psi_{,k}. \tag{6}$$

Since the considered problem is assumed to be axisymmetric, it can be reduced to 2-D, if cylindrical coordinates $\mathbf{x} = [r, \theta, z]$ are used. All physical quantities are then independent of angular coordinate θ, thus $\mathbf{x}$ becomes $\mathbf{x} = [r, z]$. Finally, for the axisymmetric piezoelectric body we can rewrite the governing equations (1), (2) into the following cylindrical coordinate form

$$\sigma_{rr,r}(r, z) + \sigma_{rz,z}(r, z) + \frac{\sigma_{rr}(r, z) - \sigma_{\theta\theta}(r, z)}{r} = 0, \tag{7}$$

$$\sigma_{rz,r}(r, z) + \sigma_{zz,z}(r, z) + \frac{\sigma_{rz}(r, z)}{r} = 0, \tag{8}$$

$$D_{r,r}(r, z) + D_{z,z}(r, z) + \frac{D_r(r, z)}{r} = 0, \tag{9}$$

where volume density of body forces and volume density of free charges are vanishing. The constitutive equations (3), (4) are rewritten for orthotropic piezoelectric materials in cylindrical coordinates as

$$\sigma_{rr} = c_{11}\varepsilon_{rr} + c_{12}\varepsilon_{\theta\theta} + c_{13}\varepsilon_{zz} - e_{31}E_z, \tag{10}$$

$$\sigma_{\theta\theta} = c_{12}\varepsilon_{rr} + c_{11}\varepsilon_{\theta\theta} + c_{13}\varepsilon_{zz} - e_{31}E_z, \tag{11}$$

$$\sigma_{zz} = c_{13}\varepsilon_{rr} + c_{13}\varepsilon_{\theta\theta} + c_{33}\varepsilon_{zz} - e_{33}E_z, \tag{12}$$

$$\sigma_{rz} = c_{44}\varepsilon_{rz} - e_{15}E_r, \tag{13}$$

$$D_r = e_{15}\varepsilon_{rz} + h_{11}E_r, \tag{14}$$

$$D_z = e_{31}\varepsilon_{rr} + e_{31}\varepsilon_{\theta\theta} + e_{33}\varepsilon_{zz} + h_{33}E_z. \tag{15}$$

The strain tensor ε_{ij} is then related to elastic displacements u_i as

$$\varepsilon_{rr} = u_{r,r}, \qquad \varepsilon_{\theta\theta} = \frac{1}{r}u_r, \qquad \varepsilon_{zz} = u_{z,z}, \qquad \varepsilon_{rz} = u_{r,z} + u_{z,r}. \tag{16}$$

Note again that angular strain $\varepsilon_{\theta\theta}$ is nonzero and independent of angular coordinate.

Governing equations (7)–(9) must be followed by essential and/or natural boundary conditions that are assumed for elastic and electrical fields as

$$u_i(\mathbf{x}, t) = \tilde{u}_i(\mathbf{x}, t) \text{ on } \Gamma_u, \quad \sigma_{ij}n_j = \tilde{T}_i(\mathbf{x}, t) \text{ on } \Gamma_t, \tag{17}$$

$$\psi(\mathbf{x}, t) = \tilde{\psi}(\mathbf{x}, t) \text{ on } \Gamma_p, \quad D_i n_i = \tilde{Q}(\mathbf{x}, t) \text{ on } \Gamma_q, \tag{18}$$

where n_i is the unit outward normal vector, Γ_u, Γ_t, Γ_p, Γ_q are parts of the global boundary Γ with prescribed displacements $\tilde{u}_i$, tractions $\tilde{T}_i$, electric potential $\tilde{\psi}$ and surface density of electric induction field flux $\tilde{Q}$ (normal component of electric displacements), respectively.

3. Local integral equations

The MLPG method is based on the local weak form of the governing equations that is written over local subdomain Ω_s. The local subdomain is a small region taken for each node inside the global domain [2]. The local subdomains could be of any geometrical shape. In this paper the local subdomains posses circular shape — just for simplicity. The local weak forms of (5)–(7) can be then written as

$$\int_{\Omega_s} \sigma_{rr,r}(r,z)p^* \, d\Omega + \int_{\Omega_s} \sigma_{rz,z}(r,z)p^* \, d\Omega + \int_{\Omega_s} \frac{1}{r} \left[\sigma_{rr}(r,z) - \sigma_{\theta\theta}(r,z) \right] p^* \, d\Omega = 0, \quad (19)$$

$$\int_{\Omega_s} \sigma_{rz,r}(r,z)q^* \, d\Omega + \int_{\Omega_s} \sigma_{zz,z}(r,z)q^* \, d\Omega + \int_{\Omega_s} \frac{1}{r}\sigma_{rz}(r,z)q^* \, d\Omega = 0, \quad (20)$$

$$\int_{\Omega_s} D_{r,r}(r,z)w^* \, d\Omega + \int_{\Omega_s} D_{z,z}(r,z)w^* \, d\Omega + \int_{\Omega_s} \frac{1}{r}D_r(r,z)w^* \, d\Omega = 0, \quad (21)$$

where $p^*(\mathbf{x})$, $q^*(\mathbf{x})$, $w^*(\mathbf{x})$ are arbitrary test functions. The Heaviside unit step functions are chosen as test functions for the present problem in the same way as in [17]. The local weak forms are the starting point for deriving local integral equations (LIEs) with the use of Gauss divergence theorem

$$\int_{\partial\Omega_s} \sigma_{rr}(r,z)n_r \, d\Gamma + \int_{\partial\Omega_s} \sigma_{rz}(r,z)n_z \, d\Gamma + \int_{\Omega_s} \frac{1}{r} \left[\sigma_{rr}(r,z) - \sigma_{\theta\theta}(r,z) \right] d\Omega = 0, \quad (22)$$

$$\int_{\partial\Omega_s} \sigma_{rz}(r,z)n_r \, d\Gamma + \int_{\partial\Omega_s} \sigma_{zz}(r,z)n_z \, d\Gamma + \int_{\Omega_s} \frac{1}{r}\sigma_{rz}(r,z) \, d\Omega = 0, \quad (23)$$

$$\int_{\partial\Omega_s} D_r(r,z)n_r \, d\Gamma + \int_{\partial\Omega_s} D_z(r,z)n_z \, d\Gamma + \int_{\Omega_s} \frac{1}{r}D_r(r,z) \, d\Omega = 0, \quad (24)$$

where $\partial\Omega_s$ represents the boundary of the local subdomain Ω_s and n_r, n_z are the unit outward normal vectors.

3.1. Meshless discretization

Traditional mesh-based approaches use mesh of finite elements to discretize the solution domain, however in the MLPG method the solution domain is discretized solely using nodal points distributed without any restrictions in mutual position. Special techniques are required to approximate unknown quantities in terms of nodal values only. The moving least-squares (MLS) approximation is used in this paper for the approximation of displacements u_r, u_z and electric potential field ψ as

$$u_r(r,z) = \sum_{i=1}^{n} \phi^i(r,z)\hat{u}_r^i, \quad (25)$$

$$u_z(r,z) = \sum_{i=1}^{n} \phi^i(r,z)\hat{u}_z^i, \quad (26)$$

$$\psi(r,z) = \sum_{i=1}^{n} \phi^i(r,z)\hat{\psi}^i, \quad (27)$$

where the nodal values $\hat{u}_r^i$, $\hat{u}_z^i$, $\hat{\psi}^i$ are so called fictitious parameters for the displacements and electric potential, and $\phi^i(r,z)$ is called the MLS shape function associated with the node $i \in \{1, 2, \ldots, n\}$ and n is the number of nodes whose support domains involve the evaluation point (r,z). As a weight function, the 4th order spline-type function [2] is used, ensuring C^1 continuity in the analyzed domain. In the similar manner also the appropriate derivatives can be obtained with use of the shape function derivative as

$$u_{j,k}(r,z) = \sum_{i=1}^{n} \phi^i_{,k}(r,z)\hat{u}_j^i, \qquad \psi_{,k}(r,z) = \sum_{i=1}^{n} \phi^i_{,k}(r,z)\hat{\psi}^i, \quad (28)$$

where $\phi^i_{,k}(r,z)$ represents the MLS shape function derivative.

Applying the trial functions given by (25)–(27) for approximation of $u_r(r, z)$, $u_z(r, z)$, $\psi(r, z)$ and their derivatives in constitutive relations (10)–(15) and their subsequent insertion into local integral equations (22)–(24) is leading to the discretized local integral equations in the form

$$\sum_{i=1}^{n} \hat{u}_r^i \int_{\partial\Omega_s} \left[c_{11}(\mathbf{x})n_r\phi_{,r}^i(r, z) + \frac{c_{12}(\mathbf{x})}{r}n_r\phi^i(r, z) + c_{44}(\mathbf{x})n_z\phi_{,z}^i(r, z) \right] \mathrm{d}\Gamma +$$

$$\sum_{i=1}^{n} \hat{u}_r^i \int_{\Omega_s} \left[\frac{c_{11}(\mathbf{x})}{r}\phi_{,r}^i(r, z) + \frac{c_{12}(\mathbf{x})}{r^2}\phi^i(r, z) - \frac{c_{11}(\mathbf{x})}{r^2}\phi^i(r, z) - \frac{c_{12}(\mathbf{x})}{r}\phi_{,r}^i(r, z) \right] \mathrm{d}\Omega +$$

$$\sum_{i=1}^{n} \hat{u}_z^i \int_{\partial\Omega_s} \left[c_{13}(\mathbf{x})n_r\phi_{,z}^i(r, z) + c_{44}(\mathbf{x})n_z\phi_{,r}^i(r, z) \right] \mathrm{d}\Gamma + \quad (29)$$

$$\sum_{i=1}^{n} \hat{\psi}^i \int_{\partial\Omega_s} \left[e_{31}(\mathbf{x})n_r\phi_{,z}^i(r, z) + e_{15}(\mathbf{x})n_z\phi_{,r}^i(r, z) \right] \mathrm{d}\Gamma = 0,$$

$$\sum_{i=1}^{n} \hat{u}_r^i \int_{\partial\Omega_s} \left[c_{44}(\mathbf{x})n_r\phi_{,z}^i(r, z) + \frac{c_{13}(\mathbf{x})}{r}n_z\phi^i(r, z) + c_{13}(\mathbf{x})n_z\phi_{,r}^i(r, z) \right] \mathrm{d}\Gamma +$$

$$\sum_{i=1}^{n} \hat{u}_r^i \int_{\Omega_s} \frac{c_{44}(\mathbf{x})}{r}\phi_{,z}^i(r, z) \,\mathrm{d}\Omega + \sum_{i=1}^{n} \hat{u}_z^i \int_{\partial\Omega_s} \left[c_{33}(\mathbf{x})n_z\phi_{,z}^i(r, z) + c_{44}(\mathbf{x})n_r\phi_{,r}^i(r, z) \right] \mathrm{d}\Gamma +$$

$$\sum_{i=1}^{n} \hat{u}_z^i \int_{\Omega_s} \frac{c_{44}(\mathbf{x})}{r}\phi_{,r}^i(r, z) \,\mathrm{d}\Omega + \sum_{i=1}^{n} \hat{\psi}^i \int_{\Omega_s} \frac{e_{15}(\mathbf{x})}{r}\phi_{,r}^i(r, z) \,\mathrm{d}\Omega + \quad (30)$$

$$\sum_{i=1}^{n} \hat{\psi}^i \int_{\partial\Omega_s} \left[e_{15}(\mathbf{x})n_r\phi_{,r}^i(r, z) + e_{33}(\mathbf{x})n_z\phi_{,z}^i(r, z) \right] \mathrm{d}\Gamma = 0,$$

$$\sum_{i=1}^{n} \hat{u}_r^i \int_{\partial\Omega_s} \left[e_{15}(\mathbf{x})n_r\phi_{,z}^i(r, z) + \frac{e_{31}(\mathbf{x})}{r}n_z\phi^i(r, z) + e_{31}(\mathbf{x})n_z\phi_{,r}^i(r, z) \right] \mathrm{d}\Gamma +$$

$$\sum_{i=1}^{n} \hat{u}_r^i \int_{\Omega_s} \frac{e_{15}(\mathbf{x})}{r}\phi_{,z}^i(r, z) \,\mathrm{d}\Omega + \sum_{i=1}^{n} \hat{u}_z^i \int_{\partial\Omega_s} \left[e_{15}(\mathbf{x})n_r\phi_{,r}^i(r, z) + e_{33}(\mathbf{x})n_z\phi_{,z}^i(r, z) \right] \mathrm{d}\Gamma +$$

$$\sum_{i=1}^{n} \hat{u}_z^i \int_{\Omega_s} \frac{e_{15}(\mathbf{x})}{r}\phi_{,r}^i(r, z) \,\mathrm{d}\Omega - \sum_{i=1}^{n} \hat{\psi}^i \int_{\partial\Omega_s} \left[h_{11}(\mathbf{x})n_r\phi_{,r}^i(r, z) + h_{33}(\mathbf{x})n_z\phi_{,z}^i(r, z) \right] \mathrm{d}\Gamma - \quad (31)$$

$$\sum_{i=1}^{n} \hat{\psi}^i \int_{\Omega_s} \frac{h_{11}(\mathbf{x})}{r}\phi_{,r}^i(r, z) \,\mathrm{d}\Omega = 0.$$

4. The patch test

If the patch test is applied to a homogeneous piezoelectric material, the linear displacement and potential field specified on the outer boundary will result into constant mechanical strains and constant electric field vector and thus also the governing equations (1) and (2) will be satisfied. However in case of continuously non-homogeneous FGM material imposed linear mechanical and electrical fields will satisfy condition of constant strain, but governing equations will not be satisfied and some external load would be required to maintain equilibrium. This is the reason why the modified patch test that does not require constant strain field is applied in the present paper. Nonlinear displacement and electric potential field are applied resulting in non-constant strains; however equilibrium is maintained for the analyzed plate.

Properties of functionally graded materials can be specified during their manufacturing process according the specific requirements of the design application. In the problem considered here the exponential variation of material properties in z direction is specified as

$$P(z) = P^{(0)} e^{\delta z/h}, \tag{32}$$

where P is graded material property, $P^{(0)}$ is material property at the bottom of the plate, h is the height of the plate and δ is a gradation constant. Change of the gradation constant is resulting to the change of graded material property along the thickness of the plate.

Considering the variation of material properties (27), the axial displacement u_z and electric potential ψ can be specified on the outer boundaries of the plate as

$$u_z = u_{z0} + A e^{-\delta z/h}, \qquad \psi = \psi_0 + B e^{-\delta z/h}, \tag{33}$$

where u_{z0}, ψ_0, A are arbitrarily chosen constants, while the constant B can be determined after inserting (33) into the governing equations with vanishing the radial displacement $u_r = 0$. Then the constant B is specified as

$$B = -\frac{c_{13}^{(0)}}{e_{31}^{(0)}} A. \tag{34}$$

Applying the relations (32)–(34) in the constitutive equation (10) for the radial stress gives

$$\sigma_{rr} = c_{11}^{(0)} e^{\delta z/h} 0 + c_{12}^{(0)} e^{\delta z/h} \frac{0}{r} + c_{13}^{(0)} e^{\delta z/h} \left(-\frac{A\delta}{h}\right) e^{-\delta z/h} + e_{31}^{(0)} e^{\delta z/h} \left(\frac{c_{13}^{(0)}}{e_{31}^{(0)}} \frac{A\delta}{h}\right) e^{-\delta z/h} =$$

$$= -c_{13}^{(0)} \left(\frac{A\delta}{h}\right) + c_{13}^{(0)} \left(\frac{A\delta}{h}\right) = 0. \tag{35}$$

In the same way we obtain the expressions of remaining constitutive equations (11)–(15) as

$$\sigma_{\theta\theta} = \sigma_{rz} = D_r = 0, \tag{36}$$

$$\sigma_{zz} = \left(-c_{33}^{(0)} + e_{33}^{(0)} \frac{c_{13}^{(0)}}{e_{31}^{(0)}}\right) \frac{A\delta}{h} = \text{const.}, \tag{37}$$

$$D_z = -\left(e_{33}^{(0)} + h_{33}^{(0)} \frac{c_{13}^{(0)}}{e_{31}^{(0)}}\right) \frac{A\delta}{h} = \text{const.} \tag{38}$$

Recall that in homogeneous media, the constant values of the secondary fields (stresses and electric displacements) and the constant values of the gradients of the primary fields (elastic displacements and the electric potential) are obeyed simultaneously. Now in FGM case, the use of the exponential variation for the prescribed axial displacement field and electric potential field (33) leads to exponential (non-constant) variation of axial strain and axial electric field vector

$$E_z = -\psi_{,z} = \frac{c_{13}^{(0)}}{e_{31}^{(0)}} \frac{A\delta}{h} e^{-\delta z/h}, \qquad \varepsilon_{zz} = u_{z,z} = -\frac{A\delta}{h} e^{-\delta z/h}. \tag{39}$$

This clearly denies the constant strain conditions. Nevertheless, the stresses as well as electric displacements are constant like in the case of patch test for homogeneous media. Moreover, the considered fields represent the exact solution of the governing equations and can be employed in the patch test involving FGM.

The patch test is suitable for testing new computational methods because it directly offers an "exact" benchmark solution that can be utilized for the error estimation and convergence analysis. Equivalently the expression (33) can be considered as the exact solution to the plate subjected to applied tension

$$\tilde{T}_z(z=0) = \left(c_{33}^{(0)} - e_{33}^{(0)} \frac{c_{13}^{(0)}}{e_{31}^{(0)}} \right) \frac{A\delta}{h}, \tag{40}$$

$$\tilde{T}_z(z=h) = \left(-c_{33}^{(0)} + e_{33}^{(0)} \frac{c_{13}^{(0)}}{e_{31}^{(0)}} \right) \frac{A\delta}{h}, \tag{41}$$

$$\tilde{T}_r(r,z) = 0 \tag{42}$$

and prescribed potential field on the outer boundary

$$\psi(r,z) = \psi_o - \frac{c_{13}^{(0)}}{e_{31}^{(0)}} A \mathrm{e}^{-\delta z/h}. \tag{43}$$

In the MLPG solution, the collocation approach can be used to impose essential boundary conditions directly using the MLS approximations (25)–(27) for primary field variables directly. In the present paper these approximations are applied to specify the expression (33) on the boundary nodes.

5. Numerical example

The patch test is performed on the circular plate with radius $r_0 = 0.3$ m and thickness $h = 0.03$ m. PZT-4 piezoelectric material is considered as bottom material with properties taken from [17]. The considered rectangular geometry is discretized with nodal points only and no restrictions are set for the mutual position of nodes. Vanishing radial displacement and exponentially varying axial displacement is prescribed on the boundary nodes according to (33). The input constants are chosen as $u_{z0} = 0.03$ m, $\psi_0 = 5 \times 10^7$ V, $A = 0.01$ together with grading exponent $\delta = 0.693\,1$. Using this exponent is leading to material properties increased two times at the top of the plate, since according to (32) one gets

$$P(z=h) = P^{(0)} \mathrm{e}^{0.693\,1\frac{h}{h}} = 2P^{(0)}. \tag{44}$$

Five nodal distributions were selected as $3 \times 3, 5 \times 5, 9 \times 5, 11 \times 7$ and 15×7 corresponding to total 9, 25, 45, 77 and 105 nodes, respectively. For each nodal distribution the radius of support domain is chosen to cover all nodal points in order to avoid the sensitivity of the model to this parameter. The radius of support domain [2] is an important parameter for MLS approximation. It determines the number of nodes n that are used for approximation of unknown quantities in (25)–(28).

Numerical results for the axial displacement field $u_z(r,z)$ and electric potential field $\psi(r,z)$ are shown in Fig. 2 for the case with 11×7 nodal distribution. According to (35)–(37), the resulting radial stress and shear stress must vanish while axial stress should be constant. Results for these stresses evaluated at nodes located in the middle of the plate shown in Fig. 3 confirm this statement.

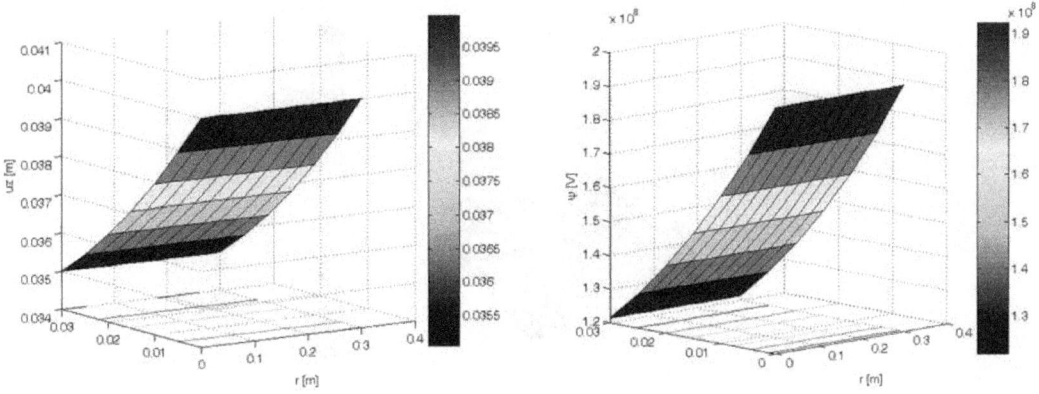

Fig. 2. Resulting axial displacement (left) and electric potential distribution (right)

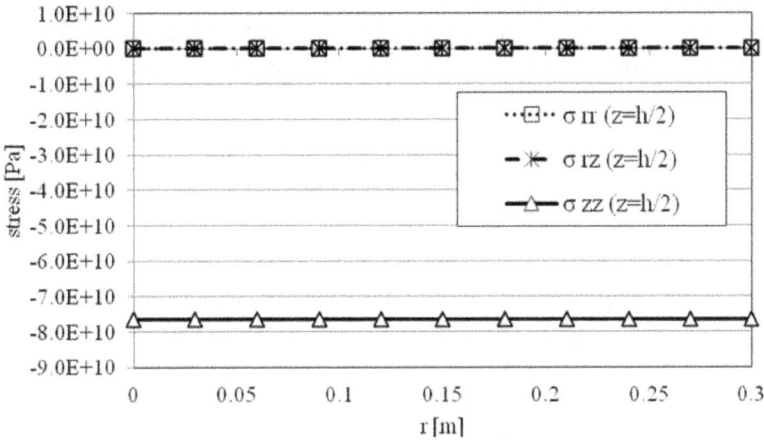

Fig. 3. Stress distribution taken at the middle of the plate ($z = h/2$)

Using the exact benchmark solution, one can compute relative errors of the numerical analysis. Relative errors are computed for axial displacements. The relative error for every node is specified in percentage as

$$e = 100 \frac{|u_z^{num} - u_z^{exact}|}{|u_z^{exact}|} \ [\%], \tag{45}$$

where u_z^{num} represents computed value of axial displacement and u_z^{exact} represents "exact" solution as obtained by straightforward use of (33). Fig. 4 shows the distribution of relative error for 11×7 nodal distribution.

As can be seen in Fig. 4, relative error is much higher for the interior nodes compared to nodes on the boundary. The axial displacement on the boundary nodes is specified by the MLS approximation of (33) while for the interior nodes the results are obtained from the discretized LIE (30). The numerical integration of domain integrals in LIEs is probably causing the errors. Spline-type weight functions used in MLS approximation may generate complicated shape over the local subdomain Ω_s, which may be difficult to integrate. Breitkopf et al. [5] proposed custom integration scheme to integrate MLS shape functions correctly. However in the present paper, the standard Gauss integration scheme is chosen, since the error is in the acceptable level.

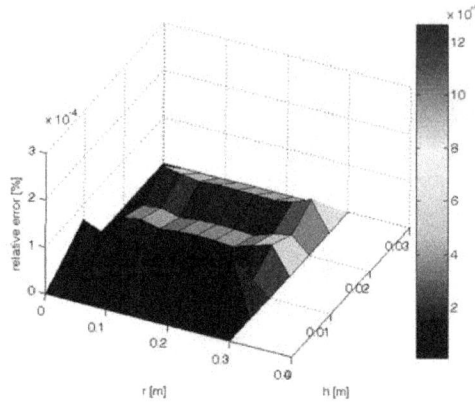

Fig. 4. Relative nodal error distribution over discretized plate

It may be convenient to specify other type of error estimation that would tackle the error of the numerical solution from the global viewpoint. Averaged percentage error [15] was introduced for this reason as

$$AE = 100 \frac{\|u_z^{num} - u_z^{exact}\|}{\|u_z^{exact}\|} \ [\%],\tag{46}$$

where the norm is given as $\|*\| = \left(\sum_{a=1}^{N_t} [* (\mathbf{x}^a)]^2 \right)^{\frac{1}{2}}$ and N_t is the total number of nodes used in the analysis. Using this error estimate the convergence of the method for varying nodal densities can be determined as shown in Fig. 5. Both the axial displacement and the electric potential convergence are analyzed. Results show that the axial displacement is reaching better accuracy than the electric potential; however the convergence rate for both quantities is almost the same. The convergence curve for the axial displacement also indicates the increase of accuracy with increasing the number of nodes in z-axis direction, since the material gradation is in the same direction.

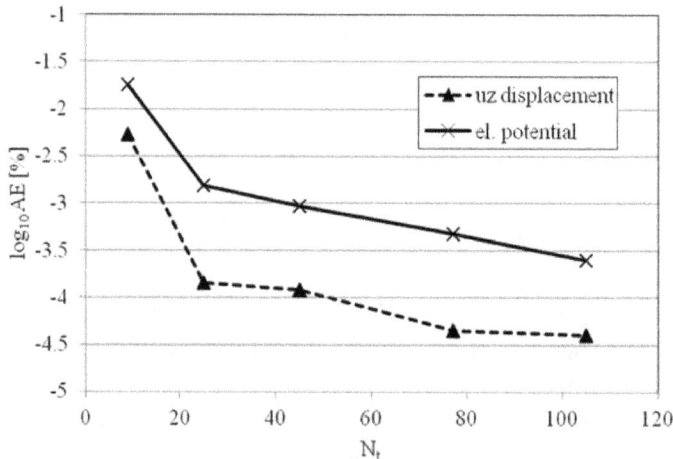

Fig. 5. Averaged relative error distribution and convergence rate

6. Conclusion

The patch test is an important tool used in development of new numerical computational techniques. In the present paper the patch test is used to examine convergence of meshless solution for functionally graded piezoelectric circular plate. The employed meshless MLPG method is a truly meshless method since no elements are required neither for approximation nor for integration of unknowns. The MLS approximation is applied to approximation of unknown elastic as well as electric quantities.

The patch test analysis of material with constant material properties requires constant strain field (produced by applied linear displacement field) to maintain the equilibrium, this is however not the case for continuously non-homogeneous material. Non-linear distribution of displacement and electrical fields were derived as the exact solution of the governing equations involving the exponential material gradation rule. The numerical example showed that the developed formulation passed the patch test with positive convergence for increased number of nodes.

The patch test was presented as a simple tool for verification of new computer codes since the data are easy to prepare and the exact benchmark results are instantly applicable for determination of accuracy and convergence of numerical results.

Acknowledgements

The present paper has been supported by the Slovak Science and Technology Assistance Agency through the grant registered under number APVV-0014-10. The support is gratefully acknowledged.

References

[1] Adachi, A., Kitamura, Y., Iwatsubo, T., Integrated design of piezoelectric damping system for flexible structure, Applied Acoustics 65 (2004) 293–310.

[2] Atluri, S. N., The Meshless Method (MLPG) For Domain & BIE Discretizations, Tech Science Press, Forsyth, GA, 2004.

[3] Atluri, S. N., Zhu, T., A new meshless local Petrov-Galerkin (MLPG) approach in computational mechanics, Computational Mechanics 22 (1998) 117–127.

[4] Belytschko, T., Liu, W. K., Moran, B., Nonlinear Finite Elements for Continua and Structures, John Willey & Sons, Chichester, England, 2000.

[5] Breitkopf, P., Rassineux, A.,Villon, P., Custom integration scheme for patch test in MLS meshfree methods, Computational Fluid and Solid Mechanics 2003 (K. J. Bathe (Ed.)), Elsevier Ltd, 2003, pp. 1 876–1 879.

[6] Lancaster, P., Salkauskas, T., Surfaces generated by moving least square methods, Mathematics of Computation 37 (1981) 141–158.

[7] Liew, K. M., Lim, H. K., Tan, M. J., He, X. Q., Analysis of composite beams and plates with piezoelectric patches using the element-free Galerkin method, Computational Mechanics 29 (2002) 486–497.

[8] Irons, B. M., Razzaque, A., Experience with the patch test for convergence of finite element method. In: The Mathematics of Finite Elements with Application to Partial Differential Equations (Aziz, A. R. (Ed.)), Academic Pres USA, 1972, p. 557–587.

[9] Semedo Garção, J. E., Mota Soares, C. M., Mota Soares, C. A., Reddy, J. N., Analysis of laminated adaptive plate structures using layerwise finite elements, Computers and Structures 82 (2004) 1 939–1 959.

[10] Sladek, J., Sladek, V., Solek, P., Saez, A., Dynamic 3D axisymmetric problems in continuously nonhomogeneous piezoelectric solids, International Journal of Solids and Structures 45 (2008) 4 523–4 542.

[11] Sladek, J., Sladek, V., Solek, P., Wen, P. H., Thermal Bending of Reissner-Mindlin Plates by the MLPG, CMES – Computer Modeling in Engineering & Sciences 28 (2008) 57–76.

[12] Sladek, J., Sladek, V., Stanak, P., Zhang, Ch., Meshless Local Petrov-Galerkin (MLPG) Method for Laminate Plates under Dynamic Loading, CMC – Computers, Materials & Continua 15 (2010) 1–26.

[13] Sladek, J., Sladek, V., Stanak, P., Pan, E., The MLPG for bending of electroelastic plates, CMES – Computer Modeling in Engineering & Sciences 64 (2010) 267–298.

[14] Sladek, J., Sladek, V., Stanak, P., Zhang, Ch., Wunshe, M., Bending of circular piezoelectric plates with functionally graded material properties. Proceedings Third International Symposium on Computational Mechanics in conjunction with Second Symp. on Comp. Structural Enginering. ((Ed) Yeong-Bin Yang, Liang-Jeng Leu, Chuin-Shan David Chen), Taipei, National Taiwan University Press, 2011, pp. 42–43.

[15] Sladek, V., Sladek, J., Tanaka, M., Local Integral Equations and two Meshless Polynomial Interpolations with Application to Potential Problems in Non-homogeneous Media, CMES – Computer Modeling in Engineering & Sciences 7 (2005) 69–83.

[16] Stanak, P., Sladek, J., Sladek, V., Krahulec, S., Bending of functionally graded circular plates with piezoelectric layer by the MLPG method. Proceedings of 18th International conference Engineering Mechanics 2012 (J. Naprstek, C. Fisher (Eds.)), Svratka Czech Republic. ITAM CAS Prague, 2012.

[17] Stanak, P., Sladek, J., Sladek, V., Krahulec, S., Composite circular plate analyzed as a 3-D axisymmetric piezoelectric solid, Building Research Journal 59, (2011) 125–140.

[18] Suresh, S., Mortensen, A., Fundamentals of Functionally Graded Materials, Institute of Materials, London, 1998.

[19] Tiersten, H. F., Linear piezoelectric plate vibrations, Plenum Press, New York, 1969.

Modeling of damage evaluation in thin composite plate loaded by pressure loading

M. Dudinský^{a,*}, M. Žmindák^{a}, P. Frnka^{a}

^{a}*Faculty of Mechanical Engineering, University of Žilina, Univerzitná 1, 010 26 Žilina, Slovak Republic*

Abstract

This article presents the results of numerical analysis of elastic damage of thin laminated long fiber-reinforced composite plate consisting of unidirectional layers which is loaded by uniformly distributed pressure. The analysis has been performed by means of the finite element method (FEM). The numerical implementation uses layered plate finite elements based on the Kirchhoff plate theory. System of nonlinear equations has been solved by means of the Newton-Raphson procedure. Evolution of damage has been solved using the return-mapping algorithm based on the continuum damage mechanics (CDM). The analysis was performed using own program created in MATLAB. Problem of laminated fiber-reinforced composite plate fixed on edges for two different materials and three different laminate stacking sequences (LSS) was simulated. Evolution of stresses vs. strains and also evolution of damage variables in critical points of the structure are shown.

Keywords: damage, finite element method, continuum damage mechanics, composite plate

1. Introduction

Composite materials are now common engineering materials used in a wide range of applications. They play an important role in the aviation, aerospace and automotive industry, and are also used in the construction of ships, submarines, nuclear and chemical facilities, etc. The meaning of the word damage is quite broad in everyday life. In continuum mechanics the term damage is referred to as the reduction of the internal integrity of material due to generating, spreading and merging of small cracks, cavities and similar defects. In the initial stages of the deformation process the defects (microcracks, microcavities) are very small and relatively uniformly distributed in the microstructure of a material. If the damage reaches the critical level (depends on type of loading and used material), subsequent growth of defects will concentrate in some of the defects already present in material [7]. Damage is called elastic, if the material deforms only elastically (in macroscopic level) before the occurrence of damage, as well as during its evolution. This damage model can be used if the ability of the material to deform plastically is low. Fiber-reinforced polymer matrix composites can be considered as such materials [11]. Commercial FEM software can perform analyses with many types of material nonlinearities, such as plasticity, hyperelasticity, viscoplasticity, etc. However, almost no commercial software (except for ABAQUS) contains a module for damage analysis of composite materials.

The goal of this paper is to present the numerical results of elastic damage analysis of thin laminated composite plate consisting of unidirectional long fiber-reinforced layers which is

---

*Corresponding author. e-mail: martin.dudinsky@fstroj.uniza.sk.

loaded by uniformly distributed pressure. The analysis was performed by own software created in MATLAB programming language. This software can perform numerical analysis of elastic damage based on continuum damage mechanics utilizing finite element method using layered plate finite elements based on the Kirchhoff plate theory. Locking effect was not removed, since this is a rather complicated issue.

2. Theoretical and numerical modeling background

A number of material modeling strategies exist to predict failure in laminated composites, subjected to static or impulsive loads. Broadly, they can be classified as [12, 15, 19]:

- strength-based failure criteria,

- fracture mechanics approach (based on energy release rates),

- plasticity or yield surface approach,

- damage mechanics approach.

Strength-based failure criteria (failure criteria approach) are commonly used with FEM to predict failure events in composite structures. These approaches are based on the equivalent stresses or strains in the critical failure areas. Numerous criteria have been derived to relate internal stresses and experimental measures of material strength to the onset of failure (maximum stress or strain, Hill, Hoffman, Tsai-Wu, etc.). These classical criteria implemented in most commercial FE codes are not able to physically capture the failure mode. Some of them cannot deal with materials having a different strength in tension and compression. The Hashin criteria are briefly reviewed in [11] and improvements were proposed by Puck and Schurmann [14] over Hashin's theories are examined.

However, few criteria can represent several relevant aspects of the failure process of laminated composites, e.g. the increase on apparent shear strength when applying moderate values of transverse compression, or detrimental effect of the in-plane shear stresses in failure by fiber kinking.

2.1. Continuum damage mechanics

From a physical point of view, damage represents surface discontinuities in form of microcracks or volume discontinuities in form of cavities in a material. They are formed as the material undergoes an increasing loading. The objective of the damage mechanics is to predict, through mechanical variables, the response of a material in the presence of damage. Damage is initiated at certain stress level and it generally increases with increasing stress from the virgin state up to a macroscopic crack initiation or failure.

Continuum Damage Mechanics (CDM) considers damaged materials as a continuum, in spite of heterogenity, micro-cavities, and micro-defects. The response to the loading conditions is determined on the basis of the constitutive relations between macroscopic variables (e.g. stress, strain) and internal variables which model, on a macroscopic scale, the irreversible changes occurring at the microscopic level.

We consider a volume of material free of damage if no cracks or cavities can be observed at the microscopic scale. The opposite state is the fracture of the volume element. Theory of damage describes the phenomena between the virgin state of material and the macroscopic

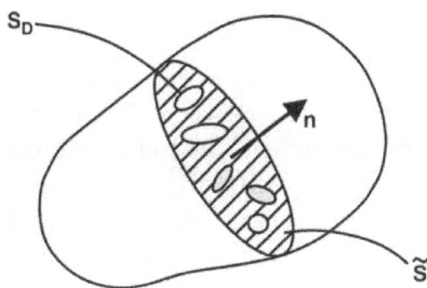

Fig. 1. Representative volume element for damage mechanics

onset of crack [6, 16]. The representative volume element must be of sufficiently large size compared to the inhomogenities of the composite material. In Fig. 1 this volume is depicted. One section of this element is related to its normal and to its area S. Due to the presence of defects, an effective area $\tilde{S}$ for resistance of load can be found. Total area of defects is therefore

$$S_D = S - \tilde{S}. \tag{1}$$

The local damage related to the direction n is defined as:

$$D = \frac{S_D}{S}. \tag{2}$$

For isotropic damage, the dependence on the normal n can be neglected, i.e.

$$D = D_n \ \forall n. \tag{3}$$

We note that damage D is a scalar assuming values between 0 and 1. For $D = 0$ a material is undamaged, for $0 < D < 1$ a material is damaged, for $D = 1$ complete failure occurs. The quantitative evaluation of damage is not a trivial issue, it must be linked to a variable that is able to characterize the phenomenon. Several papers can be found in literature where the constitutive equations of the materials are a function of a scalar variable of damage [2, 3]. For the formulation of a general multidimensional damage model it is necessary to generalize the scalar damage variables. It is therefore necessary to define corresponding tensorial damage variables that can be used for general states of deformation and damage [18].

2.2. Numerical modeling

One of the most powerful computational methods for structural analysis of composites is the FEM. The starting point should be a "validated" FE model, with a reasonably fine mesh, correct boundary conditions, material properties, etc. [1]. As a minimum requirement, the model is expected to produce stress and strains that have reasonable accuracy to those of the real structure prior to failure initiation. In spite of a great success of the finite and boundary element methods as effective numerical tools for the solution of boundary-value problems on complex domains, there is still growing interest in the development of new advanced methods. Many meshless formulations are becoming popular due to their high adaptivity and a low cost to prepare input data for numerical analysis [4, 5, 13].

2.3. FEM formulation for Kirchhoff plate

Plate models are used to study structural components which are subjected to bending loads and their thickness is smaller than the others dimensions. This characteristic allows representing the plate using the reference middle surface. Therefore the geometric domain used for the formulation of plate models is the middle surface.

A plate resists transverse loads by means of bending, exclusively. The flexural properties of a plate depend greatly upon its thickness in comparison with other dimensions. Plates may be classified into three groups according to the ratio a/t, where a is a typical dimension of a plate in a plane and t is a plate thickness. The first group is represented by thick plates having ratios $a/t \leq 8 \ldots 10$. The second group refers to plates with ratios $a/t \geq 80 \ldots 100$. These plates are referred to as membranes. The most extensive group represents an intermediate type of plates, so-called thin plates with $8 \ldots 10 \leq a/t \leq 80 \ldots 100$ [17].

One of the most widely used theory for thin plates is the Kirchhoff (classical) plate theory. The Kirchhoff (classical) laminate plate theory and the first-order shear deformation theories describe with reasonable accuracy the kinematics of most laminates [19]. The details of derivation of equations governing the behavior of thin plates are given in [17]. The equations are represented here for clarity.

In this subsection formulation for plate element based on the Kirchhoff plate theory for symmetric balanced laminate will be presented. The most widely used plate elements in FEM are linear and quadratic elements with 3 degrees of freedom (DOFs) in node: w, θ_x, θ_y. When using linear four-node elements, one element has 12 DOFs and 12 shape functions are required.

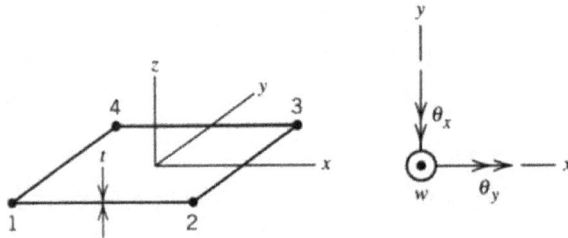

Fig. 2. Four-node Kirchhoff plate element and DOFs in node

It is worth noting that shape functions must have C^1 continuity. Displacements within the element are interpolated as

$$w = \mathbf{N}\,\hat{\mathbf{u}} \tag{4}$$

where w is displacement in given point of the element, $\mathbf{N} = [N_1, N_2, \ldots, N_{3xn}]$ is vector of values of shape functions in this point, n is number of nodes in element and
$\hat{\mathbf{u}} = \left[\hat{w}_1, \hat{\theta}_{x1}, \hat{\theta}_{y1}, \ldots, \hat{w}_n, \hat{\theta}_{xn}, \hat{\theta}_{yn}\right]^T$ is vector of nodal displacements.

Matrix $\mathbf{B}$, which in the case of plate elements gives the relation between curvatures and nodal displacements, has the form of

$$\mathbf{B} = \begin{bmatrix} \dfrac{\partial^2 N_1}{\partial x^2} & \dfrac{\partial^2 N_2}{\partial x^2} & & \dfrac{\partial^2 N_n}{\partial x^2} \\[2mm] \dfrac{\partial^2 N_1}{\partial y^2} & \dfrac{\partial^2 N_2}{\partial y^2} & \cdots & \dfrac{\partial^2 N_n}{\partial y^2} \\[2mm] 2\dfrac{\partial^2 N_1}{\partial x \partial y} & 2\dfrac{\partial^2 N_2}{\partial x \partial y} & & 2\dfrac{\partial^2 N_n}{\partial x \partial y} \end{bmatrix}. \tag{5}$$

The element stiffness matrix for unidirectional element has the form of

$$\mathbf{k} = \int_A \mathbf{B}^T \mathbf{D_K} \mathbf{B} \, dA. \tag{6}$$

Matrix $\mathbf{D_K}$ gives the relation between internal moments and curvatures. More details about this matrix are given e.g. in [9]. The element stiffness matrix is integrated numerically, most often by means of Gauss quadrature [10]. The element stiffness matrix calculation for layered rectangular element with edges parallel to x and y axis by means of Gauss quadrature is performed as follows

$$\mathbf{k} = \sum_{n=1}^{NL} \int_{x_1}^{x_2} \int_{y_1}^{y_4} \mathbf{B}^T \mathbf{D_K} \mathbf{B} \, dy \, dx \approx$$

$$\approx \sum_{n=1}^{NL} \sum_{i=1}^{n_{Gx}} \frac{x_2 - x_1}{2} \sum_{j=1}^{n_{Gy}} \frac{y_4 - y_1}{2} \mathbf{B}^T(x_{int_i}, y_{int_j}) \mathbf{D_K} \mathbf{B}(x_{int_i}, y_{int_j}) W_i W_j, \tag{7}$$

$$x_{int_i} = \frac{x_2 - x_1}{2} x_{Gi} + \frac{x_2 + x_1}{2}, \tag{8}$$

$$y_{int_j} = \frac{y_4 - y_1}{2} y_{Gj} + \frac{y_4 + y_1}{2}, \tag{9}$$

where NL is number of layers, x_1, x_2, y_1, y_4 are x and y coordinates of nodes, which are in subscript, x_{Gi}, y_{Gj} are Gauss points, W_i, W_j are corresponding weights and n_{Gx} and n_{Gy} is number of Gauss points in x- and y-axis direction.

3. Damage model used

The model for fiber-reinforced lamina mentioned next was presented by Barbero and de Vivo [2] and is suitable for fiber-reinforced composite materials with polymer matrix. On the lamina level these composites are considered as ideal homogenous and transversely isotropic. All parameters of this model can be easily identified from available experimental data. It is assumed that damage in principal directions is identical with the principal material directions (1, 2, 3) throughout the damage process. This is due to the fact that the dominant modes of damage are micro-cracks, fiber breaks and fiber-matrix debond, all of which can be conceptualized as cracks parallel or perpendicular to the fiber direction [3]. Therefore the evolution of damage is solved in the lamina coordinate system. The model predicts the evolution of damage and its effect on stiffness and subsequent redistribution of stress.

3.1. Damage surface and damage potential

Damage surface is defined by tensors $\mathbf{J}$ and $\mathbf{H}$ [3]

$$J = \begin{bmatrix} J_{11} & 0 & 0 \\ 0 & J_{22} & 0 \\ 0 & 0 & J_{33} \end{bmatrix}, \qquad H = [H_1, H_2, H_3]. \tag{10}$$

Damage surface is similar to the Tsai-Wu damage surface [6], and it is commonly used for predicting failure of fiber-reinforced lamina with respect to experimental material strength values. Damage surface and damage potential have the form of [3]

$$g(\mathbf{Y}, \gamma) = \sqrt{J_{11} Y_1^2 + J_{22} Y_2^2 + J_{33} Y_3^2} + \sqrt{H_1 Y_1^2 + H_2 Y_2^2 + H_3 Y_3^2} - (\gamma + \gamma_0), \quad (11)$$

$$f(\mathbf{Y}, \gamma) = \sqrt{J_{11} Y_1^2 + J_{22} Y_2^2 + J_{33} Y_3^2} - (\gamma + \gamma_0), \quad (12)$$

where the thermodynamic forces Y_1, Y_2 and Y_3 can be calculated by means of relations

$$
\begin{aligned}
Y_1 &= \frac{1}{\Omega_1^2} \left(\frac{\bar{S}_{11}}{\Omega_1^4} \sigma_1^2 + \frac{\bar{S}_{12}}{\Omega_1^2 \Omega_2^2} \sigma_1 \sigma_2 + \frac{\bar{S}_{66}}{\Omega_1^2 \Omega_2^2} \sigma_6^2 \right), \\
Y_2 &= \frac{1}{\Omega_2^2} \left(\frac{\bar{S}_{22}}{\Omega_2^4} \sigma_2^2 + \frac{\bar{S}_{12}}{\Omega_1^2 \Omega_2^2} \sigma_1 \sigma_2 + \frac{\bar{S}_{66}}{\Omega_1^2 \Omega_2^2} \sigma_6^2 \right), \\
Y_3 &= 0.
\end{aligned}
\quad (13)
$$

where stresses and components of matrix $\bar{S}$ are defined in the lamina coordinate system. Matrix $\bar{S}$ gives the strain-stress relations in the effective configuration [2]. Ω_1 and Ω_2 are components of a second-order tensor $\boldsymbol{\Omega} = \sqrt{\mathbf{I} - \mathbf{D}}$, called the integrity tensor. The eigenvalues D_i of damage tensor $\mathbf{D}$ describe the load-carrying area reduction on the three planes orthogonal to the principal direction of the tensor $\mathbf{D}$. Equations (11) and (12) can be written for particular simple stress states: tension and compression in fiber direction, tension in transverse direction, in-plane shear. Tensors $\mathbf{J}$ and $\mathbf{H}$ can be derived in terms of material strength values.

3.2. Hardening parameters

In the present damage model isotropic hardening is considered and hardening function was used in the form of

$$\gamma = c_1 \left[\exp \left(\frac{\delta}{c_2} \right) + 1 \right]. \quad (14)$$

where δ denotes the hardening variable. The hardening parameters γ_0, c_1 and c_2 are determined by approximating the experimental stress-strain curves for in-plane shear loading. If this curve is not available, we can reconstruct it using the function

$$\sigma_6 = F_6 \tanh \left(\frac{G_{12}}{F_6} \gamma_6 \right), \quad (15)$$

where F_6 is the in-plane shear strength, G_{12} is the in-plane initial (undamaged) shear modulus and γ_6 is the in-plane shear strain (in the lamina coordinate system). This function represents experimental data very well.

3.3. Critical damage level

Reaching of critical damage level is dependent on stress values in lamina. If in a point in lamina only normal stresses in the fiber direction and across the fibers (i.e. normal stresses in lamina coordinate system) occur, then simply comparing the values of damage variables with critical values of damage variables for given material at this point is sufficient. The damage has reached the critical level if at least one of the values of D_1, D_2 in the point of lamina is greater or equal to its critical value. The magnitude of these critical damage parameter values can be estimated from statistical models of the failure process of each type of loading. If in given point of lamina also shear stress occurs (in lamina coordinate system), it is additionally necessary to compare the value of the product of $(1 - D_1)(1 - D_2)$ with value of k_s parameter from Table 3 for given material. If the value of this product is less or equal to k_s, the damage has reached the critical level. Value of k_s is determined from the relation between damaged in-plane shear modulus G_{12}^* and undamaged in-plane shear modulus G_{12}

$$k_s = \frac{G_{12}^*}{G_{12}}. \tag{16}$$

3.4. Implementation of numerical method

The Newton-Raphson method was used for solving the system of nonlinear equations. Evolution of damage has been solved using the return-mapping algorithm described in [2]. The input values are strains and strain increments in lamina coordinate system, state variables D_1, D_2, and δ in integration point from the start of the last performed iteration, $\bar{\mathbf{C}}$ matrix (gives the stress-strain relations in the effective configuration [3]) and damage parameters related to damage model. The output variables are D_1, D_2, and δ, stresses and strains in lamina coordinate system in this integration point at the end of the last performed iteration. Another output is damaged tangent constitutive matrix $\mathbf{C}^{ed}$ in lamina coordinate system, which reflects the effect of damage on the behavior of structure. Flowchart of the return-mapping algorithm used in numerical damage analysis is described in Fig. 3.

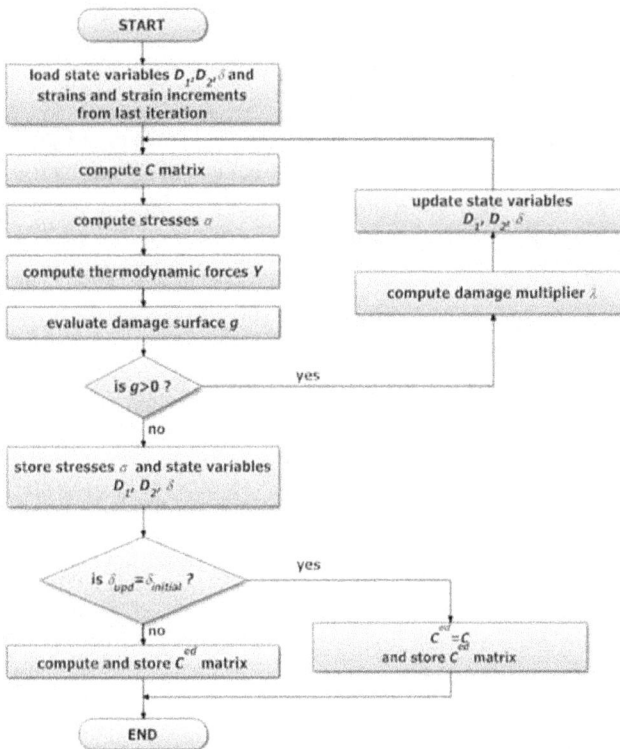

Fig. 3. Flowchart of the return-mapping algorithm used in numerical damage analysis of thin composite plates

4. Numerical example and results

One problem for two different materials and three different laminate stacking sequences (LSS) was simulated in order to study damage of laminated long fiber-reinforced composite plates consisting of unidirectional layers. Composites consist of carbon fibers embedded in epoxy matrix. Composite plate fixed on its edges with dimensions of $125 \times 125 \times 2.5$ mm and LSS of $[0, 45, -45, 90]_S$, $[0, 90, 45, -45]_S$ and $[45, 0, -45, 90]_S$ was loaded by uniformly distributed pressure $p = 0.5$ MPa perpendicular to the surface of the plate (Fig. 4). Own program created in MATLAB language was used for this analyses.

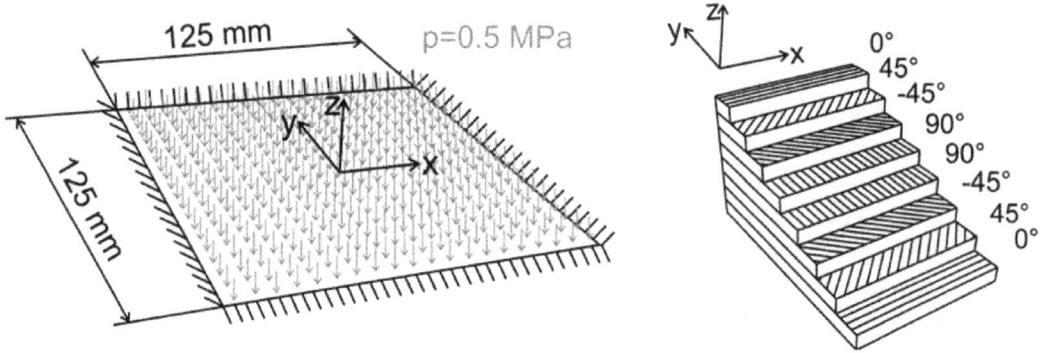

Fig. 4. Force and displacement boundary condition of the analyzed plate and schematic illustration of the LSS $[0, 45, -45, 90]_S$

Material properties, damage parameters, hardening parameters and critical values of damage parameters [2] are given in Tables 1–3. Subscripts t and c in Table 3 denote critical damage parameter values for tensile and compressive loading, respectively. Critical value for damage parameter D_2 is listed only for tensile loading because it is difficult to find accurate model for estimating the critical value of this parameter for transverse compressive loading. In this model it is assumed that critical value of this parameter for compressive loading is equal to critical value for tensile loading. Parameters J_{33} and H_3 are equal to zero. The plate model was divided into 20×20 elements and was analyzed in 50 load substeps.

Table 1. Material properties

	E_1 [GPa]	E_2 [GPa]	G_{12} [GPa]	ν_{12}
M30/949	167	8.13	4.41	0.27
M40/948	228	7.99	4.97	0.292

Table 2. Damage and hardening parameters

	J_1	J_2	H_1	H_2	γ_0	c_1	c_2
M30/949	$0.952 \cdot 10^{-3}$	0.438	$25.585 \cdot 10^{-3}$	$-21.665 \cdot 10^{-3}$	-0.6	0.30	-0.395
M40/948	$2.208 \cdot 10^{-3}$	0.214	$10.503 \cdot 10^{-3}$	$-8.130 \cdot 10^{-3}$	-0.12	0.10	-0.395

Table 3. Critical values of damage variables

	D_{1t}^{cr}	D_{1c}^{cr}	$D_{2t}^{cr} = D_{2c}^{cr}$	k_s
M30/949	0.105	0.111	0.5	0.944
M40/948	0.105	0.111	0.5	0.908

Linear static analysis of the plate with LSS of $[0, 45, -45, 90]_S$ has shown that maximum magnitudes of stresses in fiber direction and direction transverse to fibers as well as equivalent (von Mises) stress occur in the outer layers in the middle of two opposite edges of the plate and maximum magnitudes of shear stress in lamina coordinate system occur in layers 2 (2nd from the bottom) and 7. However, the results of damage analysis have shown that critical damage level will not be reached in the outer layers at first, but in layer 2 (2nd layer from the bottom) and layer 7 for both materials.

For plate from material M30/949 critical damage level has been reached between 17th and 18th load substep in several locations in layer 2 and layer 7. Critical loading for plate from this material (macrocrack will be present in the plate) is $p = 0.175$ MPa. For material M40/948 critical damage level has been reached between 46th and 47th load substeps in several locations in layer 2 and layer 7. Critical loading is $p = 0.465$ MPa. In Figs. 5–6 evolution of stress vs. strain in lamina (local) coordinate system in layer 7 in integration point where critical damage level was reached at first for LSS $[0, 45, -45, 90]_S$ for both materials are shown. In Figs. 7–8 evolution of damage variables in the same point are shown.

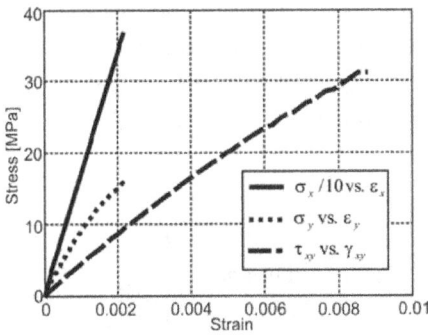

Fig. 5. Stress vs. strain evolution in lamina coordinate system in layer 7 in integration point where critical damage level was reached at first for material M30/949 and LSS $[0, 45, -45, 90]_S$

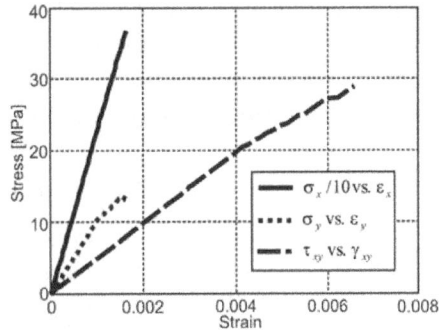

Fig. 6. Stress vs. strain evolution in lamina coordinate system in layer 7 in integration point where critical damage level was reached at first for material M40/948 and LSS $[0, 45, -45, 90]_S$

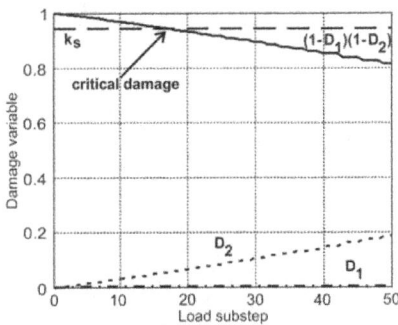

Fig. 7. Evolution of damage variables in layer 7 in integration point where critical damage level was reached at first for material M30/949 and LSS $[0, 45, -45, 90]_S$

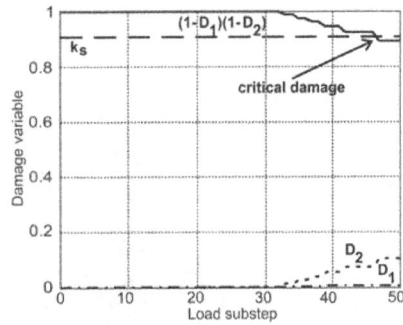

Fig. 8. Evolution of damage variables in layer 7 in integration point where critical damage level was reached at first for material M40/948 and LSS $[0, 45, -45, 90]_S$

Linear static analysis of the plate with LSS of $[0, 90, 45, -45]_S$ has shown that maximum magnitudes of stresses in fiber direction and direction transverse to fibers as well as shear stress in lamina coordinate system and equivalent (von Mises) stress occur in the outer layers. Results of damage analysis have shown that critical damage level for plate from material M30/949 will be reached in these layers. Critical damage level has been reached between 29th and 30th load substep. Critical loading for plate from this material is $p = 0.299$ MPa. For plate from material M40/948 critical damage level has not been reached.

On the other hand linear static analysis of the plate with LSS of $[45, 0, -45, 90]_S$ has shown that maximum magnitudes of stresses in fiber direction and direction transverse to fibers as well as equivalent (von Mises) stress do not occur in the outer layers, but in layers 2 and 7. Maximum magnitudes of shear stress in lamina coordinate system occur in the outer layers. Critical damage level in plate with this LSS will be reached in the outer layers at first for both materials. Critical damage level has been reached between 13th and 14th load substep in plate from material M30/949 and between 34th and 35th load substep in plate from material M40/948. Critical loadings are $p = 0.137$ MPa and $p = 0.346$ MPa.

Overall results of the damage analyses relating to the critical damage level for all LSSs and both materials are listed in Table 4. In Figs. 9–14 distribution of value of the product $(1 - D_1)(1 - D_2)$, which is required for assessing the critical damage level, in layers 1 and 2 for plate from material M30/949 with LSS of $[45, 0, -45, 90]_S$ after applying 15, 30 and 50 load substeps, which corresponds to loadings 0.15 MPa, 0.30 MPa and 0.50 MPa.

Table 4. Overall results of the damage analyses relating to critical damage level

LSS	material	layers in which the critical damage level was reached at first	layers with critical damage level after applying full loading	critical loading
$[0, 45, -45, 90]_S$	M30/949	2, 7	1, 2, 3, 6, 7, 8	0.175 MPa
	M40/948	2, 7	2, 7	0.465 MPa
$[0, 90, 45, -45]_S$	M30/949	1, 8	1, 2, 3, 6, 7, 8	0.299 MPa
	M40/948	–	–	–
$[45, 0, -45, 90]_S$	M30/949	1, 8	1, 2, 3, 6, 7, 8	0.137 MPa
	M40/948	1, 8	1, 8	0.346 MPa

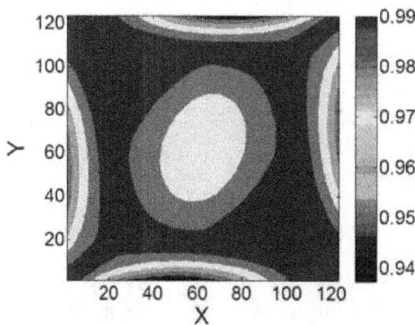

Fig. 9. Distribution of the value of $(1 - D_1)(1 - D_2)$ in layer 1, material M30/949, LSS $[45, 0, -45, 90]_S$ after load substep 15 ($p = 0.15$ MPa)

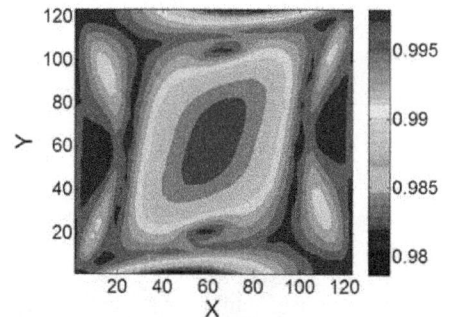

Fig. 10. Distribution of the value of $(1 - D_1)(1 - D_2)$ in layer 2, material M30/949, LSS $[45, 0, -45, 90]_S$ after load substep 15 ($p = 0.15$ MPa)

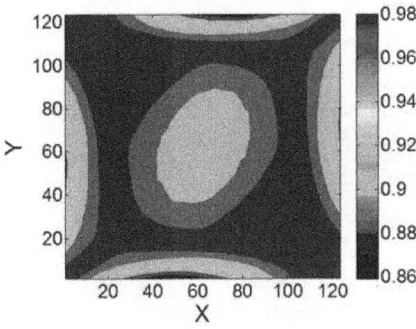

Fig. 11. Distribution of the value of $(1 - D_1)(1 - D_2)$ in layer 1, material M30/949, LSS $[45, 0, -45, 90]_S$ after load substep 30 ($p = 0.30$ MPa)

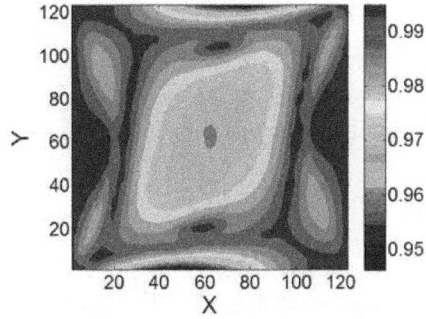

Fig. 12. Distribution of the value of $(1 - D_1)(1 - D_2)$ in layer 2, material M30/949, LSS $[45, 0, -45, 90]_S$ after load substep 30 ($p = 0.30$ MPa)

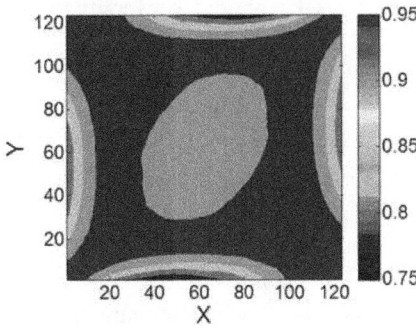

Fig. 13. Distribution of the value of $(1 - D_1)(1 - D_2)$ in layer 1, material M30/949, LSS $[45, 0, -45, 90]_S$ after load substep 50 ($p = 0.50$ MPa)

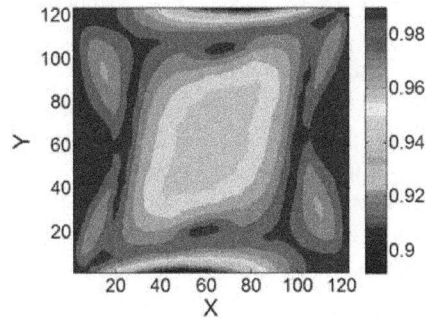

Fig. 14. Distribution of the value of $(1 - D_1)(1 - D_2)$ in layer 2, material M30/949, LSS $[45, 0, -45, 90]_S$ after load substep 50 ($p = 0.50$ MPa)

5. Conclusion

In the current study, we have focused on solving elastic damage analysis of thin laminated long fiber-reinforced composite plate consisting of unidirectional layers which is fixed on its edges and loaded by uniformly distributed pressure for different materials and different LSSs. The postulated damage surface reduces to the Tsai-Wu surface in stress space. However, presented model goes far beyond simple failure criteria by identifying a damage threshold, hardening parameters for the evolution of damage, and critical values of damage for which material failure occurs. The analysis results show that change of material, change of laminate stacking sequence as well as presence of shear stress have significant influence on the evolution of damage as well as on location of critical damage and load at which critical level of damage will be reached. Critical damage level has not necessary to be reached in places with maximum magnitude of equivalent stress, but can be reached in other places.

Acknowledgements

The authors gratefully acknowledge the support by the Slovak Grant Agency VEGA 09-015-00 and VEGA 1/1226/12

References

[1] Bathe, K. J., Finite Element Procedures, Prentice Hall, New Jersey, 1996.

[2] Barbero, E. J., de Vivo, L., A Constitutive Model for Elastic Damage in Fiber-Reinforced PMC Laminae. International Journal of Damage Mechanics 10 (1) (2001) 73–93.

[3] Barbero, E. J., Finite Element Analysis of Composite Materials, CRC Press, Boca Raton, 2007.

[4] Chen, Y., Lee, J. D., Eskandarian, A., Meshless Methods in Solid Mechanics, Springer, New York, 2006.

[5] Guiamatsia, I., Falzon, B. G., Davies, G. A. O., Iannucci, L., Element-free Galerkin modelling of composite damage, Composites Science and Technology 69 (2009) 2 640–2 648.

[6] Jain, J. R., Ghosh, S., Damage Evolution in Composites with a Homogenization-based Continuum Damage Mechanics Model, International Journal of Damage Mechanics 18 (6) (2009) 533–568.

[7] Jirásek, M., Zeman, J., Deformation and damage of materials, Czech Technical University, Prague, 2006. (in Czech)

[8] Kaw, A. K., Mechanics of Composite Materials. 2nd ed., CRC Press, Boca Raton, 2006.

[9] Kollár, L., Springer, G. S., Mechanics of Composite Structures, Cambridge University Press, New York, 2003.

[10] Kompiš, V., Žmindák, M., Kaukič, M., Computational methods in mechanics: Linear analysis, University of Žilina, Žilina, 1997.

[11] Kormaníková, E., Riecky, D., Žmindák, M., Strength of composites with fibers, In Murín, J., Kompiš, V., Kutiš, V., eds.: Computational Modelling and Advanced Simulations, Springer Science + Business Media B.V., 2011.

[12] Laš, V., Zemčík, R., Progressive Damage of Unidirectional Composite Panels, Journal of Composite Materials 42 (1) (2008) 25–44.

[13] Liu, G. R., Gu, Y. T., An Introduction to Meshfree Methods and Their Programming, Springer, Berlin, 2005.

[14] Puck, A., Schurmann, H., Failure analysis of FRP laminates by means of physically based phenomenological models, Composite Science and Technology 62 (12–13) (2002) 1 633–1 662.

[15] Tay, T. E., Liu, G., Yudhanto, A., Tan, V. B. C., A Micro Macro Approach to Modeling Progressive Damage in Composite Structures, International Journal of Damage Mechanics 17 (1) (2008) 5–28.

[16] Tumino, D., Capello, F., Catalanotti, G., A continuum damage model to simulate failure in composite plates under uniaxial compression, Express Polymer Letters 1 (1) (2007) 15–23.

[17] Ventsel, E., Krauthammer, T., Thin Plates and Shells: Theory, Analysis, and Applications, Marcel Dekker, Inc., New York, 2001.

[18] Voyiadjis, G. Z., Kattan, P. I., A Comparative Study of Damage Variables in Continuum Damage Mechanics, International Journal of Damage mechanics 18 (4) (2009) 315–340.

[19] Zhang, Y. X, Chang, C. H., Recent developments in finite element analysis for laminated composite plates, Composite Structures 88 (1) (2009) 147–157.

Permissions

All chapters in this book were first published in ACM, by University of West Bohemia; hereby published with permission under the Creative Commons Attribution License or equivalent. Every chapter published in this book has been scrutinized by our experts. Their significance has been extensively debated. The topics covered herein carry significant findings which will fuel the growth of the discipline. They may even be implemented as practical applications or may be referred to as a beginning point for another development.

The contributors of this book come from diverse backgrounds, making this book a truly international effort. This book will bring forth new frontiers with its revolutionizing research information and detailed analysis of the nascent developments around the world.

We would like to thank all the contributing authors for lending their expertise to make the book truly unique. They have played a crucial role in the development of this book. Without their invaluable contributions this book wouldn't have been possible. They have made vital efforts to compile up to date information on the varied aspects of this subject to make this book a valuable addition to the collection of many professionals and students.

This book was conceptualized with the vision of imparting up-to-date information and advanced data in this field. To ensure the same, a matchless editorial board was set up. Every individual on the board went through rigorous rounds of assessment to prove their worth. After which they invested a large part of their time researching and compiling the most relevant data for our readers.

The editorial board has been involved in producing this book since its inception. They have spent rigorous hours researching and exploring the diverse topics which have resulted in the successful publishing of this book. They have passed on their knowledge of decades through this book. To expedite this challenging task, the publisher supported the team at every step. A small team of assistant editors was also appointed to further simplify the editing procedure and attain best results for the readers.

Apart from the editorial board, the designing team has also invested a significant amount of their time in understanding the subject and creating the most relevant covers. They scrutinized every image to scout for the most suitable representation of the subject and create an appropriate cover for the book.

The publishing team has been an ardent support to the editorial, designing and production team. Their endless efforts to recruit the best for this project, has resulted in the accomplishment of this book. They are a veteran in the field of academics and their pool of knowledge is as vast as their experience in printing. Their expertise and guidance has proved useful at every step. Their uncompromising quality standards have made this book an exceptional effort. Their encouragement from time to time has been an inspiration for everyone.

The publisher and the editorial board hope that this book will prove to be a valuable piece of knowledge for researchers, students, practitioners and scholars across the globe.

List of Contributors

J. Machalová, H. Netuka and R. Šimeček
Faculty of Science, Palacký University in Olomouc, 17. listopadu 1192/12, 771 46 Olomouc, Czech Republic

M. Brandner and H. Kopincová
NTIS -- New Technologies for Information Society, University of West Bohemia in Pilsen, Univerzitni 8, 306 14 Pilsen, Czech Republic

J. Egermaier
Department of Mathematics, University of West Bohemia in Pilsen, Univerzitni 8, 306 14 Pilsen, Czech Republic

J. Rosenberg
New Technologies-Research Centre, University of West Bohemia in Pilsen, Univerzitni 8, 306 14 Pilsen, Czech Republic

A. Civín and M. Vlk
A Institute of Solid Mechanics, Mechatronics and Biomechanics, Brno University of Technology, Technická 2896/2, 616 69 Brno, Czech Republic

Y. M. Ghugal
Applied Mechanics Department, Govt. College of Engineering, Karad – 415 124, MS, India

A. G. Dahake
Applied Mechanics Department, Govt. College of Engineering, Aurangabad – 431 005, MS, India

V. Zeman and Z. Hlaváč
Faculty of Applied Sciences, University of West Bohemia, Univerzitní 22, 306 14 Plzeň, Czech Republic

J. Zapoměl and P. Ferfecki
Centre of Smart Systems and Structures, Institute of Thermomechanics – Branch at VSB – Technical University of Ostrava, Czech Academy of Sciences, 17. listopadu 15, 708 33 Ostrava-Poruba, Czech Republic

L. Čermák
Institute of Mathematics, Brno University of Technology, Technická 2, 616 69 Brno, Czech Republic

M. Tukač and T. Vampola
Faculty of Mechanical Engineering, Czech Technical University in Prague, Technická 4, 166 07 Praha 6, Czech Republic

A. S. Sayyad
Department of Civil Engineering, SRES's College of Engineering Kopargaon-423601, M.S., India

Y. M. Ghugal
Department of Applied Mechanics, Government Engineering College, Karad, Satara-415124, M.S., India

T. Michálek and J. Zelenka
University of Pardubice, Jan Perner Transport Faculty, Department of Transport Means and Diagnostics, Section of Rail Vehicles; Detached Branch of the Jan Perner Transport Faculty, Nádražní 547, 560 02 Česká Třebová, Czech Republic

A. Jonášová, J. Vimmr and O. Bublík
Faculty of Applied Sciences, University of West Bohemia, Univerzitní 22, 306 14 Plzeň, Czech Republic

J. Machalová and H. Netuka
Faculty of Science, Palacký University in Olomouc, 17. listopadu 1192/12, 771 46 Olomouc, Czech Republic

J. Vychytil, F. Moravec and P. Kochová
Faculty of Applied Sciences, University of West Bohemia, Univerzitní 22, 306 14 Plzeň, Czech Republic

J. Kuncová and J. Švíglerová
Faculty of Medicine in Pilsen, Charles University in Prague, Lidická 1, 301 66 Plzeň, Czech Republic

K. L. Verma
Department of Mathematics, Government Post Graduate College, Hamirpur, (H.P.) 177005, India

S. Seitl
Institute of Physics of Materials, Academy of Sciences of the Czech Republic, v. v. i. Žižkova 22, 616 62 Brno, Czech Republic

D. Fernández-Zúñiga and A. Fernández-Canteli
Dept. of Construction and Manufacturing Engineering, E.P.S. de Ingeniería de Gijón, University of Oviedo, Campus de Viesques, 33203 Gijón, Spain

S. Seitl and P. Hutař
Institute of Physics of Materials, Academy of Sciences of the Czech Republic, v. v. i.Žižkova 22, 616 62 Brno, Czech Republic

T. E. García and A. Fernández-Canteli
Oviedo University, Dept. of Construction and Manufacturing Engineering; E.P.S. de Ingeniería de Gijón, University of Oviedo, Campus de Viesques, 33203 Gijón, Spain

A. A. Gholampour and M. Ghassemieh
School of Civil Engineering, University of Tehran, Tehran, Iran

P. Staňák, V. Sládek, J. Sládek, S. Krahulec and L. Sátor
Institute of Construction and Architecture, Slovak Academy of Sciences, Dúbravská cesta 9, 845 03 Bratislava, Slovakia

M. Dudinský, M. Žmindák and P. Frnka
Faculty of Mechanical Engineering, University of Žilina, Univerzitná 1, 010 26 Žilina, Slovak Republic

Index